NON-STANDARD ANALYSIS

T0319464

PRINCETON LANDMARKS
IN MATHEMATICS AND PHYSICS

NON-STANDARD ANALYSIS

Revised Edition

Abraham Robinson

PRINCETON UNIVERSITY PRESS

PRINCETON, NEW JERSEY

Library of Congress Cataloging-in-Publication Data

Robinson, Abraham, 1918–1974.
Non-standard analysis / Abraham Robinson. — Rev. ed.
p. cm. — (Princeton landmarks in mathematics and physics)
Originally published: Rev. ed. Amsterdam : North-Holland Pub. Co.;
New York : American Elsevier Pub. Co., 1974, in series: Studies in logic and
the foundations of mathematics.
Includes bibliographical references and index.
ISBN 0-691-04490-2 (pb : alk. paper)
1. Nonstandard mathematical analysis. 2. Logic, Symbolic and mathematical.
I. Title. II. Series.
QA299.82.R6 1995
515'.1—dc20 95-43750

To Renée

ANALYSE
DES
INFINIMENT
PETITS

CALCUL
DIFFEREN-
TIEL

TABLE
DES
SINUS

De la Monce inv.

Enberpun Sculp.

Engraving reproduced from Leonard Euler's 'Introductio in Analysin Infinitorum', first ed.,
Lausanne, 1748. By courtesy of the Amsterdam University Library.

FOREWORD (1996)

The second edition of Abraham Robinson's monumental work, *Nonstandard Analysis*, has been out of print for some time. It gives me great pleasure, as one of the earlier followers of Robinson, to have the opportunity to welcome the reissuing of this unique mathematical work by Princeton University Press.

During this century the mathematical community has witnessed the solution of a number of long-standing problems and conjectures. One of these problems, attributed to W. G. Leibniz, concerns the foundation of the calculus. In the early history of the calculus, arguments involving infinitesimals played a fundamental role in the derivation of the basic rules of Newton's method of fluxions. The notion of an infinitesimal, however, lacked a precise mathematical definition, and soon their widespread use came under severe attack, notably by Bishop Berkeley in England, who disdainfully referred to infinitesimals as the "ghosts of departed quantities." To counteract the critics of the theory of infinitesimals, Leibniz proposed a program to conceive of a system of numbers that would include infinitesimally small as well as infinitely large numbers. The process to arrive at such a system of numbers was to adjoin ideal objects to the existing finite quantities and to define arithmetic rules of combining them in such a way that the new ideal numbers and their reciprocals could be viewed as being either infinitesimally small or infinitely large; yet, such that the entire system would have, in some well-defined sense, the same properties as the system of real numbers. This process was inspired by the manner in which the so-called imaginary numbers were introduced as ideal objects to solve equations. When Leibniz and his followers failed in their attempts to create such a system, infinitesimal reasoning gradually lost ground and survived only as a figure of speech. In fact, infinitesimals were gradually replaced by the D'Alembert-Cauchy concept of a limit to provide a firm foundation for the calculus. The creation of a mathematical theory of infinitesimals, as envisaged by Leibniz, on which to base the calculus remained an open problem.

Despite the discovery of non-archimedean ordered field extensions of the reals around the turn of the century, Leibniz' problem remained unsolved and lay dormant until the end of the 1950s. At that time the situation changed dramatically. It then became clear that a rigorous mathematical theory of infinitesimals as proposed by Leibniz was within the grasp of mathematics.

In 1958, the aerodynamicist C. Schmieden, in collaboration with D. Laugwitz, constructed a partially ordered ring extension of the reals that included elements that could be viewed to play the role of infinitesimals for the purpose of representing the Dirac delta function by a function defined point-wise on the ring which is infinitely small except at the origin where it is infinitely large. Since their ring extension contained divisors of zero and elements of indeterminate size that could not be classified, with respect to the reals, as either infinitely small or infinitely large or finite, it did not lend itself in a straightforward manner as a number system that could be used to provide a satisfactory solution to Leibniz' problem. Nevertheless their approach was the first step in the right direction.

At the end of the 1950s, however, unaware of the Schmieden-Laugwitz approach, Abraham Robinson showed that the ordered fields that are non-standard models of the theory of real numbers could be viewed in the meta-mathematical or external sense as non-archimedean ordered field extensions of the reals that externally contain numbers that behave like infinitesimals with respect to the reals and whose reciprocals are infinitely large; that is, larger in absolute value than any positive real number. As models of the reals, these ordered fields had, in a precise formal manner, the same properties of the reals. As a consequence, these new number systems obtained by using model theoretic methods, provided a complete, satisfactory solution of Leibniz' three-centuries-old problem.

Since the models employed are known as nonstandard models, Robinson announced the creation of his theory of infinitesimals in a paper that appeared in 1961 in the *Proceedings of the Royal Academy of Sciences of Amsterdam* under the title *Non-standard Analysis*. Robinson introduced his history-making paper by giving credit to the ideas of the world-renowned logician T. Skolem. In a paper published in the *Fundamenta Mathematicae* in 1934, Skolem showed that the system of natural numbers could not be characterized by any set of its arithmetical properties that were formulated in the lower predicate calculus. In view of the categorical nature of the Peano system of axioms of arithmetic, Skolem's result came as a great surprise. The proof given by Skolem consisted of a construction of a proper extension of the system of natural numbers $\mathbb{N}$ that, under a precisely defined interpretation of its symbols, satisfies all the properties of $\mathbb{N}$ that are formulated in the lower predicate calculus in terms of a given set of arithmetic operations such as addition and multiplication. We may add here that Skolem's proper extension of $\mathbb{N}$ was the first explicit example of what are now referred to as nonstandard models of arithmetic.

After recalling Skolem's result, Robinson continued by applying the same idea to the set K_0 of all sentences of the lower predicate calculus in terms of the operations of the reals expressing properties of the real number system. The new number systems were then defined as nonstandard models of K_0. It followed that, in the meta-mathematical sense, the new number systems were non-archimedean ordered field extensions of the reals and thus, in the meta-mathematical sense, contained numbers that could serve the role of infinitesimals and their reciprocals that could serve as infinitely large numbers. After making these observations and using his new and rigorous theory of infinitesimals, Robinson derived the fundamental properties of the theory of limits, and more generally of the calculus, and in doing so, presented a completely satisfactory solution of Leibniz' problem.

Robinson did not stop there, however. He continued to test—successfully—his new methods in such fields as classical differential geometry, the classical theory of Lie groups, the theory of differential equations, and the theory of distributions that includes an improved version of the Schmieden-Laugwitz representation of the Dirac delta distribution. From the area of classical applied mathematics, Robinson discussed briefly the way in which the theory of infinitesimals could be used in the theory of boundary layers and concluded by stating that the probability that the nonstandard reals may provide a better explanation of certain observable phenomena than the standard reals should be borne in mind.

Abraham Robinson is considered to be one of the prominent mathematicians of this century for his ground-breaking work in logic and various fields of mathematics and applied mathematics, but above all for his discovery of nonstandard analysis, which he used to unite ideas of mathematical logic with mathematics proper. He was born in 1918 in Waldenburg, Lower Silesia and spent the larger part of his childhood in Breslau. In 1933, his family fled Nazi Germany and settled in Palestine. He studied mathematics at the Hebrew University under the wing of the famous mathematician Abraham Fraenkel. In June of 1940, after a brief stay in Paris studying at the Sorbonne, he barely escaped to England. During World War II, Robinson worked at the Royal Aircraft Establishment in Farnborough analyzing the design of supersonic airfoils. In a short amount of time, this work made him a specialist in aerodynamics. After the war, he continued his work in mathematical logic begun with Fraenkel and in 1949 he received his Ph.D. at the University of London. His thesis appeared later in book form under the title *The Meta-Mathematics of Algebra*. During those years he also continued his aeronautical research as a staff member of the Cranfield College of Aeronautics. In 1951, he joined the Department of Applied Mathematics of the University of Toronto, where I met him for the first time in the fall of 1956. Robinson succeeded his mentor, Abraham Fraenkel, in 1957 as Professor of Pure Mathematics at the Hebrew University. The creation of the theory of infinitesimals happened during the academic year 1959–1960, while he was visiting the Institute for Advanced Study in Princeton. During his years as Professor of Mathematics and Philosophy at the University of California, Los Angeles (from 1962 through 1967), Robinson worked full force at developing nonstandard techniques for applications to various other branches of mathematics. He also wrote this book, the first edition of which appeared in 1966, and the second in 1973. From 1967 until his untimely death on April 11, 1974, Robinson was Professor of Mathematics at Yale. During those years he had started to collaborate with P. Roquette from the University of Heidelberg on a program analyzing problems of algebraic number theory with nonstandard methods.

This most welcome reissuing of Robinson's *Non-standard Analysis* follows on the heels of the recent publication of Robinson's biography by the renowned historian of mathematics, Joseph Warren Dauben, entitled *Abraham Robinson, The creation of nonstandard analysis, a personal and mathematical odyssey,* with a forward by Benoit B. Mandelbrojt (Princeton University Press, 1995). These two books present a most valuable documentation of the history and the meaning of the creation of nonstandard analysis.

Since Robinson's death in 1974, nonstandard analysis has continued to attract the attention of the mathematical community. Recently, these methods have been applied, with great success, in probability theory, notably in the theory of stochastic processes. At regular intervals conferences and workshops devoted to nonstandard analysis have been organized. Proceedings of some of them have appeared in print.

For the interested reader we mention here the following publications: *Nonstandard Analysis, Recent Developments*, Proceedings of the Second Victoria Symposium on Non-standard Analysis, edited by A. E. Hurd; *Nonstandard Analysis and Its Applications, Lecture Notes in Mathematics* 983 (Springer Verlag, 1983); Proceedings of a Conference and Workshop held at the University of Hull in 1986, edited by Nigel Cutland, London Mathematical Society Student Texts 10 (Cambridge University Press, 1988); and *Advances in Analysis, Probability and Mathematical Physics; Contributions of Nonstandard Analysis*, Proceedings of a 1992 International Conference honoring D. Laugwitz on the Occasion of his Sixtieth Birthday, edited by S. A. Albeverio, W.A.J. Luxemburg, and M.P.H. Wolff, Mathematics and its Application Series 314 (Kluwer Academic Publishers, 1995).

In addition, a number of textbooks on nonstandard analysis at various levels have appeared, each with its own flavor. Nevertheless, paraphrasing the advice Lagrange is alleged to have given his students about reading Euler, I conclude by stating, "Read Robinson; he is the master of us all."

WILHELMUS A. J. LUXEMBURG

PREFACE

Je vois plus que jamais qu'il ne faut
juger de rien sur sa grandeur apparente.
O Dieu! qui avez donné une intelligence
à des substances qui paraissent si
méprisables, l'infiniment petit vous
coûte autant que l'infiniment grand.

VOLTAIRE, Micromégas

In the fall of 1960 it occurred to me that the concepts and methods of contemporary Mathematical Logic are capable of providing a suitable framework for the development of the Differential and Integral Calculus by means of infinitely small and infinitely large numbers. I first reported my ideas in a seminar talk at Princeton University (November 1960) and, later, in an address at the annual meeting of the Association for Symbolic Logic (January 1961) and in a paper published in the Proceedings of the Royal Academy of Sciences of Amsterdam (ROBINSON [1961]). The resulting subject was called by me Non-standard Analysis since it involves and was, in part, inspired by the so-called Non-standard models of Arithmetic whose existence was first pointed out by T. Skolem.

In the intervening years, Non-standard Analysis has developed considerably in several directions. Since many of the results have been reported so far only in courses or lectures, and in mimeographed reports, it was thought that a book dedicated entirely to this subject would be in order.

Over the years, my thinking in this area has been enlivened by discussions with several colleagues, among whom I venture to mention here R. Arens, C. C. Chang, A. Erdelyi, A. Horn, G. Kreisel, I. Lakatos, and J. B. Rosser. Special thanks are due to W. A. J. Luxemburg, whose lectures and lecture notes on Non-standard Analysis have done much to make the subject known among mathematicians.

I acknowledge with thanks the support received from the National Science Foundation (Grants GP-1812 and GP-4038) in connection with much of the research incorporated in the present book. I also wish to mention that Chapter VI is based on research sponsored in part by the Air Force Office of Scientific Research and published previously in report form (ROBINSON [1962]). Finally, I wish to put on record my indebtedness to M. Machover who read the book at the proof stage and suggested various improvements.

ABRAHAM ROBINSON

University of California, Los Angeles
April, 1965

PREFACE TO THE SECOND EDITION

Seven years have passed since the publication of the first edition of this book and almost twice as long since the inception of its subject. The fact that a new edition has become necessary indicates that the ideas described here attracted a measure of attention. The least that can be said is that the intrinsic interest of the subject and its historical relevance are, by now, widely appreciated. Beyond this, numerous articles in the recent mathematical literature contain applications of non-standard analysis to contemporary problems in areas as widely apart as algebraic number theory and mathematical economics. In particular, the interested reader may wish to consult the proceedings of several symposia in this area which were held in recent years. They are contained in: Applications of Model Theory to Algebra, Analysis and Probability, edited by W. A. J. Luxemburg (Holt, Rinehart and Winston, Toronto 1969); Contributions to Non-standard Analysis, edited by W. A. J. Luxemburg and A. Robinson, in: Studies in Logic and the Foundations of Mathematics, Vol. 69 (North-Holland, Amsterdam 1972); and the proceedings of a conference on Non-standard Analysis which was held in the spring of 1972 at the University of Victoria, B.C., Canada, to be published in the Lecture Note series of the Springer-Verlag.

Although the situation may change some day, the non-standard methods that have been proposed to date are *conservative* relative to the commonly accepted principles of mathematics (e.g., the axioms of Zermelo–Fraenkel, including the axiom of choice). This signifies that a non-standard proof can always be replaced by a standard one, even though the latter may be more complicated and less intuitive. Thus, the present writer holds to the view that the application of non-standard analysis to a particular mathematical discipline is a matter of choice and that it is natural for the actual decision of an individual to depend on his early training.

A more definite opinion has been expressed in a statement which was made by Kurt Gödel after a talk that I gave in March 1973 at the Institute for Advanced Study, Princeton. The statement is reproduced here with Professor Gödel's kind permission.

'I would like to point out a fact that was not explicitly mentioned by Professor Robinson, but seems quite important to me; namely that non-standard analysis frequently simplifies substantially the proofs, not only of elementary theorems, but also of deep results. This is true, e.g., also for the proof of the existence of invariant subspaces for compact operators, disregarding the improvement of the result; and it is true in an even higher degree in other cases. This state of affairs should prevent a rather common misinterpretation of non-standard analysis, namely the idea that it is some kind of extravagance or fad of mathematical logicians. Nothing could be farther from the truth. Rather there are good reasons to believe that non-standard analysis, in some version or other, will be the analysis of the future.

One reason is the just mentioned simplification of proofs, since simplification facilitates discovery. Another, even more convincing reason, is the following: Arithmetic starts with the integers and proceeds by successively enlarging the number system by rational and negative numbers, irrational numbers, etc. But the next quite natural step after the reals, namely the introduction of infinitesimals, has simply been omitted. I think, in coming centuries it will be considered a great oddity in the history of mathematics that the first exact theory of infinitesimals was developed 300 years after the invention of the differential calculus. I am inclined to believe that this oddity has something to do with another oddity relating to the same span of time, namely the fact that such problems as Fermat's, which can be written down in ten symbols of elementary arithmetic, are still unsolved 300 years after they have been posed. Perhaps the omission mentioned is largely responsible for the fact that, compared to the enormous development of abstract mathematics, the solution of concrete numerical problems was left far behind.'

I am greatly indebted to Peter Winkler for correcting a considerable number of printing errors which appeared in the first edition of this book and for closing a gap in the original statement of Theorem 8.1.12, page 206.

Yale University, New Haven   ABRAHAM ROBINSON
October 1973

LIST OF CONTENTS

CHAPTER IV

GENERAL TOPOLOGY

CHAPTER V

FUNCTIONS OF A REAL VARIABLE

CHAPTER VI

FUNCTIONS OF A COMPLEX VARIABLE

CHAPTER VII

LINEAR SPACES

CHAPTER VIII

TOPOLOGICAL GROUPS AND LIE GROUPS

CHAPTER IX

SELECTED TOPICS

CHAPTER X

CONCERNING THE HISTORY OF THE CALCULUS

CHAPTER I

GENERAL INTRODUCTION

1.1 Purpose of this book. Underlying the fundamental notions of the branch of mathematics known as Analysis is the concept of a limit. Derivatives and integrals, the sum of an infinite series and the continuity of a function all are defined in terms of limits. For example, let $f(x)$ be a real-valued function which is defined for all x in the open interval $(0,1)$ and let x_0 be a number which belongs to that interval. Then the real number a is the *derivative of $f(x)$ at x_0*, in symbols

1.1.1
$$f'(x_0) = \left(\frac{df}{dx}\right)_{x=x_0} = a$$

if

1.1.2
$$\lim_{x \to x_0} \frac{f(x) - f(x_0)}{x - x_0} = a .$$

Suppose we ask a well-trained mathematician for the meaning of 1.1.2. Then we may rely on it that, except for inessential variations and terminological differences (such as the use of certain topological notions), his explanation will be thus:

For any positive number ε there exists a positive number δ such that

$$\left| \frac{f(x) - f(x_0)}{x - x_0} - a \right| < \varepsilon$$

for all x in $(0,1)$ for which $0 < |x - x_0| < \delta$.

Let us now ask our mathematician whether he would not accept the following more direct interpretation of 1.1.1 and 1.1.2.

For any x in the interval of definition of $f(x)$ such that $dx = x - x_0$ is *infinitely close to* 0 but not equal to 0, the ratio df/dx, where

$$df = f(x) - f(x_0),$$

is *infinitely close* to a.

To this question we may expect the answer that our definition may be simpler in appearance but unfortunately it is also meaningless. If we then try to explain that two numbers are infinitely close to one another if their distance (the modulus of their difference) is *infinitely small*, i.e., smaller than any positive number, we shall probably be faced with the rejoinder that this is possible only if the numbers coincide. And, so we may be told charitably, this obviously is not what we meant since it would make our explanation trivially wrong.

However, in spite of this shattering rebuttal, the idea of infinitely small or *infinitesimal* quantities seems to appeal naturally to our intuition. At any rate, the use of infinitesimals was widespread during the formative stages of the Differential and Integral Calculus. As for the objection quoted above, that the distance between two distinct real numbers cannot be infinitely small, G. W. Leibniz argued that the theory of infinitesimals implies the introduction of ideal numbers which might be infinitely small or infinitely large compared with the real numbers but which were *to possess the same properties as the latter*. However, neither he nor his disciples and successors were able to give a rational development leading up to a system of this sort. As a result, the theory of infinitesimals gradually fell into disrepute and was replaced eventually by the classical theory of limits.

It is shown in this book that Leibniz' ideas can be fully vindicated and that they lead to a novel and fruitful approach to classical Analysis and to many other branches of mathematics. The key to our method is provided by the detailed analysis of the relation between mathematical languages and mathematical structures which lies at the bottom of contemporary model theory.

1.2 Summary of contents. The plan of this book is as follows. In Chapter II we describe the formal tools from mathematical logic which are required for subsequent developments. Our discussion deals with first and higher order theories and includes a proof of the finiteness principle (compactness theorem), which is of central importance for our purpose.

In Chapter III we detail the basic properties of the mathematical structures which serve as frameworks for Non-standard Arithmetic and Non-standard Analysis. We show that these structures provide an adequate foundation for a theory of infinitesimals, and we develop the elements of the Differential and Integral Calculus by means of that theory. Next we introduce first and higher order differentials and give applications to some simple problems of classical Differential Geometry. Naturally, these differentials are infinitely small, as they were taken to be, naively, in the earliest texts on the Calculus published in continental Europe, such as the *Traité des infiniment petits* of de l'Hospital. It turns out that after the removal of some glaring and frequently attacked inconsistencies, the method used in these texts can be put on a firm foundation.

In Chapter VI we show that the theory of infinitesimals possesses a generalization which is applicable to (non-metric) topological spaces. Within this theory, we reformulate various basic notions of Topology. In particular, we obtain a striking characterization of compact spaces, which has several applications.

Chapter V deals with functions of a real variable. The Lebesgue measure is defined in terms of Non-standard Analysis and several standard theorems are proved within this framework. Next, we discuss the theory of Schwartz distributions. Our approach supplies a concrete realization of these generalized functions and, in particular, provides effective means for the discussion of the notion of the local value of a distribution.

In Chapter VI, we develop the non-standard theory of functions of a complex variable. Fields of application considered in detail are (i) the analytic theory of polynomials, which deals with the location of zeros of polynomials in the complex domain, and (ii) the theory of exceptional values of entire functions, including Picard's theorems and Julia's directions. However, it is even more significant that the theory of normal families is replaced by certain generalized functions provided in a natural way by our approach.

Chapter VII brings us to the theory of normed linear spaces. The Non-standard theory of compact operators is developed in several directions. In particular, it is shown that a linear operator in Hilbert space which has a compact square, possesses a non-trivial invariant subspace. It is noteworthy that the analysis given here provided the first proof of this result, which settles a problem raised by K. T. Smith and P. R. Halmos (see BERNSTEIN and ROBINSON [1966]). Another application of the theory deals with spectral analysis.

In Chapter VIII we consider topological groups and, more particularly, Lie groups. In our theory, the infinitesimal neighborhood of the unit element of a group actually exists and constitutes a group. This group realizes the intuitive notion of the infinitesimal group for a given topological group or Lie group. It can be related to the standard concepts of the theory.

Chapter IX contains applications of variational principles to several mathematical problems within the framework of Non-standard Analysis. In particular, we give adaptations of a classical proof of Riemann's mapping theorem and of the Dirichlet integral method in Potential Theory. Next we consider several topics from Hydrodynamics. We expound the basic notions of boundary layer theory by means of infinitesimals, and we give a counterpart of de Saint-Venant's principle in the Theory of Elasticity. Finally de Saint-Venant's principle itself is interpreted rationally within the framework of Non-standard Analysis.

The last chapter contains a review of certain stages in the history of the Differential and Integral Calculus that had to do with the theory of infinitesimals. The fact that the more recent writers in this field were convinced that no such theory can be developed effectively, colored their historical judgment. Thus, a revision has now become necessary.

It is natural to ask whether a non-standard method (in the technical sense in which the term non-standard is used here, i.e., a method of Non-standard Analysis) can always be replaced by a standard mathematical proof. This question presumes that the methods of Mathematical Logic are set apart from ordinary Mathematics, and we may agree for our present purposes that the distinction is meaningful in practice. The answer to the question then is that the method of ultrapowers provides a ready means for translating a non-standard proof into a standard mathematical proof in each particular case. However, in the course of doing so we may complicate the proof considerably, so that frequently the resulting procedure will be less desirable from a heuristic point of view. At the same time there may well exist a shorter mathematical proof which can be obtained independently.

The development of the theory of infinitesimals has very nearly stood still during the last one hundred and fifty years. During the same period an enormous amount of effort and ingenuity has been poured into the classical methods. Nevertheless, we believe to have shown that even at this late stage non-standard methods can supplement standard methods effectively, both in throwing new light on old theories and in order to find

new results. We hope that some of the experts, in the subjects touched
upon in this book or in subjects not touched upon here, will consider the
application of non-standard methods to their field. They may find that in
some cases the appropriate non-standard interpretation of a classical
theory is not discovered without effort. But once it has been found the
subsequent reformulation and development of the theory can be a
rewarding experience.

One might express the hope that some branches of modern Theoretical
Physics, in particular those that are afflicted with divergence problems,
might be treated with profit by Non-standard Analysis. The fact that the
present book contains only applications to classical Applied Mathematics
most probably is a testimony to the limitations of the author and not of
the method.

CHAPTER II

TOOLS FROM LOGIC

2.1 The Lower Predicate Calculus. In this chapter we introduce the reader to some formalisms whose appreciation is essential for the sequel. The text is written in such a way that it can be understood by anyone with a rudimentary knowledge of Mathematical Logic and of the elements of Abstract Set Theory.

We proceed to describe the language of the Lower, or First Order, Predicate Calculus. Disregarding variations of minor importance, this may be said to be the basic formalism of Mathematical Logic.

A first order language L is defined in the following way.

The atomic symbols of L are:

(i) Individual object symbols, or individual constants, usually small italics, with or without subscripts or primes, such as a, b', c_n^m, or occasionally, to conform with common usage, other symbols such as the numerals $0, 1, 2, \ldots$ The set of individual constants is arbitrary but fixed (and is usually transfinite).

(ii) (Individual) variables, x, y', z_{kl} (Italics from the end of the alphabet, with or without subscripts, etc.). Their number is supposed to be infinite but countable.

(iii) Relation symbols of order $n, n \geq 1$, $S(\ )$, $R(\ , \ , \ )$, where n denotes the number of places in the parentheses. The set of relation symbols is arbitrary but fixed and may be transfinite. (Relation symbols of order $n = 0$ are not required and, for ease of understanding, will be excluded here.)

(iv) The connectives, $\neg$ (negation), $\vee$ (disjunction), $\wedge$ (conjunction), $\supset$ (implication), $\equiv$ (equivalence).

(v) The quantifiers, $(\exists\ )$ – existential, and $(\forall\ )$ – universal.

(vi) The brackets, $[$, and $]$.

Atomic formulae are obtained from the atomic symbols by filling the empty places of relative symbols with either individual constants or variables, e.g., $S(a)$, $S(x_n)$, $R(x,a,y)$. From these, well-formed formulae (wff) are obtained in successive stages, beginning with the atomic formulae, by applying connectives and quantifiers in the appropriate manner to well-formed formulae obtained already. At the same time brackets are introduced in such a way that the mode of construction of a well-formed formula can be determined unambiguously from its form. In more detail –

If X is an atomic formula, $[X]$ is a well-formed formula; if X and Y are wff, then $[\neg X]$, $[X \vee Y]$, $[X \wedge Y]$, $[X \supset Y]$, $[X \equiv Y]$ are wff; and if X is a wff then $[(\exists y)X]$ and $[(\forall y)X]$ are wff (where y stands for an arbitrary variable) provided y does not already appear in X under the sign of a quantifier, i.e., provided X does not already contain $(\exists y)$ or $(\forall y)$. X is said to be the scope of the quantifier under consideration in $[(\exists y)X]$ or $[(\forall y)X]$ and in any wff which is obtained from these by the further (repeated) application of connectives or quantifiers.

An occurrence of a variable, e.g. y, in a wff X is *free* if y is not in $(\exists y)$ or $(\forall y)$ or in their scope in X. The wff X is called a *sentence* if it does not contain a free occurrence of a variable, otherwise it is called a *predicate*.

It is usual to omit brackets if this does not lead to any ambiguity, or if any ambiguity which is introduced thereby is irrelevant. Thus, we write $[X_1 \vee X_2 \vee X_3]$ indiscriminately for $[[X_1 \vee X_2] \vee X_3]$ and for $[X_1 \vee [X_2 \vee X_3]]$, since (as we shall see) it is irrelevant which of the two latter expressions is meant.

We say that the first order language L' is an *extension* of the first order language L if the set of atomic symbols of L is a subset of the set of atomic symbols of L'.

A *first order structure*, M, consists of a non-empty set of *individuals*, A, in which a set P of n-ary relations e.g., $S(\ ,\ ,\ ...)$ are *defined* in the following sense. If $S(\ ,\ ,...,)$ is an n-ary relation which belongs to P and $(a_1,...,a_n)$ is an ordered n-tuple of elements of A then it is determinate whether or not $S(a_1,...,a_n)$ *holds* in M. (In set theoretic versions of the notion of a structure, a relation is actually identified with the set of n-tuples which hold for it.) For example, we may consider that an algebraic field is given by its set of individuals and by the three relations $E(\ ,\ )$, $S(\ ,\ ,\ )$, $P(\ ,\ ,\ )$, where E is binary and S and P are ternary and where $E(a,b)$ stands for $a=b$ (equality), and $S(a,b,c)$ and

$P(a,b,c)$ stand for $a+b=c$ and $ab=c$ (addition and multiplication) respectively.

2.2 Interpretation.   Now let M be a first order structure and let L be a first order language. Suppose that we are given a one-to-one mapping C between the set of individuals of M and a subset of the set of individual constants of L and between the relations of M and a subset of the set of relation symbols of L in such a way that n-ary relations correspond to n-ary relation symbols. We shall say that a well-formed formula X of L is *defined* in M (with respect to the mapping C) if all the individual constants and relation symbols which occur in X correspond to individuals and relations of M under C.

For any sentence X which is defined in M we shall now define what it signifies that X holds in M or *is true in M*, or that *M is satisfied by X*, by means of the following conditions. If X is defined, but does not hold in M we shall say also that *X is false in M*.

(i) Let $X=[Y]$ be defined in M where Y is an atomic formula. Thus $Y=R(a_1,...,a_n)$ where R is an n-ary relation symbol and $a_1,...,a_n$ are individual constants such that Y, under C, R corresponds to a relation R' in M and $a_1,...,a_n$ correspond to individuals $a'_1,...,a'_n$ respectively. Then X *holds in M* (under C) by definition, if and only if $R'(a'_1,...,a'_n)$ holds in M.

(ii) Suppose that the sentence $X=[\neg Y]$ is defined in M. Thus, Y also is defined in M. Then X holds in M if and only if Y does not hold in M. Again, suppose that the sentence X which is defined in M is given by $X=[Y\vee Z]$. Then X *holds in M* if and only if at least one of the two sentences Y or Z holds in M. For the remaining connectives, supposing the sentence X to be defined in M, the rules which decide whether or not X holds in M are as follows.

If $X=[Y\wedge Z]$, X *holds in M* if and only if both Y and Z hold in M. If $X=[Y\supset Z]$, X *holds in M* provided Z holds in M, and also if neither Y nor Z hold in M; X does *not hold* in M if Y holds in M but Z does not hold in M. If $X=[Y\equiv Z]$, X *holds in M* if and only if both Y and Z hold in M or neither Y nor Z holds in M.

(iii) Suppose now that X has been obtained by existential quantification. Thus, X is of the form $X=[(\exists y)Z(y)]$. Then X holds in M if and only if there exists an individual constant a such that $Z(a)$ holds in M. (This implies that a corresponds to some individual of M under C). We have used the notation $Z(y), Z(a)$ in order to indicate that $Z(a)$ is obtained from the scope of $(\exists y)$ by substituting a for (every occurrence of) y.

If y does not occur in Z the correct interpretation of our rule is that X holds in M if and only if Z holds in M.

Finally, suppose that X is of the form $X = [(\forall y) Z(y)]$. Then X holds in M if for every individual constant a of L which corresponds to an individual of M under C, the sentence $Z(a)$ holds in M.

One shows by induction following the construction of a well-formed formula (or, more concretely, following the number of square brackets in a formula) that conditions (i), (ii), and (iii) jointly decide unambiguously for every X which is defined in M whether or not X holds in M, for the given C. The question whether or not a sentence X holds in a structure M depends only on the mapping C_X between the individual constants and relation symbols which actually occur in X and the appropriate individuals and relations of M, provided there are enough individual constants and relation symbols available in L to extend C_X to a mapping C in which all individuals and relations of M are mapped on corresponding entities in L. For a given M, we may at any rate extend L to a language L' in which this condition is satisfied. In order to avoid trivial but tiresome exceptions, we shall suppose from now on that such an extension of the given language has been carried out tacitly where necessary. That is to say, when stating that X holds in a structure M for a one-to-one mapping C between the individual constants and relation symbols which occur in X and certain individuals and relations in M, we shall imply that this is the case, according to our original definition, in an extension L' of L in which C_X can be extended to a mapping C as above.

Similarly, let K be a set of sentences in a first order language L and let there be given a one-to-one mapping C_K between the individual constants and relation symbols that occur in K and individuals and relations of a (first order) structure M, so that individual constants correspond to individuals and, for each $n \geq 1$, n-ary relation symbols correspond to n-ary relations. Then we shall say that K holds in M (is true in M, is satisfied by M or, that M is a *model* of K) for the given mapping C_K if in some extension L' of L, C_K can be extended to a mapping C as above, which ranges over all individuals and relations of M, such that all sentences of K hold in M under this mapping, in the sense explained previously. Here again, the answer to the question whether or not K holds in M depends only on C_K in the sense that if $C=C_1$ and $C=C_2$ are two extensions of C_K which range over all of M then the sentences of K hold in M under C_1 if and only if they hold in M under C_2.

2.3   Ultraproducts.   Let $Q=\{M_v\}$ be a set of first order structures, where v varies over a non-empty index set I. Suppose that the M_v are similar, that is to say that the same relations are defined in all of them. (Notice that this concept of similarity is very wide. For example, all algebraic fields are similar if we take them to be given by the three relations of equality, addition, and multiplication, as introduced above. We do not mean that the relations need be the same extensionally, i.e., we identify, e.g., the relations of addition in different fields although they will in general be different in the set theoretic sense.) Let D be an ultrafilter in I (or, which is the same, a maximal dual ideal in the Boolean algebra of subsets of I). That is to say, D is a set of subsets of I which satisfies the following conditions.

    (i)   The empty set does not belong to D, $\emptyset \notin D$.

    (ii)   If $A \in D$ and $B \in D$ then $A \cap B \in D$.

    (iii)   If $A \in D$ and $A \subset B \subset I$ then $B \in D$.

    (iv)   For any $A \subset I$, either $A \in D$ or $I - A \in D$ (although, by (i) and (ii) the
          sets A and $I - A$ cannot both belong to D).

By means of Q and D we now define a new first order structure Q_D, called an *ultraproduct* as follows. The set of individuals of Q_D is the set of functions $f(x)$ defined on I and satisfying the condition that $f(v) \in M_v$ for every $v \in I$. Now let $R(x_1,x_2,...,x_n)$ be a relation which is given for the structures M_v (where we have inserted the variables $x_1,...,x_2$ into the empty places of R for easier reading). Let $f_1,...,f_n$ be individuals of Q_D. Thus, $f_i=f_i(x)$, $i=1$, ..., n where $f_i(v) \in M_v$, for $v \in I$, as detailed. We determine the relation R in Q_D by defining that $R(f_1, ..., f_n)$ shall hold in Q_D if and only if the set $\{v | R(f_1(v), ..., f_n(v))$ holds in $M_v\}$ — i.e., the set of elements v of I for which $R(f_1(v), ..., f_n(v))$ holds in M — is an element of the ultrafilter D.

This completes the definition of Q_D.

2.4   Prenex normal form.   A wff X in L is said to be in *prenex normal form* if, in the construction of X from atomic formulae, the quantifiers (if any) are applied after the connectives (if any) or, which is the same, if the connectives are in the scope of all quantifiers.

For the purposes of this and the following section, two sentences X and Y will be called *equivalent* if they contain the same individual constants and relation symbols, and if $X \equiv Y$ holds in all structures in which this sentence is defined. This will be the case if and only if for any structure M in which X and Y are defined, and by the same correspondence, either both X and Y hold in M or neither X nor Y holds in M.

2.4.1 THEOREM. For every sentence X, there exists an equivalent sentence X_0 which is in prenex normal form.

In order to establish this result, we observe that if X is equivalent to Y and Y is equivalent to Z according to the above definition, then X is equivalent to Z, i.e., the (meta-mathematical) relation of equivalence between sentences is transitive. For a given X, we may therefore find an equivalent Y by a *succession* of modifications each leading from a given sentence to an equivalent sentence. The modifications required here are based on the following facts which can be verified without difficulty by means of the definitions of section 2.2.

(i) If X contains a wff of the form $[Y \equiv Z]$ and we replace it by $[[Y \supset Z] \wedge [Z \supset Y]]$ then we obtain a sentence X' which is equivalent to X, by 2.2 (ii).

(ii) If X contains a wff of the form $[Y \supset Z]$, and we replace it by $[[\neg Y] \vee Z]$ then we obtain a sentence X' which is equivalent to X, again by 2.2(ii).

(iii) The sentence $Y = [\neg [Z \vee W]]$ is equivalent to $Y' = [[\neg Z] \wedge [\neg W]]$, by 2.2(ii), and the sentence $Y = [\neg [Z \wedge W]]$ is equivalent to $Y' = [[\neg Z] \vee [\neg W]]$, by 2.2(ii) (these are 'de Morgan's laws'). Hence, if a sentence X contains a *well-formed formula* $Y = [\neg [Z \vee W]]$, and we replace Y by $Y' = [[\neg Z] \wedge [\neg W]]$ then we obtain a sentence X' which is equivalent to X. And if the sentence X contains a wff $Y = [\neg [Z \wedge W]]$ and we replace the latter by $Y' = [[\neg Z] \vee [\neg W]]$ then we obtain a sentence X' which is equivalent to X.

(iv) If the sentence Y is given by $Y = [\neg [(\exists y)Z]]$ then Y is equivalent to $Y' = [(\forall y)[\neg Z]]$, by 2.2(ii) and 2.2(iii), and if $Y = [\neg [(\forall y)Z]]$ then Y is equivalent to $Y' = [(\exists y)[\neg Z]]$, also by 2.2(ii) and 2.2(iii). Accordingly, if in any sentence we replace a *well-formed formula* $[\neg [(\exists y)Z]]$ or $[\neg [(\forall y)Z]]$ by $[(\exists y)[\neg Z]]$ or $[(\forall y)[\neg Z]]$ respectively, then we obtain an equivalent sentence.

(v) If $X = [[(\exists y)Z] \lor W]$ and if u is a variable that does not occur in X then X is equivalent to $X' = [(\exists u)[Z' \lor W]]$ where Z' is obtained from Z by substituting u for y. Similarly, if $X = [W \lor [(\exists y)Z]]$, then with u and Z' as before, X is equivalent to $X' = [(\exists u)[W \lor Z']]$. Similarly statements hold if the existential quantifiers, $(\exists y)$, $(\exists u)$ are replaced by universal quantifiers, $(\forall y)$ and $(\forall u)$; or if the connective of disjunction, $\lor$, is replaced everywhere by the connective of conjunction, $\land$ or if both of these modifications are carried out simultaneously. Finally, if this transformation is carried out for a *well-formed formula X* which is contained in a sentence Y, then the result is a sentence Y' which is equivalent to Y provided the new variable u does not appear in Y within a quantifier which includes X in its scope.

By means of (i) and (ii) we may replace any given sentence X by an equivalent sentence which contains neither $\equiv$ nor $\supset$. For example,

$$X = [[(\forall y)[R(a, y)]] \supset [\neg[(\forall y)[[Q(y)] \lor [(\exists z)[S(y, z)]...]$$

is equivalent to

$$X_1 = [[\neg[(\forall y)[R(a, y)]]] \lor [\neg[(\forall y)[[Q(y)] \lor [(\exists z)[S(y, z)]...].$$

Next, employing (iii) and (iv), we may transform every sentence X which does not contain $\equiv$ or $\supset$ into an equivalent sentence in whose construction every introduction of the connective of negation precedes the introduction of the connectives of disjunction and conjunction and of the quantifiers. Thus, continuing the above example, we may replace X_1 by the equivalent

$$X_2 = [[(\exists y)[\neg[R(a, y)]]] \lor [\neg[(\forall y)[[Q(y)] \lor. [(\exists z)[S(y, z)]...].$$

Continuing this operation, we obtain in successive stages the sentences

$$X_3 = [[(\exists y)[\neg[R(a, y)]]] \lor [(\exists y)[\neg[[Q(y)] \lor [(\exists z)[S(y, z)]...],$$

$$X_4 = [[(\exists y)[\neg[R(a, y)]]] \lor [(\exists y)[[\neg[Q(y)]] \land [\neg[(\exists z)[S(y, z)]...],$$

and

$$X_5 = [[(\exists y)[\neg[R(a, y)]]] \lor [(\exists y)[[\neg[Q(y)]] \land [(\forall z)[\neg S(y, z)]...].$$

The last sentence, X_5, has the required form.

Using (v) above we may now modify the sentence obtained so far so that, finally, all connectives precede all quantifiers in the construction of the sentence from atomic formulae. Thus, in the particular example considered above, X_5 is equivalent to

$$X_6 = [(\exists u)[[\neg[R(a,u)]] \vee [(\exists y)[(\forall z)[[\neg[Q(y)]] \wedge [\neg[S(y,z)]...]$$

and to

$$X_0 = [(\exists u)[(\exists w)[(\forall x)[[\neg[R(a,u)]] \vee [[\neg[Q(w)]] \wedge [\neg[S(w,x)]]]]]]]]$$

The last sentence, X_0, is in prenex normal form.

Observe, that for a given sentence X, the equivalent sentence in prenex normal form is not unique. This can be shown by means of the above example. More simply, if $X = [[(\exists y)[Q(y)]] \wedge [(\forall z)[R(z)]]]$ then both the sentence $[(\exists y)[(\forall z)[[Q(y)] \wedge [R(z)]...]$ and the sentence $[(\forall z)[(\exists y)[[Q(y)] \wedge [R(z)]...]$ are equivalent to X and in prenex normal form.

Observe that the sentences cannot be obtained from each other merely by renaming variables.

From now on, we shall permit ourselves to simplify our notation by omitting

(i)   brackets around atomic formulae
(ii) the outermost brackets around any wff
(iii) brackets between successive quantifications.

Making use of these rules we may, for example, write the above X_0 as

$$(\exists u)(\exists w)(\forall x)[[\neg R(a,u)] \vee [[\neg Q(w)] \wedge [\neg S(w,x)]]].$$

2.5 The Finiteness Principle. In this section we shall prove a central result of the Lower Predicate Calculus, which is of fundamental importance for our theory.

A set of sentences will be called *consistent* if it possesses a model.

2.5.1 FINITENESS PRINCIPLE (COMPACTNESS THEOREM). Let K be a set of sentences such that every finite subset of K is consistent. Then K is consistent.

For the proof, we may suppose that all sentences of K are in prenex normal form. For if this is not the case from the outset we consider a set K_0 which is obtained from K by substituting for every sentence X in K an equivalent sentence X_0 which is in prenex normal form. Suppose that every finite subset of K is consistent and let K_0' be a finite subset of K_0. Then K_0 is obtained from a finite subset K' of K by substituting for every sentence of K' an equivalent sentence in prenex normal form. Since K' is consistent it possesses a model and this is also a model of K_0', i.e., K_0' also is consistent. Hence, assuming that we have already proved our assertion for K_0, we may conclude that K_0 possesses a model, and this is also a model of K. Accordingly, we shall suppose from now on that the sentences of K, as given, are in prenex normal form.

We shall require a mode of interpretation of sentences in prenex normal form which was conceived by Skolem and Hilbert and developed further by Herbrand. To introduce it, let us consider the sentence $X = (\forall x)$ $(\exists y) Q(x,y)$ where $Q(x,y)$ is a wff. According to 2.2, the assertion that X holds in a structure M signifies that for every individual constant a, denoting (i.e., in correspondence with) an individual of M, there exists an individual constant b such that $Q(a,b)$ holds in M. Applying the axiom of choice, we may select such a b for every a, and we may then say that there exists a function $\phi(x)$ (called a *Skolem function* or *functor*) defined on the individual constants which denote individuals of M, and taking values in the same set, such that for every a in the set, $Q(a,\phi(a))$ holds in M. In this expression, ϕ is a function from individual constants (i.e., entities of the language L) to individual constants, but we shall permit ourselves to use the same symbol also in order to denote the function from and to the corresponding individuals. Thus if a and b denote individuals a' and b' of M respectively, and $b = \phi(a)$, then we write also $b' = \phi(a')$. The fact that such a function exists implies, conversely, that the sentence $(\forall x)(\exists y) Q(x, y)$ holds in M. The expression $Q(x,\phi(x))$, which is outside our formal language L, is called a *(Skolem) open form sentence*. By saying that $Q(x,\phi(x))$ holds in M we mean that $Q(a,\phi(a))$ holds in M for every individual constant which denotes an individual of M.

In order to exemplify the general case, consider the sentence

2.5.2 $(\forall x)(\exists y)(\forall z)(\exists u)(\exists v)(\forall w)(\exists t) Q(x, y, z, u, v, w, t)$

where Q is a wff without quantifiers. 2.5.2 holds in a structure M if and only if the *open form sentence*

2.5.3 $Q(x, \phi(x), z, \psi(x, z), \chi(x, z), w, \varrho(x, z, w))$

holds in M for an appropriate choice of the Skolem functions

$$\phi(x), \psi(x, z), \chi(x, z), \varrho(x, z, w),$$

i.e., if

2.5.4 $Q(a, \phi(a), b, \psi(a, b), \chi(a, b), c, \varrho(a, b, c))$

holds in M for all a,b, and c which denote (correspond to) individuals in M. Here again, the symbols $\phi, \psi, \chi, \varrho$, which are outside the formal language L will be used both for functions from individual constants which denote individuals of M to individual constants, and for the functions from the corresponding individuals to the corresponding individuals. Note that the Skolem functions correspond to the existential quantifiers in the given sentence X, and their arguments correspond to the universally quantified variables which precede the respective existential quantifier in X. If X begins with an existential quantifier then the corresponding Skolem function reduces to a constant, i.e., it takes a single value (in M).

In order to prove 2.5.1. let K be a set of sentences in prenex normal form such that every finite subset of K is consistent. We shall construct a model for K in the form of an ultraproduct. The index set I is defined as the set of all finite subsets of K (so that I is not empty). In order to agree with the notation of section 2.3 we shall denote these subsets of K by small Greek letters, $v, \mu, \ldots$. For every $v \in I$ we select a corresponding structure M_v such that M_v is a model of v under some specified mapping C_v. Such a structure exists for every $v \in I$, by assumption, but the determination of an indexed set $Q = \{M_v\}$ where v varies over I involves the axiom of choice (except in the trivial case where I is finite). For every v, C_v is a mapping from the individual constants and relation symbols of the sentences of v onto corresponding individuals and relations of M_v. We may then identify relations in different M_v if they are denoted by the same relation symbols in K. If, for some v, a relation symbol R of K does not correspond to any relation in M_v (because R does not occur in the sentences of v) then we define a corresponding relation R', with the same number of variables, arbitrarily in M_v, e.g., by stipulating that R' does not hold for any individuals of M_v. In this way we ensure that the structures of $Q = \{M_v\}$ are all similar. Having chosen Q it remains to select an appropriate ultrafilter D on I.

For every $v \in I$, let d_v be the set of elements μ of I such that $v \subset \mu$, v is a subset of μ, and let D_0 be the set of all these d_v, in symbols $D_0 = \{d_v\}_{v \in I}$.

Then D_0 has the following properties.

(i) $\emptyset \notin D_0$, D_0 does not contain the empty set.

For if $A \in D_0$ then $A = d_\nu$ for some $\nu \in I$ and so $\nu \in A$, A is not empty.

(ii) If $A \in D_0$ and $B \in D_0$ then $A \cap B \in D_0$.

Indeed if $A \in D_0$ and $B \in D_0$ then $A = d_\nu$, $B = d_\mu$ for certain elements ν and μ of I. Let $\sigma = \nu \cup \mu$. Then σ is a finite subset of K and hence, is an element of I. We claim that $d_\sigma = A \cap B$. For if $\varrho \in A \cap B$ then $\varrho \supset \nu$ and $\varrho \supset \mu$ and so $\varrho \supset \mu \cup \nu$. Thus $\varrho \in d_\sigma$. While $\varrho \in d_\sigma$ implies $\varrho \supset \sigma$, $\varrho \supset \nu$, $\varrho \supset \mu$, hence $\varrho \in A$, $\varrho \in B$, $\varrho \in A \cap B$. This shows that $A \cap B = d_\sigma$ and so $A \cap B \in I$.

Consider the sets of subsets of I which are extensions of D_0 and which satisfy conditions (i) and (ii) above. Among these sets there are maximal ones, by Zorn's lemma. Let D be one of them. We are going to show that, in addition to (i) and (ii) above, D possesses also the following properties.

(iii) If $A \in D$ and $A \subset B \subset I$ then $B \in D$.

For let $D' = \{B | B \subset I$ and $A \subset B$ for some $A \in D\}$. Thus, D' consists of all subsets of I which contain elements of D as subsets. Hence, $D \subset D'$. Also, if $B \in D'$ then $B \neq \emptyset$ since B contains a subset A which belongs to D and hence, is not empty. Finally if B and B' belong to D' then $B \supset A, B' \supset A'$ for some A and A' in D. Hence, $B \cap B' \supset A \cap A'$, where $A \cap A' \in D$, by (ii) above. But then $B \cap B' \in D'$. This shows that D' is a set of subsets of I which satisfies conditions (i) and (ii) above. But D is a maximal set of this kind and so $D = D'$. Now it follows immediately from the definition of D' that if $A \in D'$ and $A \subset B \subset I$ then $B \in D'$. This proves (iii) since $D = D'$.

(iv) If $A \subset I$ then either $A \in D$ or $I - A \in D$.

Observe that A and $I - A$ cannot both belong to D since their intersection is empty. Suppose that, for some $A \subset I$, neither A nor $I - A$ belongs to D. Define D_1 by

$$D_1 = \{B | B = C \cap F \text{ where } A \subset C \subset I \text{ and } F \in D\}.$$

Then $D \subset D_1$ for if $F \in D$ then $B = F$ for $C = I$. Also, if $B \in D_1$ and $B' \in D_1$ so that $B = C \cap F$, $B' = C' \cap F'$ for some C, F, C', F' satisfying $A \subset C \subset I$, $F \in D$, $A \subset C' \subset I$, $F' \in D$, then $B \cap B' = (C \cap C') \cap (F \cap F')$ where $A \subset C \cap C' \subset I$ and $F \cap F' \in D$, so that $B \cap B'$ also belongs to D_1. Moreover $A \in D_1$ as we see by taking $C = A$, $F = I$ in the definition of D_1. Thus, D_1 is a proper extension of D which satisfies (ii) above. It then follows from the maximum property of D that D_1 cannot satisfy (i). Accordingly $\emptyset \in D_1$, and there exists a set $F_1 \in D$ such that $A \cap F_1 = \emptyset$. Similarly, defining D_2 by

$$D_2 = \{B | B = C \cap F \text{ where } I - A \subset C \subset I \text{ and } F \in D\} \text{ and arguing as before,}$$

we find that there exists an $F_2 \in D$ such that $(I - A) \cap F_2 = \emptyset$. Now let $F_3 = F_1 \cap F_2$, so that $F_3 \in D$. Then

$$F_3 = I \cap F_3 = (A \cup (I - A)) \cap F_3 = (A \cap F_3) \cup ((I - A) \cap F_3)$$
$$= (A \cap (F_1 \cap F_2)) \cup ((I - A) \cap (F_1 \cap F_2)) \subset (A \cap F_1) \cup ((I - A) \cap F_2) = \emptyset.$$

But this is impossible, by (i). This proves (iv).

Comparing the conditions (i) to (iv) in this section with conditions (i) to (iv) in section 2.3 above, we see that D is an *ultrafilter* on I. We are going to show that the ultraproduct Q_D, $Q = \{M_v\}$ is a model of K for an appropriate interpretation C_K of the individual constants and relation symbols of K. Recall that we have identified the relations in the different M_v which correspond to a given relation symbol R in K. Thus we consider that R denotes a relation R' which occurs in all the M_v. By the definition of an ultraproduct R' then occurs also in Q_D and we define that it correspond to R under C_K. As for the interpretation of the individual constants of K under C_K, we proceed as follows. Let a be an individual constant which occurs in K, i.e., which occurs in at least one sentence of K. We have to determine an element of Q_D which corresponds to K, i.e., we have to define an appropriate function $f(v)$ on I such that $f(v) \in M_v$. Now if some sentence of v contains a then M_v contains an individual a' which corresponds to a under C_v. We then put $f(v) = a'$. In the alternative case, if a does not occur in v, we choose $f(v)$ as an arbitrary element of M_v.

Now let X be any sentence of K. X is in prenex normal form, by assumption. We shall suppose that it is exemplified by 2.5.2 above. We have to show that X holds in Q_D or, equivalently, that the Skolem functions $\phi(x), \psi(x,z), \chi(x,z), \varrho(x,z,w)$ can be defined in Q_D in such a way that 2.5.3 holds in Q_D. For this purpose, we may suppose that we have already chosen appropriate Skolem functions $\phi, \psi, \chi, \varrho$ in all the M_v for which $X \in v$, and that these functions have been defined arbitrarily in the remaining M_v (subject only to the condition that the functional values are in the same structures M_v as the arguments, respectively).

Let f, g, and h be any elements of Q_D. Thus $f = f(v)$, $g = g(v)$, $h = h(v)$ where v ranges over I. We then define $k = \psi(f,g)$ by

2.5.5 $k(v) = \psi(f(v), g(v))$ for all $v \in I$.

Then $k \in Q_D$ as required. We define the functions ϕ, χ, and ϱ in Q_D in a similar way.

We are now going to show that, with these definitions, the sentence

2.5.6 $R(f, \phi(f), g, \psi(f,g), \chi(f,g), h, \varrho(f,g,h))$

(compare 2.5.4) holds in Q_D. R does not contain any quantifiers. By 2.2. (i) and 2.2 (ii), the question whether or not 2.5.6 holds in Q_D is determined entirely by the *truth values* ('holds in Q_D' or 'does not hold in Q_D') of the atomic formulae which occur in 2.5.6. Let these atomic formulae be $R_i(f_1^i, \ldots f_{k_i}^i)$, $i = 1, \ldots, j$ where f_l^i stands for one of the individual constants which occur in X, or for one of f, g, h, or for one of the functional values $\phi(f)$, $\psi(f,g)$, etc. By the definition of Q_D, $R_i(f_1^i, \ldots, f_{k_i}^i)$ holds in Q_D if and only if the set

2.5.7 $A_i = \{v \mid R_i(f_1^i(v), f_2^i(v), \ldots, f_{k_i}^i(v))$ holds in $M_v\}$

belongs to D. Let $A_i' = A_i$ if A_i belongs to D and $A_i' = I - A_i$ if A_i does not belong to D and hence $I - A_i$ belongs to D. Then A_i' belongs to D for $i = 1, \ldots, j$ and so the set $A = A_1' \cap A_2' \cap \ldots \cap A_j'$ belongs to D and is not empty. Also, for every $v \in A$, an atomic formula $R_i(f_1^i, \ldots, f_{k_i}^i)$ holds in Q_D if and only if $R_i(f_1^i(v), \ldots, f_{k_i}^i(v))$ holds in M_v.

On the other hand, let B be the subset of I which is defined by

$$B = \{v \mid R(f(v), \phi(f(v)), g(v), \psi(f(v), g(v)), \chi(f(v), g(v)),$$
$$h(v), \varrho(f(v), g(v), h(v))) \text{ holds in } M_v\}.$$

Let $\lambda = \{X\}$, where we recall that X is the sentence given by 2.5.2, and let $\mu \in d_\lambda$, so that μ is any finite subset of K which contains X. Then X holds in M_μ, and so

$$R(f(\mu), \phi(f(\mu)), g(\mu), \psi(f(\mu), g(\mu)), \ldots, \varrho(f(\mu), g(\mu), h(\mu)))$$

holds in M_μ.

We conclude that μ belongs to B and hence that d_λ is a subset of B. It then follows from the definition of D that B belongs to D and hence that $A \cap B$ is not empty. If we now choose any element η of $A \cap B$ then, by the definition of A, $R_i(f_1^i, \ldots, f_{k_i}^i)$ holds in Q_D if and only if $R_i(f_1^i(\eta), \ldots, f_{k_i}^i(\eta))$ holds in M_η, $i = 1, \ldots, j$ and, by the definition of B,

$$R(f(\eta), \phi(f(\eta)), g(\eta), \psi(f(\eta), g(\eta)), \ldots, \varrho(f(\eta), g(\eta), h(\eta)))$$

holds in M_η. We conclude that $R(f, \phi(f), g, \psi(f,g), \ldots, \varrho(f,g,h))$ holds in Q_D, and hence, that X holds in Q_D, as required.

Notice that the proof procedure applies equally if $R(x,y,z,u,v,w,t)$ contains individual constants, or if X begins with an existential quantifier. Thus, if a is an individual constant which occurs in X, and $p(v)$ is the element of Q_D which is denoted by a, then $p(\mu)$ is the element of M_μ which is denoted by a for all $\mu \in d_\lambda$ (see above). Accordingly, we are able to show that X holds in Q_D under the specified correspondence in all cases. This completes the proof of Theorem 2.5.1.

2.6 Higher order structures and corresponding languages. The axiomatic systems for many algebraic concepts such as groups or fields are formulated in a natural way within a first order language such as described in the preceding sections. However, interesting parts of the *theory* of such a concept may well extend beyond the resources of a first order language. Thus, in the theory of groups, statements regarding subgroups, or regarding the existence of subgroups of certain types will, in general, involve quantification with respect to sets of individuals, and this cannot be expressed within a first order language. In other cases, e.g., in the theory of natural numbers or in the theory of real numbers, some of the axioms themselves are outside the language of the Lower Predicate Calculus. Thus, the axiom of induction involves 'all properties' or alternatively 'all subsets' of the natural numbers. The straightforward formulation of this phrase again requires quantification of sets. The same is true of Dedekind's axiom, which involves quantification with respect to Dedekind cuts, i.e., ordered pairs of sets (of real numbers). The following framework for higher order structures and higher order languages copes with these and similar cases. It is rather straightforward and suitable for our purposes.

Let A be an abstract set of specified cardinal. We consider, in addition to A, all sets and n-ary relations of A (where we may regard sets as singulary relations), all sets of relations, relations between relations and between individuals and relations, etc. As in the Lower Predicate Calculus, we do not introduce symbols for functions since a function $y = f(x_1,...,x_n)$ can always be represented by a relation $R(x_1,...,x_n,y)$ such that $R(x_1,...,x_n,y)$ holds if and only if y is equal to $f(x_1,...,x_n)$.

Notice that in the above enumeration of sets and relations we have excluded 'abnormal' entities, such as the sets which contain both individuals and sets of individuals. More specifically, we shall consider only sets or relations of definite *type* τ. *Types* are entities which are built up in a finite number of steps, beginning with an arbitrarily chosen object 0 (which will be identified with the natural number 0), according to the following rules.

(i) 0 is a type.

(ii) If n is a positive integer and $\tau_1,\ldots,\tau_n$ are types then the sequence $(\tau_1,\ldots,\tau_n)$ is a type.

The set of all types will be denoted by T. Evidently, the number of types is countable.

We now assign types to all individuals, sets, and relations, which will be included in our considerations (and at the same time exclude the 'abnormal' entities mentioned above) by means of the following definitions.

The entities of type 0 are the individuals of A, and we set $A_0 = A$. We shall refer to the individuals also as *relations* of type 0. An n-ary relation whose arguments are of type 0, is of type $(0,\ldots,0)$ – where 0 appears n times –, $n \geqslant 1$. For any fixed n, the set of these relations will be denoted by $A_{(0,\ldots,0)}$. Suppose that we have already defined the relations of types $\tau_1,\ldots,\tau_n$, (including the possibility $\tau_i = 0$) and let $A_{\tau_i}, i = 1,\ldots,n$ be the sets of these relations. We then assign the type $\tau = (\tau_1,\ldots,\tau_n)$ to any n-ary relation whose ith argument varies over relations of type τ_i, $i = 1,\ldots,n$. Thus, an n-ary relation of type $\tau = (\tau_1,\ldots,\tau_n)$ is a subset of the cartesian product $A_{\tau_1} \times A_{\tau_2} \times \ldots \times A_{\tau_n}$. The set of these relations will be denoted by A_τ, or alternatively, by $(A_{\tau_1},\ldots,A_{\tau_n})$. Thus, the set A_τ itself is of type (τ). We adopt the convention that for each $\tau \neq 0$ there is a distinct empty relation of that type. The relations introduced above satisfy the restriction of the simple theory of types.

Let S be any set. Informally, we may regard a function $\sigma(v)$ from an index set $N = \{v\}$ into S as a *subset of S with repetitions* if we count only the *number of times* that an element of S appears as a functional value in $\sigma(v)$. Or to put it more accurately, we consider functions $\sigma(v)$ as above and we regard two such functions as equivalent, $\sigma(v) \sim \sigma'(v')$, $v \in N$, $v' \in N'$ where N and N' are any two index sets if there is a one-to-one mapping from N onto N', $v' = \phi(v)$ such that $\sigma(v) = \sigma'(\phi(v))$ for all $v \in N$. Then a *subset S' of S with repetitions*, (write $S' < S$) is, by definition, an equivalence class with respect to the relation $\sim$ just defined. If $a \in S$ appears as a functional value of $\sigma(v)$ in any one of the representation of S' we shall permit ourselves to write $a \in S'$ and say that a is *contained* in S', and if every element of S appears in S' just once we shall write, somewhat loosely, $S' = S$.

By a *higher order structure* M we mean any set $\{B_\tau\}$ which is indexed in T (in other words, a function which is defined on T) such that for some non-empty set of individuals A, every B_τ is a subset of A_τ with repetitions, but $B_0 = A_0 = A$. (Thus, every element of A_0 appears in a function $\sigma(v)$ which represents B_0 just once, as explained above). In addition, $M = \{B_\tau\}$ is supposed to satisfy the following condition. For any type $\tau \neq 0$, $\tau = (\tau_1,...,\tau_n)$, if $R \in B_\tau$ and the n-tuple $(R_1,...,R_n)$ satisfies R then $R_i \in B_{\tau_i}$. Thus, every $R \in B_\tau$, $\tau = (\tau_1,...,\tau_n)$ which from the set-theoretical point of view is given by a subset of the cartesian product $A_{\tau_1} \times A_{\tau_2} \times ... \times A_{\tau_n}$, coincides with its restriction to $B_{\tau_1} \times B_{\tau_2} \times ... \times B_{\tau_n}$. In particular, if $\tau = (\tau_1)$ so that the elements of B_τ are singulary relations, i.e., sets, then our condition states that all the elements of these sets are contained in B_{τ_1}.

If, for every type τ, B_τ contains all relations of A_τ then we shall say that M is *full*. If no B_τ contains any relation repeatedly (so that the corresponding $\sigma(v)$ takes each value at most once) then we say that the structure is *normal*. Putting it loosely, we may say that a structure is full and normal if $B_\tau = A_\tau$ for all τ.

In order to simplify our language we shall from now on regard individuals as relations of type 0.

There are various formalisms for the description of higher order structures. In the classical language of Principia Mathematica, or in the simpler version given by Hilbert-Ackermann, the notation is chosen so as to reflect the distinction between individuals and relations. Here we shall find it more convenient to adopt a language which, syntactically, belongs to the family of languages introduced in section 2.2 above. This language, to be denoted by Λ, contains a relation symbol $\Phi_\tau(\ ,...,)$ for each type $\tau \neq 0$ and if $\tau = (\tau_1,...,\tau_n)$ then Φ_τ is supposed to be $(n+1)$-ary. No other relation symbols are included in Λ. Because of their changed interpretation (see below) we shall refer to the individual object symbols or individual constants simply as *object symbols or constants*. In any given situation, we shall suppose that a sufficient supply of these is available in our language.

Among the well-formed formula of Λ we distinguish a subclass, whose elements will be called *stratified formulae*. For any $\Phi_{\tau_0}(x, y_1,...,y_n)$ $\tau_0 = (\tau_1,...,\tau_n)$, we call τ_i the type of the $(i+1)$th place of Φ_{τ_0}, $i = 0, 1,...,n$. Then a wff X will be said to be stratified, if any given constant or variable occurs in X only at places of the same type. Thus, the wff $[\Phi_{(0)}(x,y)]$ is stratified, but $(\forall x)[\neg[\Phi_{(0)}(x,x)]]$ is not stratified. The sentence

$[(\exists x)[\varPhi_{(0)}(a,x)]] \wedge [(\forall x)\varPhi_{(0)}(x,b)]]$ is not stratified, according to our definition. Our condition of stratification might be relaxed so as to admit the last formula but this will not be necessary for our purpose.

For a given higher order structure $M = \{B_\tau\}$, there is supposed to be specified a one-to-one mapping C from a set of constants (object symbols) of $\varLambda$ onto the totality of individuals *and* relations of M (elements of the various B_τ). We shall say, as before, that a constant of $\varLambda$ denotes (or, *is the name of*) the corresponding individual or relation of M under C. Let X be a stratified sentence which is a bracketed atomic formula, $X = [\varPhi_\tau (a,b_1,\ldots,b_n)]$. We say that X is *admissible* in M (under C) if $a, b_1,\ldots,b_n$ are in the domain of C and denote relations (or individuals) $R, R_1 \ldots, R_n$ in M such that $R \in B_\tau$, $R_i \in B_{\tau_i}$, $i = 1,\ldots,n$. Under these conditions, we say that X is *true* in M or holds in M if $(R_1,\ldots, R_n)$ *satisfies* the relation R in M, i.e., in the set theoretic approach if the sequence $(R_1,\ldots, R_n)$ belongs to R. If X is admissible but not true in M then we say that X is *false* in M.

More generally if X is any stratified sentence then we say that X is admissible in M (under C) if every constant a which occurs in X is mapped under C on a relation R such that the type of R is also the (joint) type of the places in which a occurs in X.

We now consider the set K_0 of all stratified sentences of $\varLambda$ which are admissible in M under a given correspondence C. We have already defined under what conditions a bracketed atomic formula is said to hold in M. For a sentence X which is obtained by the application of a connective, we define whether or not X holds in M as in 2.2(ii) above, where we observe that if $X = [\neg Y]$ and X is admissible then Y is admissible and if $X = [Y \cup Z]$, etc., and X is admissible then Y and Z are admissible.

Suppose now that X is a stratified sentence of the form $X = [(\exists y)Z]$, which is admissible in M under C. If y does not occur in Z at all then Z is a stratified sentence which is admissible in M and we then say that X is true or false in M accordingly as Z is true or false in M. If Z contains y, $Z = Z(y)$, let τ be the (joint) type of the places in which y occurs within (the atomic formulae of) $Z(y)$. We define that X holds in M if and only if there is a constant a in $\varLambda$ which is mapped by C on a relation R of type τ in M such that $Z(a)$ holds in M. (Notice that at any rate $Z(a)$ is admissible in M if the type of R is the type of the places of y.) Similarly, if $X = [(\forall y)Z]$ and y does not occur in Z then X is said to hold in M, by definition if and only if Z holds in M. And if y occurs in Z, $Z = Z(y)$, then X is said to hold in M, by definition, if and only if $Z(a)$ holds in M for *all* constants a such that the type of the corresponding relation R in M is also the type of the places of y in $Z(y)$.

We see (as in section 2.2) that the decision whether or not a sentence X is admissible in M and whether or not X holds in M depends only on the restriction of the mapping C to the constants of X. Or, putting it in a more accurate way, suppose we are given a one-to-one mapping C_X from the constants of a stratified sentence X onto a set of relations (including individuals) of M such that the type of the place in which any constant appears in X is also the type of the relation on which it is mapped. If then we extend C_X to a one-to-one correspondence C between constants and relations whose range consists of all relations of M then X is admissible in M under C and the question whether or not X holds in M depends only on C_X and not on the particular choice of the extension C.

Now let K be a set of sentences in Λ such that (i) the elements of K are stratified and (ii) for any constant a which occurs in K, the type of the places in which a appears is the same throughout K (and will be called *the type of a in K*). We shall then say that K is stratified.

For a stratified K as described, let C_K be a one-to-one mapping from the set of constants which occur in K onto a set of relations in M such that the type of any constant which occurs in K coincides with the type of the relation on which it is mapped. Supposing that Λ contains enough constants (as is usually assumed implicitly) we may then extend C_K to a mapping C from a set of constants in Λ to all relations of M (including individuals, as before) such that all sentences of K are admissible in M under C. Whether any particular sentence of K holds or does not hold in M is then again dependent only on C_K.

2.7 Type symbols. Given a language Λ as in the preceding section, we extend Λ to a language Λ' by adding a set of singularly (one-place) relation symbols $\Theta_\tau(\ )$ where τ varies over all types. The Θ_τ will be called *type symbols*.

We map the stratified wff of Λ on certain wff of Λ' by means of a function $X = \lambda(X)$ which is defined inductively, following the construction of a wff in Λ, as follows.

(i) If X is a bracketed atomic formula then $\lambda(X) = X$.

(ii) If $X = [\neg Y], \lambda(X) = [\neg\lambda(Y)]$; if $X = [Y \vee Z]$,

$\lambda(X) = [\lambda(Y) \vee \lambda(Z)]$; if $X = [Y \wedge Z], \lambda(X) = [\lambda(Y) \wedge \lambda(Z)]$, and so on, for the remaining connectives.

(iii) If $X = [(\exists y) Z]$, $\lambda(X) = [(\exists y)[\Theta_\tau(y) \wedge \lambda(Z)]]$ where τ is the joint type of the places at which y occurs in Z. If y does not occur in Z at all then we put $\tau = 0$.

If $X = [(\forall y) Z]$, $\lambda(X) = [(\forall y)[\Theta_\tau(y) \supset \lambda(Z)]$ where, again, τ is the joint type of the places at which y occurs in Z, and $\tau = 0$ if y does not occur in Z at all.

We might simplify the definition under (iii) by ruling out empty quantifications.

$X_\lambda = \lambda(X)$ will be called the *type transform* of X.

Let M be a higher order structure. A first order structure M_λ will be associated with M in the following way. The individuals of M_λ are the individuals and relations of M. The relations of M_λ consist of a set of singulary relations $S_\tau(x)$ where τ ranges over the set of types, T, together with a set $F_\tau(x, y_1, \ldots, y_n)$ where τ ranges over $T - \{0\}$ and F_τ has the same number of variables as Φ_τ. If R is an individual of M_λ i.e., an individual or relation of M, then we define that $S_\tau(R)$ holds in M_λ if and only if τ is the type of R in M. And if F_τ is $(n+1)$-ary and $\tau = (\tau_1, \ldots, \tau_n)$ and $R, R_1, \ldots, R_n$ are individuals of M_λ then we define that $F_\tau(R, R_1, \ldots, R_n)$ holds in M_λ if and only if τ is the type of R and $\tau_1, \ldots, \tau_n$ are the types of $R_1, \ldots, R_n$ respectively, and $(R_1, \ldots, R_n)$ satisfies the relation R.

For a given one-to-one mapping C from a set of constants of $\varLambda$ onto the individuals and relations of a higher order structure M, let C' be the following mapping from constants and relation symbols of $\varLambda'$ onto the individuals and relations of the corresponding M_λ. For the individuals of M_λ (which are the individuals and relations of M), C' coincides with C. And as for the relation symbols, Θ_τ is mapped on S_τ for all τ and Φ_τ is mapped on F_τ for all types $\tau \neq 0$.

In these circumstances we are going to prove the following theorem.

2.7.1 THEOREM.   Let X be a stratified sentence which is admissible in M under C. The X holds in M if and only if $X_\lambda = \lambda(X)$ holds in M_λ (under C').

The proof is by induction following the construction of the sentence X. We assume throughout that X is stratified and admissible in M.

Suppose first that X is a bracketed atomic formula, $X = [\Phi_\tau(a, b_1, \ldots, b_n)]$

where $\tau = (\tau_1,\ldots,\tau_n)$ and where $a,b_1,\ldots,b_n$ denote $R,R_1,\ldots,R_n$ respectively under C. In this case, X holds in M if and only if the n-tuple $(R_1,\ldots,R_n)$ satisfies the relation R in M. But $a,b_1,\ldots,b_n$ denote $R,R_1,\ldots,R_n$ also under C', where these relations are now taken as individuals of M_λ. Then X holds in M_λ if and only if $F_\tau(R,R_1,\ldots,R_n)$ holds in M_λ, i.e., if and only if $X = [\Phi_\tau(a,b_1,\ldots,b_n)]$ holds in M_λ. But, in the case under consideration, $X = X_\lambda$, and so X holds in M if and only if X_λ holds in M_λ, as required.

Suppose next that X has been obtained by the application of a connective. If $X = [\neg Y]$ then Y also is stratified and admissible in M. Moreover, $X_\lambda = [\neg Y_\lambda]$ and so X holds in M if and only if Y does not hold in M: X_λ holds in M_λ if and only if Y_λ does not hold in M_λ: and Y holds in M if and only if Y_λ holds in M_λ by the assumption of induction. Combining these we conclude that X holds in M if and only if X_λ holds in M_λ.

If $X = [Y \vee Z]$ then $X_\lambda = [Y_\lambda \vee Z_\lambda]$. In this case, Y and Z both are stratified and admissible. Also, if X holds in M, then at least one of the two sentences Y or Z holds in M and we may suppose without loss of generality that Y holds in M. Then Y_λ holds in M_λ by the assumption of induction, so X_λ holds in M_λ. Conversely, if X_λ holds in M_λ then we may suppose that Y_λ holds in M_λ. Hence Y holds in M by the assumption of induction and so X holds in M, as required. Similar procedures apply to the remaining connectives.

Lastly, suppose that X has been obtained by quantification. If $X = [(\exists y)Z]$ then we distinguish two cases. (i) y does not occur in Z. In that case X holds in M if and only if Z holds in M, and Z holds in M if and only if Z_λ holds in M_λ by the assumption of induction. Also, since $B_0 = A_0 = A$ is not empty, there exists a constant a such that $\Theta_0(a)$ holds in M_λ and hence, such that $\Theta_0(a) \wedge Z_\lambda$ holds in M_λ if and only if Z_λ holds in M_λ. Hence, if X holds in M then $\Theta_0(a) \wedge Z_\lambda$ holds in M_λ and so $X_\lambda = (\exists y)[\Theta_0(y) \wedge Z_\lambda]$ holds in M_λ. If X_λ does not hold in M_λ, then Z_λ cannot hold in M_λ and so X cannot hold in M. (ii) y occurs in Z, $Z = Z(y)$. In that case, X holds in M provided $Z(a)$ holds in M for some constant a such that $Z(a)$ is stratified and admissible in M. Hence, by the assumption of induction $\lambda(Z(a))$ holds in M_λ. But $\lambda(Z(a))$ can also be obtained by substituting a in $\lambda(Z(y))$ and since $Z(a)$ is stratified, $\Theta_\tau(a)$ holds in M_λ for the type τ of the places at which a occurs in $Z(a)$. It follows that $\Theta_\tau(a) \wedge \lambda(Z(a))$, and hence $X_\lambda = (\exists y)[\Theta_\tau(y) \wedge \lambda(Z)]$ holds in M_λ. On the other hand, if X_λ holds in M_λ, then $\Theta_\tau(a) \wedge \lambda(Z(a))$ holds in M_λ for some a. Now the fact that $\Theta_\tau(a)$ holds in M_λ shows that $Z(a)$ is stratified and admissible in M. Accordingly, $Z(a)$ is either true or false in M. If $Z(a)$ were false in M then, by the assumption of induction, $\lambda(Z(a))$ would be false in M_λ, contrary to assumption. We conclude that $Z(a)$ holds in M and hence that $X = [(\exists y) Z(y)]$ holds in M.

If $X = [(\forall y) Z]$ and y does not occur in Z then $X_\lambda = (\forall y)[\Theta_0(y) \supset Z_\lambda]$. Suppose that X holds in M, then Z holds in M hence, by the assumption of induction, Z_λ holds in M_λ, and so X_λ holds in M_λ. Conversely, if X_λ holds in M_λ then $\Theta_0(a) \supset Z_\lambda$ holds in M_λ for all a in the domain of C. Choosing an a which denotes an element of $A_0 = A$ – and there are such a – we have that $\Theta_0(a)$ holds in M_λ and may then conclude that Z_λ holds in M_λ. Hence, by the assumption of induction, Z holds in M and, furthermore, X holds in M.

If $X = [(\forall y) Z]$ and y occurs in Z, $Z = Z(y)$, then $X_\lambda = (\forall y)[\Theta_\tau(y) \supset \lambda(Z(y))]$. Suppose that X holds in M and let a be any constant in the domain of C' (or, which is the same, of C). If a denotes a relation of type $\tau' \neq \tau$ then $\Theta_\tau(a)$ does not hold in M_λ and so $\Theta_0(a) \supset \lambda(Z(a))$ holds in M_λ. If a denotes a relation of type τ then $Z(a)$ is stratified and admissible in M since τ is the type of the places in which a appears. Also, $Z(a)$ holds in M since X holds in M. Hence, by the assumption of induction, $\lambda(Z(a))$ holds in M_λ and so $\Theta_\tau(a) \supset \lambda(Z(a))$ holds in M_λ. We conclude that X_λ holds in M_λ.

Again, if $X = [(\forall y) Z]$ and y occurs in Z and $X_\lambda = (\forall y)[\Theta_\tau(y) \supset \lambda(Z(y))]$ holds in M_λ, let a be any constant which denotes a relation of type τ in M. Then $\Theta_\tau(a)$ holds in M_λ and so $\lambda(Z(a))$ holds in M_λ. But, by the definition of type transforms, τ is the type of the places of y in $Z(y)$ and so $Z(a)$ is stratified and admissible in M under C. Hence, by the assumption of induction, $Z(a)$ holds in M. Since a was arbitrary subject only to the condition that the relation denoted by a has the type of the places of y in $Z(y)$, i.e., subject to the condition that $Z(a)$ be admissible in M, we conclude that $[(\forall y) Z]$ holds in M. This completes the proof of 2.7.1.

2.8   Finiteness principle for higher order theories.   Let K be a set of sentences in a language Λ as defined in section 2.6. If there exists a higher order model M such that for some mapping C as above all sentences of K are admissible and hold in M, then we say that K is consistent in the sense of higher order theory. It will be our main purpose in this section to show that the finiteness principle, Theorem 2.5.1, is equally true for higher order theories. This is expressed by the following theorem in which consistency is to be interpreted in the sense of higher order theory.

2.8.1 THEOREM.   Let K be a set of sentences in a language Λ as introduced in section 2.6 above. Suppose that every finite subset of K is consistent. Then K is consistent.

We prove this theorem by reducing it to 2.5.1. As a first step, we extend the language Λ to a language Λ' as in section 2.7. We then introduce the set K_λ which is obtained from K by replacing the sentences of K by their type transforms. We claim that K_λ is consistent in the sense of first order theory.

Indeed, let H' be any finite subset of K_λ. Then there is a finite subset of K, which will be called K', such that H' is obtained from K' by replacing the sentences of K' by their type transforms. By 2.5.1 the consistency of K_λ will be established if we can show that H' is consistent.

Now, by assumption, K' is consistent in the sense of higher order theory and possesses a model M'. Define a corresponding first order structure M'_λ as in section 2.7, then the sentences of H' hold in M'_λ, by 2.7.1 and so H' is consistent, as required, showing that K_λ is consistent.

Moreover, for every H', K', M', and M'_λ as considered here, it is not difficult to verify that M'_λ satisfies also the following sentences (2.8.2–2.8.6).

2.8.2 $$(\exists x)\,\Theta_0(x).$$

2.8.3 $$(\forall x)\left[\,\neg\Theta_\tau(x)\vee\neg\Theta_\sigma(x)\right],$$

where τ and σ range over all types such that $\tau\neq\sigma$.

2.8.4 $$(\forall x)(\forall y_1)...(\forall y_n)\left[\Phi_\tau(x,y_1,...,y_n)\right.$$
$$\left.\supset\left[\Theta_\tau(x)\wedge\Theta_{\tau_1}(y_1)\wedge\cdots\wedge\Theta_{\tau_n}(y_n)\right]\right],$$

where τ ranges over all types except 0, $\tau=(\tau_1,...,\tau_n)$.

2.8.5 $$(\forall x)(\forall x_1)...(\forall x_{i-1})(\forall x'_i)(\forall x''_i)(\forall x_{i+1})...(\forall x_n)\left[\left[\Theta_{\tau_i}(x'_i)\right.\right.$$
$$\wedge\Theta_{\tau_i}(x''_i)\wedge\left[(\forall y_1)...(\forall y_k)\left[\Phi_{\tau_i}(x'_i,y_1,...,y_k)\right.\right.$$
$$\left.\left.\equiv\Phi_{\tau_i}(x''_i,y_1,...,y_k)\right]\right]\supset\left[\Phi_\tau(x,x_1,...,x_{i-1},x'_i,x_{i+1},...,x_n)\right.$$
$$\left.\left.\equiv\Phi_\tau(x,x_1,...,x_{i-1},x''_i,x_{i+1},...,x_n)\right]\right],$$

where $\tau = (\tau_1, ..., \tau_n)$, $\tau_i = (\sigma_1, ..., \sigma_k)$ for some i, $1 \leq i \leq n$. These sentences express conditions of extensionality. They hold in M'_λ because of the condition that the elements R_i of an n-tuple $(R_1, ..., R_n)$ which is contained in a relation of M are all contained in M (i.e., in the appropriate B_τ).

Finally, if a is any constant which occurs in K' and hence in H', M'_λ satisfies the sentence.

2.8.6 $$Z_a = \Theta_\tau(a)$$

where τ is the type of the places at which a occurs in K'.

Let H_0 be the set of all sentences enumerated under 2.8.2–2.8.5 and let H_c be the set of sentences 2.8.6 as a ranges over *all* constants of K. We claim that the set $K_\lambda \cup H_0 \cup H_c$ is consistent.

Indeed, let $\bar{H}$ be any finite subset of $K_\lambda \cup H_0 \cup H_c$. Then $\bar{H}$ may be written as $\bar{H} = H'_1 \cup H'_0 \cup H'_c$ where H'_1, H'_0, H'_c are finite and $H'_1 \subset K_\lambda$, $H'_0 \subset H_0$ and $H'_c \subset H_c$. We choose a finite subset of K_λ, H'_2 such that the sentences of H'_2 contain all the constants a which occur in H_c, then the set $H' = H'_1 \cup H'_2$ possesses a model M'_λ as above, and this structure satisfies the sentences of H_0 and, a fortiori, of H'_0 and at the same time satisfies all sentences Z_a for the constants a which occur in H', and in particular, the sentences of H'_c. Thus, M'_λ is a model of $\bar{H}$, the set $\bar{H}$ is consistent, and so the set $K_\lambda \cup H_0 \cup H_c$ is consistent, by 2.5.1. Let N be a model of $K_\lambda \cup H_0 \cup H_c$ in the sense of first order theory. From N, we define a higher order structure M in the following way.

Let A be the set of all individuals R of N such that $S_0(R)$ holds in N. A is not empty, for the sentence which is given by 2.8.2 asserts precisely the existence of such an R, and that sentence holds in N. For every type τ, define A_τ as the set of all relations of type τ which can be obtained beginning with A as set of individuals, as in section 2.6. Thus, $A_0 = A$, $A_{(0)}$ is the set of all subsets of A, $A_{(0,0)}$ is the set of all binary relations on A, etc. In order to define the higher order structure $M = \{B_\tau\}$, we identify the relations of M, of type τ, with the individuals of N which satisfy S_τ in M. Thus, if R is an individual of N then R is an element of B_τ if and only if $S_\tau(R)$ holds in N. This yields, in particular, $B_0 = A_0 = A$. To complete our definition, we still have to identify the elements of B_τ with specific elements of A_τ for $\tau \neq 0$, i.e., with specific set-theoretic relations of type τ. We do this by induction on τ, following the construction of τ. Thus, let $\tau = (\tau_1, ..., \tau_n)$ and suppose that we have already identified each of the elements of $B_{\tau_1},...,B_{\tau_n}$ with a relation in $A_{\tau_1},...,A_{\tau_n}$ respectively. Now let R be any element of B_τ then we identify R with the relation which is satisfied just by the n-tuples $R_1,...,R_n$ such that $R_i \in B_{\tau_i}$ and $F_\tau(R,R_1,...,R_n)$ holds in N. This determines the structure M. Observe that by 2.8.3 every R belongs to not more than one B_τ. Also, by 2.8.4, an element of B_τ is always identified with an element of A_τ, although there may well be individuals of N which do not satisfy any S_τ and so do not belong to M. Finally, if R_i, R'_i are elements of B_{τ_i} which have been identified with the same relation in A_{τ_i}, $i=1,...,n$, then $(R_1,...,R_n)$ and $(R'_1,...,R'_n)$ satisfy the same relation R in B_τ, $\tau=(\tau_1,...,\tau_n)$, by 2.8.5.

By construction, N is a model of $K_\lambda \cup H_0 \cup H_c$ under some one-to-one mapping C' under which Θ_τ corresponds to S_τ for all τ, Φ_τ corresponds to F_τ for $\tau \neq 0$, and a set of constants which includes the constants of K_λ corresponds to the individuals of N. We propose to show that M is a model of K in the sense of higher order theory under the mapping C under which any relation R of M is the image of the same constant as it is under C' when regarded as an individual of N. (Since we are now dealing with a higher order structure it is taken for granted that Φ_τ denotes F_τ for all $\tau \neq 0$).

Let J be the set of stratified sentences which are admissible in M under C. Then J includes K. Indeed, since the sentences of every finite subset of K are stratified, all sentences of K must be stratified. Also, if a constant a occurs in a sentence X of K at places of type τ, then $Z_a = \Theta_\tau(a)$ holds in N, since Z_a belongs to H_c. It follows that the relation R which is denoted by a belongs to B_τ and so X is admissible in M. This proves the assertion that $K \subset J$. Denoting by J_λ the set of type transforms of J, we conclude that $K_\lambda \subset J_\lambda$. In order to prove 2.8.1 it is now sufficient to show that, for any $X \in J$, X holds in M if and only if X_λ holds in N. For, since K_λ holds in N, we may then conclude that K holds in M, K is consistent.

The interdependence of the structures M and N is not quite the same as that of the structures M and M_λ in section 2.7 since N may contain individuals that do not become individuals or relations of M (i.e., those individuals of N that do not satisfy any S_τ). However, it is still true that an individual R of N is a relation of type τ in M if and only if $S_\tau(R)$ holds in N. And if $R, R_1,...,R_n$ are relations of types $\tau, \tau_1,...,\tau_n$ in M which are denoted by constants $a, b_1,...,b_n$, where $\tau=(\tau_1,...,\tau_n)$, then $\Phi_\tau(a,b_1,...,b_n)$ holds in M if and only if it holds in N. Using these facts we may prove, following the construction of the sentences $X \in J$ that X holds in M if and only if X_λ holds in N, almost exactly as we proved in 2.7 that a sentence X which is stratified and admissible in M, holds in M if and only if X_λ holds in M_λ. This completes the proof of 2.8.1.

It is important to realize that we have been able to prove the finiteness principle for higher order structures only by means of a deviation from what might be called the standard explication of that notion. Thus, in the standard explication, one would expect *all* subsets of the set of individuals to be present within the structure, and similarly, *all* relations, *all* relations between sets, etc. In other words, in a more intuitive interpretation of the notion of a higher order structure, we would expect the structure to be *full*. However, as we shall see in the sequel, other higher order structures also can be of great interest.

Another, less essential, point which should be kept in mind is that according to our definition a relation in A_τ may occur repeatedly in B_τ. The same possibility exists for first order structures. If one holds that any given relation should occur in a structure only once since it is a purely set theoretical, extensional, entity, one may modify our notions accordingly. It then becomes natural to assume that the mapping C is many-one i.e., that an individual or relation in a given structure may have several names.

Finally, we observe that except for notational differences we may regard a first order structure as a higher order structure $M = \{B_\tau\}$ in which the B_τ are empty unless $\tau = 0$ (for the individuals) or $\tau = (0,0,...,0)$ where 0 appears n times, $n = 1,2,...$ (for the n-ary relations).

2.9   Enlargements.   Let K and X be a set of sentences and a single sentence, respectively, in a first order language. Then we say that X is *defined* in K if every constant or relation symbol which occurs in X occurs also in K (i.e., in some sentence of K). We say that K is *contradictory*, or inconsistent, if K is not consistent and we say that X is *deducible* from K if $K \cup \{\neg X\}$ is contradictory, in symbols $K \vdash X$.

Similarly, consider a stratified set of sentences K and a stratified sentence X in a higher order language. Then X is *defined* in K if all constants that occur in X occur also in K, X is *admissible* in K if $K \cup \{X\}$ is stratified; and X is *deducible* from K if $K \cup \{\neg X\}$ is stratified and *contradictory* (i.e., not consistent, inconsistent), in symbols $K \vdash X$.

Let K be a stratified set of sentences and let Γ be the set of constants which occur in K. The *type of $b \in \Gamma$* is the type of the places at which b occurs within sentences of K. Suppose that the type of b is $\tau = (\tau_1, \tau_2)$ so that b denotes a binary relation (in any interpretation). Let Δ_b be the set of all elements g of Γ such that

2.9.1 $K \vdash (\exists y)\, \Phi_\tau(b, g, y)$.

(Thus, in any interpretation, the constants of Δ_b denote the elements of the domain of the first argument of the relation denoted by b.) b will be called *concurrent* if for every finite subset $\{g_1, \ldots, g_n\}$ of Δ_b,

2.9.2 $K \vdash (\exists y)\big[\Phi_\tau(b, g_1, y) \wedge \Phi_\tau(b, g_2, y) \wedge \cdots \wedge \Phi_\tau(b, g_n, y)\big]$.

Let Γ_0 be the set of concurrent elements of Γ. For every $b \in \Gamma_0$ we select a distinct constant a_b which does not belong to Γ and we assume as usual that there are enough constants available in the language for this purpose. Let Γ_b be the set of all these a_b.

For every $b \in \Gamma_0$, define K_b as the set of sentences

2.9.3 $X_{bg} = \Phi_\tau(b, g, a_b)$,

where g ranges over Δ_b. Let K_0 be the union of the sets K_b as b ranges over Γ_0. The set $H = K \cup K_0$ will be called *the enlargement* of K. Evidently H is unique except for 'typographical' variations (different choice of constants a_b).

To continue, we require some auxiliary considerations.

Let J be a set of sentences in a first order language and let G be the set of individual constants which occur in J. Let D be a mapping from G onto a set of individual constants G_1, and let J_1 be obtained by replacing the individual constants in the sentences of J by their images under D. Under these conditions we may show that –

2.9.4   THEOREM.   If J_1 is consistent then J is consistent.

Indeed, since J_1 is consistent there exists a structure M_1 which is a model of J_1 under a mapping E_1 under which the set P_1 of individuals of M_1 is the map of a set of constants that include G_1. We now define a structure M in the following way. (i) If a is an individual of P_1 which is not mapped on a constant of G_1 under E_1^{-1} then a belongs to the set of individuals of M, P. (ii) If $a \in P_1$ is mapped on any element $a' \in G_1$ under E_1^{-1}, let $\{a'_\nu\}$ be the set of constants which are mapped on a' by D. For every a'_v in that set we then introduce into P a distinct individual a_v. P shall be the totality of individuals defined in one of these two ways. Thus, to every element b of P corresponds a unique element a of P_1, where b is identical with a for an individual a which is not mapped on G_1, and a is mapped on the same $a' \in G_1$ by E_1^{-1} as a'_v is by D, where $b = a_v$, in the alternative case. Let this mapping be denoted by E_0. In general, E_0 is many-one.

The relations of M are the same as the relations of M_1 (but not in the extensional, set-theoretic sense, see the beginning of section 2.3) and are defined thus. If R is an n-ary relation in M, and $b_1,...,b_n$ are elements of P, let $a_1,...,a_n$ be the corresponding elements of P_n under the mapping E_0. We then define that $(b_1,...,b_n)$ satisfies R in M if and only if $(a_1,...,a_n)$ satisfies R in M_1. This completes the definition of M. We claim that M is a model of J under a mapping E which is defined as follows. E coincides with E_1 as far as relations are concerned as well as for individuals of M_1 which were introduced under (i). For individuals $b = a_v$ which were introduced under (ii), these are to be the images by E of the corresponding a'_v.

Let F be a mapping from individual constants to individual constants which is defined on the domain of E in the following way. If b' is mapped on b by E and b is mapped on a by E_0 and a' is mapped on a by E_1 then a' shall be the image of b' by F. In other words, F is the composition of the mappings E_1, E_0 and E_1^{-1}, $F = E_1^{-1} * E_0 * E$. It will be seen that D is the restriction of F to G.

F induces a mapping from the set of sentences which are defined in M under E to the set of sentences which are defined in M_1 under E_1, and we write $X_1 = F_1(X)$ where X_1 is obtained from X by replacing the constants of X by their images under F. Then J_1 is the image of J under F_1. We claim that any sentence X which is defined in M, holds in M under E if and only if $F_1(X)$ holds in M_1 under E_1. For bracketed atomic sentences, $X = [R(b'_1,...,b'_n)]$, this follows immediately from our definition of M and E and for all other sentences X which are defined in M under E it can then be verified by induction following the construction of these sentences (compare section 2.2). Since M_1 is a model of J_1, by assumption, we conclude that M is a model of J. This shows that J is consistent, as asserted by 2.9.4.

A corresponding result holds in higher order theory. Thus, if J is a set of sentences in higher order theory, and D is a many-one mapping from the set of constants G which occur in J, onto a set of constants G_1, and J_1 is defined from J by means of D as before, then if J_1 is stratified, so is J, and if J_1 is consistent, so is J. Accordingly, we shall suppose from now on that 2.8.4 represents the statement of this fact for both first and higher order theories.

We return to the notion of an enlargement H of a set of sentences K, $H = K \cup K_0$ where K_0 is the set of sentences given by 2.9.3.

2.9.5 THEOREM. Let K be a stratified set of sentences. If K is consistent then its enlargement, H, is consistent as well.

PROOF. By 2.8.1, we only have to show that every finite subset of H is consistent or, more strongly, that the union J of K and of any finite subset K_0' of K_0, $J = K \cup K_0'$, is consistent. Let

$$K_0' = \{\Phi_{\tau_1}(b_1, g_{11}, a_{b_1}), ..., \Phi_{\tau_1}(b_1, g_{1k_1}, a_{b_1}),$$
$$\Phi_{\tau_2}(b_2, g_{21}, a_{b_2}), ..., \Phi_{\tau_2}(b_2, g_{2k_2}, a_{b_2}),$$
$$.................................$$
$$\Phi_{\tau_n}(b_n, g_{n1}, a_{b_n}), ..., \Phi_{\tau_n}(b_n, g_{nk_n}, a_{b_n})\},$$

where the constants $b_1, b_2, ..., b_n$ are distinct. Let M be a model of K. By 2.9.2 there exist constants $a_1, ..., a_n$ (denoting certain individuals of M) such that the set of sentences

$$K_1' = \{\Phi_{\tau_1}(b_1, g_{11}, a_1), ..., \Phi_{\tau_1}(b_1, g_{1k_1}, a_1),$$
$$\Phi_{\tau_2}(b_2, g_{21}, a_2), ..., \Phi_{\tau_2}(b_2, g_{2k_2}, a_2),$$
$$\Phi_{\tau_n}(b_n, g_{n1}, a_n), ..., \Phi_{\tau_n}(b_n, g_{nk_n}, a_n)\}$$

holds in M. We put $J_1 = K \cup K_1'$ and apply 2.9.4, where G is the set of constants which occur in J, G_1 is the set $G \cup \{a_{b_1}, ..., a_{b_n}\}$ and is thus the set of constants which occur J_1 and the mapping D maps the elements of G on themselves (reduces to the identity on G) and maps a_{b_i} on a_i, $i = 1, ..., n$. Since K and K_1 hold in M simultaneously, J_1 holds in M and is therefore consistent. We conclude that J is consistent, proving 2.9.5.

Let B be a set of concurrent elements of Γ, so that B is a subset of Γ_0. Then the union of the sets K_b (see 2.9.3) as b ranges over B is a subset of K_0 which will be denoted by K_B. The set $H_B = K \cup K_B$ (uniquely determined by B except for typographical variations) is called the B-enlargement of K. Evidently, $H_B \subset H$, so that H_B is consistent if K is stratified and consistent. In particular $H_{\Gamma_0} = H$.

A model M of K will be called *a B-model* of K if for every element b of B, there exists a constant a such that the set of sentences

2.9.6　　　　　　　　　　　$X_{bg} = \Phi_\tau(b,g,a)$

holds in M as g ranges over Δ_b. Thus, a model of H_B is a B-model of K. The converse is not true since it is compatible with the notion of a B-model that the same (relation denoted by) a in 2.9.6 serves for different concurrent b.

2.10　Examples of enlargements.　We consider the following examples of B-enlargements.

2.10.1　Let A be a countable set of individuals. Then there are among the relations on A ternary relations S and P, both of type $(0,0,0)$ which, as the relations of addition and multiplication, together turn A into the system of natural numbers. That is to say, both S and P determine functions, in the sense that for any given a and b in A there are unique individuals c and d in A such that the triples (a,b,c) and (a,b,d) satisfy S and P respectively; and if we write $c = a+b$, $d = ab$, then we obtain A as (a system isomorphic to) the system of real numbers. For simplicity, we shall take the relation of equality in the system to coincide with the relation of identity of type $(0,0)$. It is evident that we can choose the pair of relations S and P in many different ways. Supposing that we have made our choice, we denote S and P in the formal language by s and p respectively, and we denote the relation of identity of type $(0,0)$ on A by e.

We define the structure $M = \{B_\tau\}$ by $B_0 = A_0 = A$, as usual, by $B_\tau = A_\tau$ for $\tau = (0,0,...,0)$ – where 0 appears n times, $n = 1,2,3,...$ – and by $B_\tau = \emptyset$ for all other τ. Thus, S and P are included in $B_{(0,0,0)}$. Although we have presented M in the notation of higher order structures it contains only relations on individuals and with some slight conceptual modifications may therefore be regarded as a first order structure.

For a given one-to-one mapping from a set of constants of a language Λ onto the individuals and relations of M – including the mapping of s on S, p on P and e on the identity relation of type $(0,0)$ – let K be the set of stratified sentences of Λ which are admissible in M and hold in M. Consider the binary relation on A which in ordinary mathematical notation is expressed by $x < y$, in words 'y is greater than x'. This relation belongs to M since it is included in $B_{(0,0)}$. Let it be denoted by q (in the given mapping), so that $\Phi_{(0,0)}(q,x,y)$ means $x < y$. Then the domain of the first argument of q (compare 2.9.1) consists of all elements of A (i.e., of all natural numbers). Suppose that $g_1, \ldots, g_n$ denote elements of A (in the given mapping) then the sentence

$$X = (\exists y)\left[\Phi_{(0,0)}(q, g_1, y) \wedge \cdots \wedge \Phi_{(0,0)}(q, g_n, y)\right]$$

is true in M – since it asserts merely that there is a natural number which is greater than the natural numbers (denoted by) $g_1, \ldots, g_n$ – and so belongs to K. But if so then X is certainly deducible from K, i.e., 2.9.2 applies, the relation symbol g is concurrent for the specified K. Putting $B = \{q\}$, we shall call any B-model of K an *elementary* non-standard model of *Arithmetic*.

2.10.2 Let A be a set of individuals of cardinal $2^{\aleph_0}$, let S and P be ternary relations of type $(0,0,0)$ which as the relations of addition and multiplication turn A into the *field of real numbers*. Thus, given a and b in A there are unique elements c and d of A such that (a,b,c) and (a,b,d) satisfy S and P respectively and if we write $c = a + b$, $d = ab$ we now obtain A as the field of real numbers. We again take the relation of equality in the field to coincide with the relation of identity on A, to be denoted by e and define the structure M as in 2.10.1. Introducing K as the set of sentences which are admissible and hold in M, exactly as before and again taking $B = \{q\}$ where q denotes the relation of order in M, we call any B-model of K an *elementary* non-standard model of *Analysis*.

2.10.3 With A as in 2.10.1 and with the same interpretation for the relation symbols e, s, and p, choose M as a full structure, e.g., suppose that every relation of A_τ is represented just once in B_τ, briefly $B_\tau = A_\tau$ for all τ. (Recall that according to our definition of a higher order structure a relation of A_τ *may* be represented repeatedly in B_τ, so that the assumption that any relation of A_τ is represented just once in B_τ constitutes a restriction to a special case. We express this assumption, somewhat loosely, by $B_\tau = A_\tau$). With $B = \{q\}$, q as in 2.10.1, and with K as the set of all sentences which hold in M and which are formulated in terms of constants for *all* relations of M (individuals, first order *and higher order* relations) we shall call any B-model of K a *higher order* non-standard model of *Arithmetic*.

2.10.4 Similarly take A of cardinal $2^{\aleph_0}$ as in 2.10.2, and suppose that $M = \{B_\tau\}$ is a full structure, e.g., $B_\tau = A_\tau$ for all τ. Extending the set of constants used in K (the *vocabulary* of K) correspondingly while retaining the meaning of e,s,p,q, and $B = \{q\}$, we call any B-model of K a *higher order* non-standard model of *Analysis*.

In all these cases (2.10.1–2.10.4) the existence of a B-model is guaranteed by Theorem 2.9.5. It is evident, that if we ignore the higher order relations in a higher order non-standard model of Arithmetic (or, of Analysis) we obtain an elementary non-standard model of Arithmetic (or, of Analysis, respectively).

2.10.5 Finally, let A be a non-empty set of individuals of arbitrary cardinal, and let $M = \{B_\tau\}$, be a full structure with set of individuals $A = A_0$, e.g., $B_\tau = A_\tau$ for all τ. For a given one-to-one mapping from a set of constants of $\varLambda$ onto the set of relations of M, let K be the set of stratified sentences which are admissible and hold in M. Also, let $B = \varGamma_0$ be the set of all concurrent constants (relation symbols) for the given K. Theorem 2.9.5 shows that there exist higher order structures which are B-models of K. They will be called *enlargements* of M.

Let $*M$ be an enlargement of M and let $*A$ be its set of individuals. We propose to show that, in a sense to be made precise, we may regard $*A$ as an extension of A and $*M$ as an extension of M.

Let a be any element of A, and suppose that a is denoted in K by the constant a'. Since $*M$ is a model of K, a' denotes an individual or relation in $*M$, $*a$ say. However, our type restrictions ensure that $*a$ is actually an individual, i.e., an element of $*A$. For let b denote an arbitrary relation of type (0) in M. Then $\varPhi_{(0)}(a,b)$ is stratified and admissible in M and either $\varPhi_{(0)}(a,b)$ or $\lnot \varPhi_{(0)}(a,b)$ holds in M and hence, belongs to K. The fact that one of these sentences holds in $*M$ then shows that a denotes an element of $*A$ in $*M$, as asserted. The mapping $a \rightarrow {*a}$ from A into $*A$ which is obtained in this way must be one-to-one, since different elements of A are denoted by different constants of $\varLambda$ which in turn denote different constants of $*A$.

More generally, let $M = \{B_\tau\}$ and $*M = \{*B_\tau\}$. For any $R \in B_\tau$, let r be the constant of $\varLambda$ which denotes R (in the given mapping) and let $*R$ be denoted by r in the given mapping from constants of $\varLambda$ to relations of $*M$. The postulated type restrictions show again that $*R$ is of the same type as R, so that $*R\in *B_\tau$. The correspondence $\phi: R\to *R$ now yields a one-to-one mapping from B_τ into $*B_\tau$ for all τ (including $\tau=0$, the case considered separately above). Suppose now that $R\in B_\tau$, $\tau=(\tau_1,...,\tau_n)$ and that $R_i\in B_{\tau_i}$, $i=1,...,n$. Let $*R, *R_1,..., *R_n$ be the corresponding relations in $*M$ under the mapping ϕ just defined, so that $*R\in *B$, and $*R_i\in *B_{\tau_i}$, $i=1,...,n$. Then we claim that the n-tuple $(*R_1,...,*R_n)$ satisfies $*R$ in $*M$ if and only if $(R_1,...,R_n)$ satisfies R in M. Indeed, suppose that $R, R_1,..., R_n$ are denoted by $r,r_1,...,r_n$ in Λ, so that $r,r_1,...,r_n$ denote $*R, *R_1,...,*R_n$ in the other mapping. Then $X=\Phi_\tau(r,r_1,...,r_n)$ is a stratified sentence such X of $\neg X$ belongs to K according as $(R_1,...,R_n)$ does or does not satisfy R. But $*M$ is a model of K so $(*R_1,...,*R_n)$ satisfies $*R$ if $X\in K$ and does not satisfy $*R$ if $\neg X\in K$. This proves our assertion.

We may regard $*M$ as an extension of M in the special sense that if we restrict the elements of $*B_\tau$ for all τ to the images B'_τ of B_τ under ϕ, then there results a structure $M'=\{B'_\tau\}$ which is isomorphic to $M=\{B'_\tau\}$ and which can therefore be replaced by M. However, it should be observed that, for $\tau\neq 0$, the transition from $*M=\{*B_\tau\}$ to $M'=\{B'_\tau\}$ may actually change the extensional meaning of some $*R\in *B_\tau$. Suppose for example, that $*R$ is the one place relation of type (0) which is satisfied by the elements of $*A=*B_0$ in $*M$. Then $*R$ belongs to $B'_{(0)}$ as well as to $*B_{(0)}$. Suppose, as may well be the case, that B'_0 is a proper subset of $*B_0$. Then $*R$, regarded as an element of M' is satisfied only by the elements of B'_0. Thus, in carrying out the restriction from $*M$ to M' we have actually changed the extensional meaning of $*R$. Or, to put it more accurately, on restricting $*M$ to M' we pass from a relation $*R$ in $*M$ to a corresponding, but different relation R' in M'.

Still considering the enlargements described under 2.10.5, let E be the identity relation of type $(0,0)$ on A. According to our conventions, E may occur repeatedly in $B_{(0,0)}$, but must occur in it at least once. Choosing one of these occurrences suppose that it is denoted by e in Λ and that e in turn denotes the relation $*E$ in $*M$. Then $*E$ is a relation of equivalence on the domain of individuals of $*M, *A$, since the properties of an equivalence relation on the domain of individuals can be expressed as sentences of K. For example, the fact that E satisfies the transitive law is expressed by the sentence

$$X=(\forall x)(\forall y)(\forall z)\left[\left[\Phi_{(0,0)}(e,x,y)\wedge \Phi_{(0,0)}(e,y,z)\right]\supset \Phi_{(0,0)}(e,x,z)\right].$$

Thus, X belongs to K and, accordingly, holds in $*M$. That is to say, $*E$ is transitive.

More generally, for any type τ let E_τ be the identity relation on A_τ, so that E_τ is of type (τ, τ). Suppose that a particular occurrence of E_τ is denoted by e_τ in Λ while e_τ denotes $*E_\tau$ in $*M$. Then $*E_\tau$ is an equivalence relation on the set of elements of type τ in $*M$, $*B_\tau$. This includes the previous example as a special case, for $E = E_0$, $e = e_0$, $*E = *E_0$.

The relations $*E_\tau$ satisfy also certain conditions of substitutivity, as follows. If $\tau = (\tau_1, \ldots, \tau_n)$, then since $e_\tau, e_{\tau_1}, \ldots, e_{\tau_n}$ denote relations of identity on $A_\tau, A_{\tau_1}, \ldots, A_{\tau_n}$, respectively, the sentence

2.10.6 $(\forall x)(\forall x')(\forall y_1)\ldots(\forall y_n)(\forall y_1')\ldots(\forall y_n')[[\Phi_{(\tau,\tau)}(e_\tau, x, x')$
$\wedge \Phi_{(\tau_1, \tau_1)}(e_{\tau_1}, y_1, y_1') \wedge \cdots \wedge \Phi_{(\tau_n, \tau_n)}(e_{\tau_n}, y_n, y_n')$
$\supset [\Phi_\tau(x, y_1, \ldots, y_n) \equiv \Phi_\tau(x', y_1', \ldots, y_n')]]$

holds in M and accordingly, belongs to K. It follows that the sentence holds also in $*M$. Thus, if the relations $*E_\tau, *E_{\tau_1}, \ldots, *E_{\tau_n}$, hold between pairs $(R, R'), (R_1, R_1'), \ldots, (R_n, R_n')$, then the n-tuple $(R_1, \ldots, R_n)$ satisfies R, if and only if $(R_1', \ldots, R_n')$ satisfies R'.

For $\tau \neq 0$, $\tau = (\tau_1, \ldots, \tau_n)$, the relations $*E_\tau$ also satisfy conditions of extensionality. Consider the sentence

2.10.7 $(\forall x)(\forall x')[[(\forall y_1)\ldots(\forall y_n)[\Phi_\tau(x, y_1, \ldots, y_n)$
$\equiv \Phi_\tau(x', y_n, \ldots, y_n)]] \supset \Phi_{(\tau, \tau)}(e_\tau, x, x')].$

This sentence holds in M and, accordingly, holds also in $*M$. Thus, the relations $*E_\tau$ possess the formal properties of identity relations.

For a given type τ, suppose that e_τ and e_τ' denote different occurrences of the identity relation on A in M. Let $*E_\tau$ and $*E_\tau'$ be the relations denoted by e_τ and e_τ' respectively in $*M$. Then we claim that $*E_\tau$ and $*E_\tau'$ are co-extensive, i.e., that they are occurrences of the same relation. Indeed, the following sentence holds in M and $*M$, as an immediate consequence of 2.10.6.

2.10.8 $(\forall y)(\forall y')[\Phi_{(\tau, \tau)}(e_\tau, y, y') \supset [\Phi_{(\tau, \tau)}(e_\tau', y, y) \equiv \Phi_{(\tau, \tau)}(e_\tau', y, y')]].$

At the same time, $(\forall y)\Phi_{(\tau, \tau)}(e_\tau', y, y)$ holds in M and $*M$, which shows, together with 2.10.8, that

$$(\forall y)(\forall y')[\Phi_{(\tau, \tau)}(e_\tau, y, y') \supset \Phi_{(\tau, \tau)}(e_\tau', y, y')]$$

holds in M and $*M$. Similarly,

$$(\forall y)(\forall y')\left[\Phi_{(\tau,\tau)}(e_\tau',y,y')\supset\Phi_{(\tau,\tau)}(e_\tau,y,y')\right]$$

holds in M and $*M$ and so, finally, does

2.10.9 $(\forall y)(\forall y')\left[\Phi_{(\tau,\tau)}(e_\tau,y,y')\equiv\Phi_{(\tau,\tau)}(e_\tau',y,y')\right].$

The interpretation of this sentence in $*M$ is that $*E$ and $*E'$ are co-extensive.

For $\tau\neq0$, the fact a sentence $\Phi_{(\tau,\tau)}(e_\tau,r,r')$ holds in $*M$ implies that the elements of B_τ denoted by r and r' respectively, are co-extensive. Conversely, if r and r' denote co-extensive relations in $*M$ then 2.10.7 shows that $\Phi_{(\tau,\tau)}(e_\tau,r,r')$ holds in $*M$, i.e., that the relation $*E_\tau$ is satisfied by the pair of relations denoted by r and r'. Thus, $*E_\tau$ is the relation of identity on B in the usual sense, provided it is understood that elements of B_τ, $\tau=(\tau_1,...,\tau_n)$ are regarded as co-extensive, if they are satisfied by the same n-tuples $(R_1,...,R_n)$ such that R_1 occurs in B_{τ_1}, R_2 occurs in $B_{\tau_2},...,$ and R_n occurs in B_{τ_n}.

For $\tau=0$, there is no condition of extensionality, and it is indeed possible that $\Phi_{(0,0)}(e_0,a,b)$ holds in $*M$ for distinct constants a and b which denote distinct individuals in $*A=*B_0$. This cannot happen if both a and b occur already in K; for in that case $\neg\Phi_{(0,)}(e_0,a,b)$ holds in M and hence, belongs to K and holds also in $*M$. However, it may happen if only one of a and b or none of these constants occurs in K. Thus, the equivalence classes into which $*A$ is divided by the equivalence relation $*E_0$ may contain more than one element. If we wish to discard such cases, we may proceed as follows. We amend $*A$ by deleting in each equivalence class with respect to $*E_0$ all but one element. Moreover, if the class in question contains an individual a which is denoted by a constant in K – and as we have seen there can be no more than one such individual – then we retain a. Let the set of surviving individuals be A' so that A' is a subset of $*A$. The restriction from $*A$ to A' induces, in the first place, a mapping ϕ from the n-ary relations of type $\tau=(0,0,...,0)$, $n>1$ on $*A$ to corresponding relations on A', where a relation R' on A' is obtained as the image of a relation $*R$ on $*A$ simply by restricting the domains of variation of the arguments of $*R$ to A'. For any $\tau=(0,0,...,0)$ this yields a mapping from $*B_\tau$ onto a set (with repetitions) B_τ' of relations on A'. It may well happen that different relations on $*A$ are mapped into the same relation on A', but in that case, we shall count the images separately in B'_τ. Or, to put it in terms of the definition of $*B'_\tau$ as an indexed set, if $*B = \{*R_v\}$ where v varies over some index set N, then $B'_\tau = \{R'_v\}$ is indexed in the same set N and $R'_v = \phi(*R_v)$ for all $v \in N$.

Similarly, we may map the higher order relations based on the set of individuals $*A$ by corresponding restrictions on higher order relations based on A'. This yields, for every type $\tau \neq 0$, a mapping of $*B_\tau$ on a set B'_τ such that $M' = \{B'_\tau\}$ is a higher order structure with set of individuals $A' = B'_0$. Beginning with the sentences of K which are bracketed atomic formulae, it is now not difficult to verify that M' is a model of K.

Now let b be concurrent with respect to K, $b \in \Gamma_0$, and suppose for the sake of simplicity that b denotes a relation of type $(0,0)$. By the construction of $*M$ there exists an individual in $*A$, denoted by a_b, such that the sentences $\Phi_{(0,0)}(b,g,a_b)$ hold in $*M$ for all $g \in \Delta_b$ (see 2.9.3). By the construction of M', there exists an individual in A', denoted by a' say, such that $\Phi_{(0,0)}(e_0,a_b,a')$ hold in $*M$. Then the sentences $\Phi_{(0,0)}(b,g,a')$ all hold in M'. Similar arguments apply for b of other types. We may therefore conclude that M' also is an enlargement of M.

Previously, we called a structure $M = \{B_\tau\}$ *normal* if there are no repetitions in any B_τ. Still dealing with the case 2.10.5, suppose that M is normal, and consider an enlargement $*M$ of M. The normality of M entails that the relations of identity E_τ in M are unique. Let $*E_\tau$ be the corresponding relations in $*M$, then the $*E_\tau$ are equivalence relations which satisfy conditions of substitutivity 2.10.6, and, except for $\tau = 0$, the condition of extensionality 2.10.7. We divide the (indexed) relations of $B\tau$ into equivalence classes with respect to the corresponding $*E_\tau$. Let $R, R_1, \ldots, R_n$ be relations of $*M$ which belong to the same equivalence classes as the relations $R', R'_1, \ldots, R'_n$, respectively. Then 2.10.6 shows that the n-tuple $(R_1, \ldots, R_n)$ satisfies R if and only if $(R'_1, \ldots, R'_n)$ satisfies R'. It follows that if we restrict every $*B_\tau$ to a set B'_τ by deleting in every equivalence class all relations but one then we obtain a normal structure $M' = \{B'_\tau\}$ which satisfies all sentences of K that do not contain any constants. In order to take care also of sentences of K that contain constants, let r and r' be two distinct constants which occur in K and denote relations R and R' of the same type τ in $*M$. Then the sentence $\Phi_{(\tau,\tau)}(e_\tau,r,r')$ cannot belong to K since e_τ denotes the relation of identity, E_τ, in M. We conclude that R and R' belong to different equivalence classes with respect to $*E_\tau$. Thus, on passing from M to M' we may retain in every equivalence class just the relation which is denoted by a constant in K, if such a relation happens to be contained in the equivalence class in question. With this special provision, M' becomes a model of K. Moreover, we only have to repeat a previous argument in order to see that M' is an enlargement of M. The relations $*E_\tau$ are retained in M' since they are denoted by constants e_τ which occur in K, and are now genuine relations of identity like the E_τ.

If A is countable then any enlargement $*M$ as defined in 2.10.5 may be regarded as a higher order non-standard model of Arithmetic, for an appropriate choice of the relations S and P (see 2.10.3). Thus, there exist higher order non-standard models of Arithmetic which are normal. Similarly, if we take A in 2.10.5 as a set of cardinal $2^{\aleph_0}$ then our construction shows that there exist higher order non-standard models of Analysis (see 2.10.4) which are normal. If, in a higher order non-standard model of Arithmetic (Analysis) we replace all B_τ for τ other than any $(0,0,...,0)$ by the corresponding empty set then we obtain a first order non-standard model of Arithmetic (Analysis) as defined in 2.10.1 (2.10.2). Thus there exist first order models of Arithmetic (Analysis) which are normal. However, although it is now apparent that the notions introduced in 2.10.1 to 2.10.4 can be subsumed under 2.10.5 we have introduced them separately because parts of the subsequent theory can be developed on the basis of 2.10.3, 2.10.4 or even on the basis of 2.10.1, 2.10.2, alone.

2.11 General properties of enlargements. Let M be the full and normal structure for a set of individuals A, and let $*M$ be a normal enlargement of M. Let $*A$ be the set of individuals of $*M$. In a sense explained previously, we may regard $*M$ as an extension of M (and A as a subset of $*A$) by identifying the individuals and relations of M with the individuals and relations of $*M$ that are denoted by the same constants in K, respectively. This embedding procedure (injection) preserves extensionality since the constants e_τ denote identity in both M and $*M$.

The notion of concurrency was introduced originally for constants (i.e., elements of the formal language) with respect to a given set of sentences K. In the present context, when K is the set of sentences satisfied by M, we shall transfer the notion to the relations of M. Thus, a binary relation R in M, of type $\tau = (\tau_1, \tau_2)$, will be called concurrent, if the constant b which denotes R is concurrent with respect to K. Or in more detail, R is concurrent if the following condition is satisfied. Let $R_1,\dots,R_n$ be any set of relations, all of type τ_1, such that for some set of relations $R'_1,\dots,R'_n$ the pairs $(R_1, R'_1),\dots,(R_n, R'_n)$ satisfy R. Then there exists a relation S such that the pairs (R_1,S), $(R_2,S),\dots,(R_n, S)$ all satisfy R.

Let $M=\{B_\tau\}$, ${}^*M=\{{}^*B_\tau\}$, $B_0=A$, ${}^*B_0={}^*A$, as before. We shall see presently that it is, in general, not true that ${}^*M$ is a full structure based on ${}^*A$. Thus, if ${}^*A_0={}^*A$ and, for $\tau\neq0$, ${}^*A_\tau$ is the set of relations of type τ based on ${}^*A$, as explained in section 2.6, then ${}^*B_\tau$ may well be a proper subset of ${}^*A_\tau$. We shall say that the relations which belong to ${}^*B_\tau$ are *internal* while the relations of ${}^*A_\tau-{}^*B_\tau$ are *external*. This terminology is relevant only for $\tau\neq0$ since ${}^*A_0={}^*B_0$, there are no external individuals. Any internal relation which belongs already to M will be called a *standard* relation, (or standard individual, or standard number, standard point, as the case may be). Thus a standard relation is a relation which is denoted by a constant of K.

A function of n variables will be represented in our formal system by an $(n+1)$-ary relation in which the first n arguments determine the $(n+1)$th argument uniquely. Thus, if the relation in question is of type

$$\tau=(\tau_1,\dots,\tau_{n+1})$$

and denoted by r then it represents a function if the sentence

$$(\forall x_1)\dots(\forall x_n)(\forall y_1)(\forall y_2)\big[\big[\Phi_\tau(r,x_1,\dots,x_n,y_1)\wedge$$
$$\Phi_\tau(r,x_1,\dots,x_n,y_2)\big]\supset\Phi_{\tau'}(e_{\tau_{n+1}},y_1,y_2)\big]$$

holds in the structure in question, where $\tau'=(\tau_{n+1},\tau_{n+1})$. Note that the relation in question determines the function together with a specific domain of definition, which accordingly has to be regarded as part of the notion. A function will be called a *standard* function, an *internal* function, or an *external* function, if the relation which determines it is standard, or internal, or external, respectively.

2.11.1 THEOREM. Let B be a non-empty set of sets in M, of type $\tau'=((\tau))$, where τ is arbitrary. Thus B is contained in $B_{\tau'}$. Let b be the constant in K which denotes B and let ${}^*B$ be the standard set which is denoted by b in ${}^*M$. Suppose that the intersection of every finite subset of B is non-empty. Then there exist an internal relation F, of type τ in ${}^*M$

(i.e., in B_τ) such that every standard set $*G$ which is contained in $*B$
contains F.

PROOF. Observe that any standard set $*G$, denoted by g in K, is contained in $*B$ if and only if the set G in M which is denoted by the same g,
is contained in B. For the sentence $\Phi_{\tau'}(b,g)$ does or does not belong to K
according as G is or is not contained in B. In the former case, the interpretation of the sentence in $*M$ states that $*G$ belongs to $*B$. In the latter
case, $\neg\Phi_{\tau'}(b,g)$ belongs to K and this means, for $*M$, that $*G$ is not contained in $*B$.

Now let R be the relation of M, of type $\mu = ((\tau),\tau)$, and denoted by r,
which is satisfied by a pair (G, S) if and only if G is of type (τ) and is contained in B, and S is of type τ and is contained in G. Then the assumptions
of 2.11.1 show that the relation R is concurrent. It follows that there is a
relation F in $*M$ such that $\Phi_\mu(r,g,f)$ holds in $*M$ for all g which denote
standard set $*G$ in $*B$, and for the constant f which denotes F. This shows
that F is contained in all standard sets $*G$ which belong to $*B$.

Given any relation or function in M, of arbitrary type $\tau \neq 0$, e.g., R, we
shall frequently indicate the *corresponding relation* in $*M$, i.e., the relation
which is denoted by the same constant in K, by affixing an asterisk as left
superscript, so – $*R$. However, in cases when there can be no misunderstanding we shall omit the asterisk, thus denoting the corresponding
entities by the same symbol, as is in fact done in the formal language Λ.

Let R be a set (singulary relation) in M, denoted by r in K, then for
every element S of R in M, $*S$ is an element of $*R$ in $*M$. For if s denotes S
in K then the fact that $\Phi_\tau(r,s)$ holds in M, for the appropriate τ, implies
that it holds also in $*M$, i.e., that $*S$ is an element of $*R$. Similarly, if S is
an individual or relation in M which is not an element of R then $*S$ is not
an element of $*R$.

We may ask under what conditions the set $*R$ contains in addition to
the standard relation $*S$ considered above also at least one internal
relation which is not a standard relation (i.e., if $*R$ is of type (0), at least
one individual which is not contained in A). The answer to this question is
given by

2.11.2 THEOREM. The set $*R$ contains an internal relation which is not
a standard relation, if and only if R is infinite.

Indeed, if R is empty then it is easy to see that $*R$ also is empty. If R contains precisely n elements, n a positive integer, let these be $S_1, \ldots, S_n$, denoted by $s_1, \ldots, s_n$ in K. Let the type of $S_1, \ldots, S_n$ be σ so that the type of R is (σ) and let $\tau = (\sigma, \sigma)$. Then the sentence

2.11.3 $(\forall x)[\Phi_{(\sigma)}(r,x) \supset [\Phi_\tau(e_\sigma, s_1, x) \vee \Phi_\tau(e_\sigma, s_2, x) \vee \cdots \vee \Phi_\tau(e_\sigma, s_n, x)]]$

(where e_σ denotes the identity relation on A_σ, as usual) holds in M and hence, belongs to K. It follows that 2.11.3 holds also in $*M$. But e_σ denotes a relation of identity also in $*M$, and so $*R$ cannot possess any elements other than $*S_1, *S_2, \ldots, *S_n$. This shows that the condition of the theorem is necessary.

Suppose now that R contains an infinite number of elements. Let Q be the binary relation in M which is satisfied by a pair (S, S') if and only if S is an element of R and S' is an element of R which is different from S (or, to put it in a different way (S, S') satisfies Q if both S and S' satisfy R and S' is different from S). Then it is not difficult to see that Q is concurrent. Accordingly, there is a relation F in $*M$ such that $(*S, F)$ satisfies $*R$ for all standard relations $*S$ in $*R$. Thus, F is an internal relation which belongs to $*R$ although it is different from all standard elements of $*R$. This proves that the condition of the theorem is also sufficient.

Now let A' be any non-empty subset of the set of individuals A on which the complete structure M is based and let a' be the constant of K which denotes A'. From among the n-ary relations of M, of types $\tau = (0, 0, \ldots, 0) - n$ zeros, $n = 1, 2, 3, \ldots$ — we select those which hold only for arguments in A', i.e., set-theoretically, whose n-tuples consist of elements of A'. Call the resulting sets of relations A'_τ. More generally, having determined the sets $A'_{\tau_1}, \ldots, A'_{\tau_n}$, for types $\tau_1, \ldots, \tau_n$, we define the set A'_τ, $\tau = (\tau_1, \ldots, \tau_n)$ as the set of relations which are satisfied only by n-tuples $(R_1, \ldots, R_n)$ such that R_i belongs to A'_{τ_i}, $i = 1, \ldots, n$. It is not difficult to see that $M' = \{B'_\tau\}$, $B'_\tau = A'_\tau$ for all types τ is then a full extensional structure with set of individuals $B'_0 = A'_0$. M' is taken to be normal, by definition.

Passing to the structure $*M$, which is an enlargement of M, let $*A'$ be the set of individuals of $*M$ which is denoted by a' in K, in agreement with the notation introduced previously. From among the n-ary relations of $*M$ of types $\tau = (0, 0, \ldots, 0)$, i.e., among the elements of $*B_\tau$ we now select those whose n-tuples $(R_1, \ldots, R_n)$ consist of elements of $*A'$, and we denote the resulting subsets of the $*B_\tau$ by $*B'_\tau$. This notation would seem to indicate that $*B'$ is an internal set which is denoted by the same constant in K as B'_τ. Some reflection shows that this is indeed the case, for the defining property of every such B'_τ can be represented as a sentence of K. In general, we find that the sets $*B'_\tau$, which are denoted by the same constants in K as the sets B'_τ, respectively, are characterized by the property that if $\tau = (\tau_1, \ldots, \tau_n)$ then $*B'_\tau$ consists of all relations of $*B_\tau$ which are satisfied only by n-tuples $(R_1, \ldots, R_n)$ such that $R_i \in *B'_{\tau_i}$, $i = 1, \ldots, n$. We put $*M' = \{*B'_\tau\}$ where $*M'$ shall be normal by construction.

2.11.4 THEOREM. $*M'$ is an enlargement of M'.

PROOF. Let K' be the set of sentences of Λ which are admissible and hold in M', if we use the specified mapping C between the constants K and the relations of M and if we disregard the remaining relations of M and the constants denoting them. Let X be any sentence of K'; then we have to show that X holds in $*M'$. Notice that, in general, X will not hold in M. For example, if a'_0 denotes $A' = A'_0$ then $(\forall x)\Phi_{(0)}(a'_0, x)$ holds in M' but not in M. Nevertheless we can determine a sentence $Y \in K$ which expresses the fact that X holds in M' as a proposition about M. The procedure by means of which this can be achieved involves the formalization of the phrase 'which belongs to M'' in the context of quantification. Thus, let a'_τ be the constants that denote the sets A'_τ (regarded as elements of $A_{(\tau)}$) respectively. We then define a mapping μ from stratified well-formed formulae of Λ into such formulae by means of the following stipulations.

(i) $\mu(X) = X$ for (bracketed) atomic formulae;

(ii) $\mu(\neg X) = \neg \mu(X)$; $\mu(X \wedge Y) = \mu(X) \wedge \mu(Y)$; etc., for the remaining connectives;

(iii) $\mu((\exists y)Z(y)) = (\exists y)[\Phi_{(\tau)}(a'_\tau, y) \wedge \mu(Z(y))]$
        $\mu((\forall y)Z(y)) = (\forall y)[\Phi_{(\tau)}(a'_\tau, y) \supset \mu(Z(y))]$

where τ is the type of the places of y if y occurs in Z and $\tau = 0$ otherwise.

Considering the set H of all sentences X which are admissible in M' under the mapping defined above and proceeding by induction following the construction of a well formed formula, it is now not difficult to show that any $X \in H$ holds in M' if and only if $\mu(X)$ holds in M. Now if K' is the set of all $X \in H$ which hold in M', as introduced above, then K'_μ holds in M where K'_μ is the image of K' by the mapping μ. It follows that K' holds also in *M. At this point, we show (similarly as before) that a sentence $X \in H$ holds in *M' if and only if $\mu(X)$ holds in *M. From this fact we may then conclude that K' holds in *M'.

Now let b be a concurrent relation symbol with respect to K'. Thus, b denotes a concurrent relation R in M'. The same relation is concurrent also in M. Moreover, R is satisfied only by pairs (R_1, R_2) sucht hat R_1 and R_2 belong to sets A'_{τ_1}, A'_{τ_2} for the appropriate types τ_1 and τ_2, and this fact can be expressed as a sentence X of K. We conclude that there is in *M a relation S such that *R is satisfied by all pairs (R_1, S) in which R_1 is in the domain of the first argument of R (i.e., such that for some R_2, (R_1, R_2) satisfies R). But X belongs to K and so is true in *M. This shows that S belongs to the appropriate *B'_τ, and so *R is concurrent in *M'. We conclude that *M' is an enlargement of M', as asserted. The reader who is familiar with the method of relativization of quantifiers in the Lower Predicate Calculus will appreciate its connection with the above procedure. However, in the present case the situation is complicated by the fact that there is in M no single set which consists of the individuals and relations of M, because of the restrictions of type theory.

Applications of 2.11.4 arise naturally, for example in the following connection. If M represents the system of real numbers, including the higher order relations definable on it, then the system of rational numbers, M' (including its higher order relations) may be regarded as a *partial structure* of M as explained. 2.11.4 now tells us that as we pass to an enlargement *M of M, M' is carried into an enlargement *M' of M.

Two structures may be related to one another also in the following somewhat more general fashion. Let M be a full structure, and let A' be a set of relations, of specified type τ, aid denoted by a' in Λ, $A' \subset A_\tau$. We regard A' as the set of *individuals* of another full structure, M', and we obtain M' similarly as before, by selecting certain subsets of the various A_σ. Thus, as the n-ary relations of type $(0, 0, \ldots, 0)$ in M' we select the n-ary relations of type $(\tau, \tau, \ldots, \tau)$ in M which are satisfied only by n-tuples from A', and we then proceed to select n-ary relations of higher types just as in the previous case. And if *A' is the set of relations of *M which is denoted by the same constant as A', a', then the same procedure, applied to the elements of *A' as the individuals of a new structure, leads to a structure *M'. Theorem 2.11.4 still holds for this more general definition of M' and *M'. This can be shown by the method employed for the proof of the original theorem except that in passing from a sentence X which refers to M' or $*M'$ to the corresponding sentence Y which refers to M or $*M'$ we have to raise the types which appear as subscripts of the symbol 0 according to the rule $0 \to \tau$ and if $\tau_i \to \tau_i'$, $i = 1, \ldots, n$, then $(\tau_1, \ldots, \tau_n) \to (\tau_1', \ldots, \tau_n')$.

For example, M may represent the real number system, including relations of higher order, so that real functions are represented in M by relations of type $(0,0)$. Then the set of relations which represent functions of class L^2 on the interval $(-\infty, \infty)$, to be denoted by A', may be used as the set of individuals in a separable Hilbert space M'. Here again, it is important to know that as we pass to an enlargement $*M$ of M, M' is carried into an enlargement $*M'$ of M'.

This completes our exposition of the required logical setting. There is no single framework for first or higher order theories which is clearly superior to all others and one's preference may depend on the purpose that one has in mind or may be even only a matter of taste. Among the possible minor variations to the framework adopted here, we have already mentioned the ruling out of repetitions among the relations of a structure. In our terminology this amounts to the exclusion of all but normal structures and in actual fact we may confine ourselves to such structures from now on. On the contrary, it is also possible to permit repetitions among the individuals, i.e., to represent the set of individuals of a structure by a function that varies over some index set. Differing from our approach more radically, one might use axiomatic set theory rather than type theory for the development of higher order Non-standard Analysis. However, it will become apparent in due course that our methods are not affected by such variations.

Recall that we proved the finiteness principle for the Lower Predicate Calculus (2.5.1) by means of a particular ultraproduct, and we then reduced the finiteness principle for higher order theories (2.8.1) to 2.5.1. It is possible to prove 2.8.1 somewhat more directly by the introduction of ultraproducts of higher order structures. The method of ultraproducts provides apparently concrete realizations of non-standard models of Arithmetic and Analysis, and of general enlargements, in conformity with the taste of most contemporary mathematicians. However, as far as the theory which will be developed in this book is concerned, the fact that, for any given structure, an enlargement *may be assumed to exist* is all that will be required.

Our proof of the finiteness principle 2.5.1 involved the use of the axiom of choice including the use of Zorn's lemma. There are other proofs of 2.5.1 which employ the maximal ideal theorem for Boolean algebras instead. It follows that any mathematical theorem that will be proved in this book subsequently, and which does not involve the axiom of choice in other ways can be proved by means of the maximal ideal theorem for Boolean algebras alone. This is relevant since it is known that, in the appropriate axiomatic setting, the maximal ideal theorem for Boolean algebras is strictly weaker than the axiom of choice (HALPERN [1964]).

2.12 Remarks and references. The finiteness principle of the Lower Predicate Calculus is due to A. I. Malcev (MALCEV [1936]). It is closely related to the completeness theorem (GÖDEL [1930]) and its generalizations (for details, see ROBINSON [1963]). The name 'compactness theorem' was suggested by A. Tarski (TARSKI [1952]).

The notion of an ultrapower was introduced by J. Łoś (ŁOŚ [1955]). The theory of this and of related concepts has been developed further in several papers (CHANG and MOREL [1958], KOCHEN [1961], KEISLER [1962], FRAYNE et al. [1962]). The ultrapower construction overlaps with E. Hewitt's construction of hyperreal fields (HEWITT [1948], compare GILMAN and JERISON [1960]).

The extension of completeness (or compactness) arguments to higher order languages is due to L. Henkin (HENKIN [1950]).

CHAPTER III

DIFFERENTIAL AND INTEGRAL
CALCULUS

3.1 Non-standard Arithmetic. Let $*N$ be a higher order non-standard model of Arithmetic. Thus, $*N$ is a mathematical structure which possesses the following properties:

(i) Every mathematical notion which is meaningful for the system of Natural Numbers is meaningful also for $*N$. In particular, addition, multiplication, and order are defined for $*N$.

(ii) Every mathematical statement which is meaningful and true for the system of Natural Numbers is meaningful and true also for $*N$: provided that we interpret any reference to entities of any given type, e.g., sets, or relations, or functions, in $*N$ not in terms of the totality of entities of that type, but in terms of a certain subset, called the set of *internal* entities of that type. For example, if the statement contains a phrase 'for all *sets* of numbers', we interpret this as 'for all *internal* sets of numbers'. Similarly the phrases 'there exists a *set* of numbers', 'there exists a *function*', as 'there exists an *internal* set', 'there exists an *internal function*'. However, all individuals of $*N$ are internal: the phrase 'for all numbers' is interpreted in $*N$ as 'for *all* individuals of $*N$'.

(iii) The system of internal entities in $*N$ has the following property. If S is an internal set of relations, then all elements of S are internal. More generally, if S is an internal n-ary relation, $n \geq 1$, and the n-tuple $(S_1, ..., S_n)$ satisfies (belongs to) S then $S_1, ..., S_n$ all are internal relations.

(iv) $*N$ properly contains the system of natural numbers, N; there is an individual in $*N$ which, according to the relation of order defined in N and $*N$, is greater than all numbers of N.

See chapter II, in particular 2.10.3, for a more precise, formal, description of the properties of $*N$. We may, and shall, suppose that $*N$ is normal so that the constants e_τ, which denote the relations of identity in N for the various types τ, denote relations of identity also in $*N$.

Let K be the set of all stratified sentences which are admissible and hold in N as defined formally in chapter II and informally in (ii) above. In K, let e denote the relation of identity for individuals, so that $e = e_0$, while q denotes the relation of order and s and p denote the relations of addition and multiplication, respectively. Thus, the sentences $[\Phi_{(0,0)}(e,a,b)]$, $[\Phi_{(0,0)}(q,a,b)]$, $[\Phi_{(0,0,0)}(s,a,b,c)]$ and $[\Phi_{(0,0,0)}(p,a,b,c)]$ have the conventional translations $a = b$, $a < b$, $a + b = c$, and $ab = c$, respectively.

From now on, we shall refer to *all* individuals of $*N$ as natural numbers.

It is not difficult to see that the familiar laws of Arithmetic which are true for N are true also for $*N$. For example, the commutative law of addition holds in N and this can be formulated as a sentence of K, so

3.1.1 $(\forall x)(\forall y)(\forall z)\left[\Phi_{(0,0,0)}(s,x,y,z) \supset \Phi_{(0,0,0)}(s,y,x,z)\right].$

Since any sentence of K holds also in $*N$ it follows that $*N$ satisfies the commutative law of addition.

There are many other elementary assertions which can be *transferred* in this way from N to $*N$. For example, an argument which is quite similar to that just given shows that every number of $*N$ is the sum of the squares of four numbers of $*N$ (since this is known to be true for N).

At this point one might be tempted to conclude that all that can be said about $*N$ is deducible in this fashion, rather trivially, from facts about N. However, this impression would be mistaken.

q denotes the relation of order in N. Thus, q denotes a relation in N which is irreflexive, transitive, and satisfies the law of trichotomy (i.e., for any two *distinct* numbers a and b in N the relation denoted by q holds either between a and b or between b and a). This can be expressed by a sentence of K and accordingly, is true also in $*N$. Thus, q denotes a relation of order also in $*N$. We write $a < b$, as usual, if q holds between numbers a and b in $*N$. The relation of order $<$ in $*N$ is compatible with the ordinary relation of order in N since, for any numbers a and b in N, $[\Phi_{(0,0)}(q,a,b)]$ holds in $*N$ if and only if it holds in N. Thus, the numbers of N constitute an ordered subset of the ordered set of numbers of $*N$.

Moreover, the following statements also are true for N are expressible as sentences of K.

'0 is smaller then all other numbers'.

'1 is smaller than all other numbers except 0'.

'2 is smaller than all other numbers except 0 and 1', and so on.

Indeed, sentences as required are provided by

3.1.2 $(\forall x)\left[\Phi_{(0,0)}(e,0,x) \vee \Phi_{(0,0)}(q,0,x)\right]$

3.1.3 $(\forall x)\left[\Phi_{(0,0)}(e,0,x) \vee \Phi_{(0,0)}(e,1,x) \vee \Phi_{(0,0)}(q,1,x)\right]$

3.1.4 $(\forall x)\left[\Phi_{(0,0)}(e,0,x) \vee \Phi_{(0,0)}(e,1,x) \vee \Phi_{(0,0)}(e,2,x) \vee \Phi_{(0,0)}(q,2,x)\right].$

and so on, respectively, where 0, 1, 2 have the obvious denotation. This shows that the ordinary natural numbers (numbers of N) constitute an initial segment of $*N$. That is to say, any ordinary natural number (number of N) is smaller than every natural number (individual of $*N$) that is not contained in N. From now on, we shall call all natural numbers that belong to N *finite*, while all other natural numbers will be called *infinite*. Thus any finite natural number is smaller than any infinite natural number. The finite natural numbers are the *standard* natural numbers according to a general classification introduced previously.

There is no smallest infinite number. For if a is infinite then $a \neq 0$, hence $a = b + 1$ (the corresponding fact being true in N). But b cannot be finite, for then a would be finite. Hence, there exists an infinite numbers which is smaller than a.

The function $z = |x - y|$ – 'z is the absolute value of $x - y$' – is defined in $*N$ as a function of two variables which is denoted in $*N$ by the same ternary relation as the function $z = |x - y|$ in N. It follows that $|x - y|$ has the same basic properties in $*N$ as in N. Thus, for any a and b in $*N$, $|a - b|$ is equal to $a - b$ if b is smaller than or equal to a and $|a - b|$ is equal to $b - a$ if a is smaller than b.

We define a binary relation $\sim$ for the numbers of $*N$, by the condition that $a \sim b$ if and only if $|a - b|$ is finite. Observe that, by the manner of its definition, the relation $\sim$ is not necessarily internal, in fact, we shall show presently that it is *external*.

It is not difficult to see that $\sim$ is an equivalence relation, i.e., that it is reflexive, symmetrical and transitive. Moreover, the equivalence classes determined by this relation are *intervals* of the ordered set of numbers of $*N$. That is to say, if $a \sim b$ and $a < c < b$ then $a \sim c$ (and hence, $c \sim b$). It is not difficult to see that all finite numbers of $*N$ (i.e., all numbers of N) constitute an equivalence class with respect to the relation $\sim$. Now let a be an infinite natural number, and let D_a be the equivalence class of a with respect to $\sim$. Then the numbers $a - 1, a - 2, \ldots, a - n, \ldots, n$ a number in N, all exist since a is infinite and the numbers $a, a \pm 1, a \pm 2, \ldots, a \pm n$, n a number in N, all belong to D_a since $|(a \pm n) - a| = n$ is finite. Moreover, the numbers just mentioned constitute the entire class D_a. Thus, D_a possesses the order type of the integers, which may be written as $\omega^* + \omega$ (ω^* the inverse order type to ω). Since the order type of the finite natural numbers is ω we then obtain for the order type of the numbers of $^*N$,

3.1.5 $$\alpha = \omega + (\omega_* + \omega)\theta.$$

In this formula, θ is the order type of the set of equivalence classes $\{D_a\}$, a infinite, which are ordered according to the rule that $D_a < D_b$ if $a < b$. It is not difficult to see that this definition is independent of our particular choice of a and b. Thus, suppose that $D_a = D_{a'}$, and $D_b = D_{b'}$, where $a' \sim a$ and $b' \sim b$ while $|a - b|$ is infinite. We wish to show that if $a < b$ then $a' < b$ and vice versa. Suppose on the contrary that $a < b$ but $b < a'$. (If $a < b$ then $a' = b$ is excluded by the fact that this would imply $a' \sim b$. At the same time, $a \sim a'$, and so we should draw the conclusion that $a \sim b$, contrary to assumption). Then $a < b < a'$ and so b belongs to the interval between a and a'. But we have already accepted the fact that the equivalence classes constitute intervals. Hence, $a < b < a'$ together with $a \sim a'$ implies $a \sim b$, contrary to the assumption that $|a - b|$ is infinite. We conclude that $a' < b$. Similarly, if $a < b$ then $b' < a$ would imply that a belongs to the interval from b' to b, which is again impossible. This proves that the definition of the ordering in $\{D_a\}$ is legitimate. Let its order type be θ. The definition of the product of two order types now leads directly to 3.1.5.

Of any two successive numbers in $^*N$, a and $a + 1$, one at least (and at most) is divisible by 2 (i.e., even). For this statement holds it we write N for $^*N$ and can be formulated as a sentence of K. We conclude that it holds also in $^*N$. Thus, either $a = 2a'$ for some number a' or else $a + 1 = 2a'$. Accordingly, every equivalence class D_a contains (many) numbers that are divisible by 2. We shall make use of this fact in order to show that θ is dense without first or last elements. Indeed, let D_a be one of our equivalence classes, for infinite a. We are going to show that there exists an infinite number b such that $D_b < D_a$. As we have just seen, D_a contains even numbers, so we may assume that a is itself even, $a = 2b$. b is infinite for if b were finite a would be finite. Also, $b < a$, for $b \geq 2b$ only for $b = 0$ in N and this is a fact which can be denoted by a sentence of K and hence, is true also in $^*N$. Thus, either $D_b = D_a$ or $D_b < D_a$. But $D_b = D_a$ is possible

only if $|a-b|=b$ is finite, which is not the case. Hence, $D_b<D_a$. This shows
that θ does not contain a first element. A similar argument shows that θ
does not contain a last element. Now let $D_a<D_b$ where both a and b are
infinite. As we have seen, we may suppose that both a and b are even without changing the equivalence classes under consideration. Thus, $a=2a'$,
$b=2b'$. Let $c=a'+b'$, then we claim that $D_a<D_c<D_b$. Indeed,
$2c-2a=a+b-2a=b-a$ exists and is infinite by assumption, so
$c-a=b'-a'$ exists and is infinite. This proves $D_a<D_c$. Similarly,
$2b-2c=2b-(c+b)=b-a$, so $b-c=b'-a'$ exists and is infinite. This
shows that $D_c<D_b$. It follows that between any two elements of the
ordered set $\{D_a\}$, a infinite, there is another element of the set. Hence, θ is
dense. Summing up we have

3.1.6 THEOREM. The order type of $*N$ is of the form

$$\alpha=\omega+(\omega*+\omega)\theta,$$

where θ is a dense order type without first or last element.

Observe that any prime number in N remains a prime number in $*N$
according to the definition that a number $a>1$ is prime if its only divisors
are 1 and a. It follows that the number of primes in $*N$ is infinite. More
strongly, it is true that *for every number $a\in*N$ there exists a prime number p
which is greater than a*. For the statement just written in it is true in N and
can be expressed as a sentence of K. Accordingly, it holds in $*N$.

It is natural to ask whether every natural number (in $*N$) can be expressed as the product of a finite number of powers of primes. If we
understand the word *finite* here in the absolute sense then this signifies
that the primes that occur in the product are in one-to-one correspondence
with the set of natural numbers which are smaller than a number $n\in N$.
With this interpretation of the word *finite* the answer to our question is
negative. For let $\{p_n\}$ be the sequence of primes in N in ascending order of
magnitude, $p_0=2,p_1=3$, and so on. For every natural number n define

$$q_n=\prod_{k=0}^{n}p_n$$

so that $q_0=2,q_1=2\cdot3=6,q_2=2\cdot3\cdot5=30$, and so on. Then $\{q_n\}$ may be
regarded as a function which maps the finite natural numbers into the
finite natural numbers and which, accordingly is given by a relation Q of

type $(0,0)$ in N. The corresponding relation $*Q$ in $*N$, i.e., the relation of $*N$ which is denoted by the same constant of K as Q, defines a function $\{r_n\}$ from the numbers of $*N$ into the numbers of $*N$, such that $r_n = q_n$ for all finite n. In N, q_n is divisible by the first power of all primes up to p_n, by no higher power of these primes, and by no other prime. The same can therefore be said of r_n within $*N$. Thus, for infinite n, the number of prime divisors of r_n is infinite.

On the other hand, as can be seen by formalization within K, it is still true that for every natural number a in $*N$ other than 0 there exists a natural number m and a sequence of natural numbers j_n, $n=0,1,...,m$, such that a is divisible by $p_n^{j_n}$, $n=0,1,...,m$, and not by any higher power of these primes, and not by any other prime.

The following results show that there exists external sets of natural numbers in $*N$.

3.1.7 THEOREM.   The set of infinite natural numbers is an external set in $*N$.

PROOF.   Let N_i be the set of infinite natural numbers. As we have seen, N_i does not possess a smallest element. On the other hand it is true in N that every non-empty set of natural numbers possesses a smallest element. This can be expressed as a sentence of K, so

3.1.8 $(\forall x)[[(\exists y)\,\Phi_{(0)}(x,y)] \supset [(\exists y)[\Phi_{(0)}(x,y) \wedge [(\forall z)$
$$[\Phi_{(0)}(x,z) \supset [\Phi_{(0,0)}(e,y,z) \vee \Phi_{(0,0)}(q,y,z)]...],$$

where we recall that $\Phi_{(0,0)}(q,y,z)$ stands for $y<z$. Since 3.1.8 holds also in $*N$, it follows that every non-empty *internal* set of natural numbers possesses a smallest element. This shows that N_i cannot be an internal set, proving the theorem.

3.1.9 THEOREM.   The set of finite natural numbers is an *external* set in $*N$.

PROOF.   the following sentence holds in N and belongs to K.

3.1.10 $(\forall x)(\exists y)(\forall z)[[\neg\Phi_{(0)}(x,z)] \equiv \Phi_{(0)}(y,z)]$.

This sentence states that for every set of type (0) there exists a set which is

its complement, i.e., which consists of all individuals that are not contained in the former set. Since the sentence belongs to K it must be true
also in $*N$ where we may recall that the interpretation of 'set' in $*N$ is
'internal set'. It follows that if the set of finite natural numbers were an
internal set in $*N$ then the set of infinite natural numbers would be an
internal set in $*N$. But we have seen already that the set of infinite natural
numbers is not an internal set in $*N$. This proves 3.1.9.

Similarly, we may prove that the relation $\sim$, which was defined earlier
in this section by the condition that $a \sim b$ if $|a-b|$ is finite, is an external
relation in $*N$. For the sentence

$$(\forall x)(\exists y)(\forall z)\left[\Phi_{(0,0)}(x,0,z) \equiv \Phi_{(0)}(y,z)\right]$$

holds in N and $*N$. This shows that if the relation $\sim$ were internal, then
the set of natural numbers z such that $0 \sim z$, i.e., the set of finite natural
numbers would be internal, contrary to 3.1.9.

3.2 Non-standard Analysis. Let $*R$ be a higher order non-standard
model of Analysis as explained in 2.10.4. Thus, $*R$ satisfies conditions (i) –
(iv) specified at the beginning of section 3.1, where we have to replace
'Natural Numbers' everywhere by 'Real Numbers'. At the same time, we
have to substitute R and $*R$ for N and $*N$, where R is the system (higher
order structure) of real numbers, in the ordinary sense. Let K be the set of
stratified sentences which are admissible and hold in R.

$*R$ constitutes an ordered field, for the fact that R constitutes an ordered
field can be expressed by means of sentences of K. From now on, we shall
refer to all individuals of $*R$ as real numbers, reserving the name of
standard real numbers to the individuals of R. In order to simplify our
notation we shall write $a \in R$ (or $a \in *R$) in order to indicate that a is a real
number (a standard real number) although strictly speaking R and $*R$ are
not sets of individuals.

$*R$ is non-archimedean since it contains numbers which are greater then
all numbers of R, which constitutes a subfield of $*R$. Thus, for some
number a in $*R$,

$$1 < a, 1+1 < a, \ldots, 1+1+\cdots+1 < a, \ldots, \quad \text{and so on}$$

where we add 1 any (finite) number of times to itself. It is now easy to
appreciate the following facts and definitions about $*R$.

If we define the absolute value $|a|$ of a number $a\in {}^*R$ by $|a|=a$ if $a\geq 0$ and by $|a|=-a$ if $a<0$ then $|x|$ may be regarded as a standard function in $^*R$ which is an extension of the corresponding standard function in R. A real number $a\in {}^*R$ will be called *finite* if there exists a standard number $m\in R$ such that $|a|\leq m$, while any other number of $^*R$ will be called *infinite*. The finite numbers constitute a ring in R, which will be denoted by M_0.

A number $a\in {}^*R$ will be called *infinitesimal* or *infinitely small* if $|a|<m$ for *all* positive numbers m in R. By this definition, 0 is infinitesimal. The set of infinitesimal numbers constitutes a ring M_1 in $^*R$. M_1 is a subring of M_0. More particularly, M_1 is an ideal in M_0. A number $r\in {}^*R$, $r\neq 0$, is infinitesimal if and only if r^{-1} is infinite. If $a-b$ is infinitesimal, then we say that b is *infinitely close* to a, and we write $a\simeq b$.

Consider the quotient ring M_0/M_1. We claim that it is isomorphic to the field of standard real numbers.

Let A be any equivalence class in M_0 modulo M_1. Then A cannot contain two *distinct* standard real numbers r_1 and r_2. For in that case, $|r_1-r_2|$ would not be infinitesimal (not being smaller than the standard positive real number $m=|r_1-r_2|$), which contradicts the assumption that r_1-r_2 belongs to M_1.

On the other hand, A must contain at least one standard real number. For let a be an arbitrary element of A. If a is a standard real number then we have finished. If a is not a standard then we divide the standard real numbers into sets D_1 and D_2 where D_1 contains all standard real numbers that are smaller than a and D_2 contains all standard real numbers which are greater than a. Then D_2 is not empty since $a\leq |a|<m$ for some standard m and D_1 is not empty since $|a|<m$ implies $a\geq -|a|>-m$, for the same standard m. However, if $d_1\in D_1$ and $d_2\in D_2$, then $d_1<d_2$ since $d_1<a$ and $a<d_2$. Thus, the pair (D_1,D_2) constitutes a Dedekind cut. The set determines a standard real number r which is either the greatest element of D_1 or the smallest of D_2. We claim that in either case r belongs to A, since $a-r$ is infinitesimal.

Suppose first that r is the greatest element of D_1, so that $a-r<0$. If $a-r$ is not infinitesimal then there exists a positive standard real number δ such that $a-r\geq \delta$. Hence $a>r+\frac{1}{2}\delta$, which shows that the standard real number $r+\frac{1}{2}\delta$ belongs to D_1. But this contradicts the assumption that r is the greatest element of D_1 and proves that $a-r$ is infinitesimal in this case.

Suppose next that r is the smallest element of D_2. Then $r - a < 0$ and, if $a - r$ is not infinitesimal, $r - a \geq \delta$ for some standard positive real number δ. Hence $r - \frac{1}{2}\delta > a$, so that $r - \frac{1}{2}\delta$ belongs to D_2. This contradicts the assumption that r is the smallest element of D_2 and completes the proof that $a - r$ is infinitesimal in all cases. It follows that the correspondence $A \to r$, under which every equivalence class in M_0 modulo M_1 is mapped on the standard real number contained in it, provides the required isomorphism between M_0/M_1 and the field of standard real numbers R.

For any finite real number a in $*R$ we call the uniquely determined standard real number which is infinitely close to a (r in the above proof) the *standard part* of a, in symbols $r = \mathrm{st}(a)$ or, more briefly, $r = {}^0a$. Moreover, given *any* real number in $*R$ we call the set of real numbers which are infinitely close to a, the *monad* of a, to be denoted by $\mu(a)$. Thus $\mu(a)$ is the equivalence class of a in $*R$ modulo M_1, where both $*R$ and M_1 are regarded as additive groups. The elements of M_0/M_1, are the monads which belong to M_0.

Let N be the higher order structure of the natural numbers, regarded as a partial structure of R in the manner explained in section 2.11, and let $*N$ be the corresponding structure within $*R$. If $*R$ is an enlargement of R then theorem 2.11.3 shows that $*N$ is an enlargement of N. If $*R$ is only known to be a higher order non-standard model of Analysis then we can still show that all sentences of Λ that hold in N hold also in $*N$. Moreover, we claim that under the relation of order in $*R$ which is also the relation of order in $*N$, there exist individuals in $*N$ which are greater than all individuals of N. Indeed, let q be the constant of K which denotes the order relation, as before, and let v denote the sets of individuals in N and $*N$. Then the sentence,

3.2.1 $(\forall x)(\exists y)\left[\Phi_{(0,0)}(q,x,y) \wedge \Phi_{(0)}(v,y)\right]$

holds in R since it states that for every real number x there exists a natural number y which is greater than x. Accordingly, 3.2.1 holds also in $*R$. Now let a be a number of $*R$ which is greater than all numbers of R. Such an a exists since $*R$ is a non-standard model of Analysis. By 3.2.1, as interpreted in $*R$, there exists a number b which is an individual of $*N$ and which is greater than a and hence, is greater then all numbers of R and in particular, is greater than all individuals of N (which constitute a subset of the numbers of R). This shows that $*N$ is a non-standard higher-order model of Arithmetic, so that the results of 3.1 are applicable.

We shall use the symbols N and $*N$, without danger of confusion, also in order to denote the sets of individuals of the respective structures. In this sense N is the set of ordinary, or *standard* natural numbers while $*N$ is the set of natural numbers in $*R$. The numbers of N are the finite natural numbers, both in the sense of section 3.1 and in the sense introduced above for the real numbers, while the numbers of $*N-N$ are the infinite natural numbers in both senses.

3.2.2 THEOREM.   The set R is an external set in $*R$.

For if R were internal in $*R$, then the intersection of R and of $*N$, which is N, would be an internal set in $*R$ and hence in $*N$. This contradicts 3.1.9.

We show by the same method –

3.2.3 THEOREM.   The set of finite real numbers, M_0, is an external set in $*R$. So is the set of infinite real numbers.

Next we prove –

3.2.4 THEOREM.   Let a be any number in $*R$. Then $\mu(a)$ (the set of numbers which are infinitely close to a) is an external set in $*R$.

PROOF.   Suppose that $\mu(a)$ is an internal set for some a. Then the set $\mu_0 = \{b | b = c - a, c \in \mu(a)\}$ also is internal in $*R$. μ_0 consists of all numbers which are infinitely close to 0, i.e., it is the monad of 0. But if μ_0 were internal, then the set of reciprocals of elements of μ_0 other than 0 also would be internal. But this is precisely the set of infinite real numbers. This proves the theorem.

3.3   Convergence.   Let $\{s_n\}$, $n=0,1,2,...$, be an infinite sequence of real numbers in the ordinary sense. $\{s_n\}$ is represented in R by some binary relation F, which in turn is denoted by a constant f within the formal language Λ. Since F defines a function from the natural numbers into the real numbers, the following sentences hold in R.

3.3.1 $(\forall x)[\Phi_{(0)}(v,x) \equiv [(\exists y)\,\Phi_{(0,0)}(f,x,y)]]$

3.3.2 $(\forall x)(\forall y)(\forall z)[[\Phi_{(0,0)}(f,x,y) \wedge \Phi_{(0,0)}(f,x,z)] \supset \Phi_{(0,0)}(e,y,z)],$

where v denotes the set of natural numbers and e denotes the relation of identity for individuals, as before. The two sentences together state that F defines a real-valued function whose domain is the set of natural numbers. Reinterpreted in $*R$, the same sentences state that the relation $*F$ (which is denoted by f in Λ) defines a real-valued function in the sense of $*R$ whose domain is the set of natural numbers $*N$. We may regard this function also as a sequence, $\{s_n'\}$, where n varies over all natural numbers in $*N$. Now any finite natural number n determines the corresponding $s = s_n$ uniquely so that $\Phi_{(0,0)}(f,n,s)$ in both R and $*R$. It follows that $s_n = s_n'$ for all finite n. Thus, it cannot give rise to any misunderstanding if we denote the sequence determined by f in $*R$ by $\{s_n\}$ rather than by $\{s_n'\}$. The sequence $\{s_n\}$ in R then is the restriction of the sequence $\{s_n\}$ in $*R$ to the finite natural numbers n. We observe that this ambiguity in the case of the symbol $\{s_n\}$ conforms to common usage in the theory of functions where, for example, $\sin x$ may first be defined for real arguments x only, but the same symbol is retained after the definition is extended to complex values of the argument. If necessary, we shall use the phrases 'in R' or 'in $*R$' in order to indicate whether the original sequence $\{s_n\}$ is intended, or its extension to $*R$.

3.3.3 THEOREM. In order that a sequence $\{s_n\}$ in R be bounded in R, it is necessary and sufficient that the elements of $\{s_n\}$ in $*R$ are all finite.

PROOF. Suppose that $\{s_n\}$ is bounded in R. Thus, there exists a standard real number m such that $|s_n| < m$ for all $n \in N$. It follows that if $\{s_n\}$ is represented in R by a binary relation F, which in turn is denoted by a constant f in Λ, as above, then the sentence

3.3.4 $(\forall x)(\forall y)(\forall z)\left[\left[\Phi_{(0,0)}(f,x,y) \wedge \Phi_{(0,0)}(a,y,z)\right] \supset \Phi_{(0,0)}(q,z,m)\right]$

holds in R, where $\Phi_{(0,0)}(a,y,z)$ stands for 'z is the absolute value of y' and q and f have their previous meanings. But 3.3.4 belongs to K and so, holds also in $*R$. Hence $|s_n| < m$ for all natural numbers n, finite or infinite. This proves that the condition of the theorem is necessary.

The condition of the theorem is also sufficient. For suppose that it is satisfied. Then the following sentence holds in $*R$, for any positive infinite real number r.

3.3.5 $(\forall x)(\forall y)(\forall z)\left[\left[\Phi_{(0,0)}(f,x,y) \wedge \Phi_{(0,0)}(a,y,z)\right] \supset \Phi_{(0,0)}(q,z,r)\right].$

We conclude by existential quantification that the sentence

3.3.6 $X = (\exists w)(\forall x)(\forall y)(\forall z)$
$$[[\Phi_{(0,0)}(f,x,y) \wedge \Phi_{(0,0)}(a,y,z)] \supset \Phi_{(0,0)}(q,z,w)]$$

also holds in *R. While 3.3.5 is not admissible in R since the constant r does not have any interpretation in R, the sentence X, which is given by 3.3.6 is admissible in R. Hence, if X did not hold in R, then $\neg X$ would hold in R, and would belong to K and thus, would hold also in *R. This contradicts the conclusion just reached that X holds in *R and shows that X holds also in R. Accordingly, there exists a number m in R such that $|s_n| < m$ for all natural numbers in N. This completes the proof of 3.3.3.

In the proof we have made use of the general principle that if X is admissible in R and holds in *R then X holds also in R. For if not, then $\neg X$ is admissible and holds in R and hence, holds also in *R, contrary to assumption. The same argument will be applied repeatedly in the sequel.

The realization that certain statements can be formulated within the language Λ and in terms of the vocabulary of K constitutes an important element in our proof procedure. However, it would be quite impracticable actually to write down the formalized sentences in each case. Generally speaking the *possibility* of such a formulation is of paramount importance while the details are likely to be irrelevant. Accordingly, it will usually be sufficient to *state* that the formalization in question is possible, without carrying it out in practice. On occasion, a semi-formal notation may be helpful. For example, in place of X in 3.3.6 we may write

$$(\exists w)(\forall x)(\forall y)(\forall z)[[y = s_x \wedge z = |y|] \supset z < w]$$

or even more briefly,

$$(\exists w)(\forall x)[|s_x| < w].$$

3.3.7 THEOREM. Let $\{s_n\}$ be a standard infinite sequence and let s be a standard real number. Then s is the limit of $\{s_n\}$ within R, $\lim_{n \to} s_n = s$ in the classical sense if and only if $s_n \simeq s$ (i.e., s_n is infinitely close to s) for all infinite subscripts n.

PROOF. Suppose $\lim_{n \to \infty} s_n = s$ in the classical sense. Then we have to show that $s_n - s$ is infinitesimal for all infinite n. That is to say, for any standard $\varepsilon > 0$ and for any infinite natural number n we have to prove $|s_n - s| < \varepsilon$.

Now we know that for any given $\varepsilon > 0$ in R there exists a natural number ν in N such that

$$|s_n - s| < \varepsilon \qquad \text{for} \qquad n > \nu, n \in N.$$

Thus, it is true in R that

3.3.8 $(\forall x)[[x \in N \wedge x > \nu] \supset |s_x - s| < \varepsilon.$

But 3.3.8 can be expressed as a sentence of K and hence, is true also in $*R$. Since any *infinite* natural number is greater than ν we deduce that $|s_n - s| < \varepsilon$ for all infinite n. This shows that the condition of the theorem is necessary.

To see that the condition is also sufficient suppose that it is satisfied and let ε be any standard positive real number. Then for any infinite natural ν, $|s_n - s| < \varepsilon$ for $n > \nu$ since $|s_n - s|$ is then even infinitesimal. It follows that 3.3.8 holds in $*R$. As it stands, 3.3.8 cannot now be formulated within the vocabulary of K since ν is not in R. However,

3.3.9 $(\exists w)(\forall x)[[x \in N \wedge x > w] \supset |s_x - s| < \varepsilon]$

no longer contains ν, and holds in $*R$. A corresponding formal sentence can be expressed in the vocabulary of K. Accordingly, 3.3.9 holds also in R, for some natural number ν in R, $|s_n - s| < \varepsilon$ for $n > \nu$. This completes the proof of 3.3.7.

From 3.3.7 it is easy to deduce some standard theorems on limits in R. For example, if $\lim_{n \to \infty} s_n = s$, $\lim_{n \to \infty} t_n = t$, then $s_n \simeq s$, $t_n \simeq t$ for all infinite n, by 3.3.7. Hence, $s_n + t_n \simeq s + t$, $\lim_{n \to \infty}(s_n + t_n) = s + t$.

Let $\{s_n\}$ be a standard infinite sequence. A standard real number s is by definition a *limit point* of $\{s_n\}$ in R if for every standard $\varepsilon > 0$ and for every finite natural number ν there exists a finite natural number $n > \nu$ such that $|s_n - s| < \varepsilon$. Thus, if s is a limit point of $\{s_n\}$ in R then the following semi-formalized statement is true in R.

3.3.10 $(\forall x)(\forall y)[[x > 0 \wedge y \in N] \supset [(\exists z)[z \in N \wedge z > y \wedge |s_z - s| < x]]]$.

But 3.3.10 can be expressed in the vocabulary of K and, accordingly, holds also in $*R$. Hence, if we take $x = \varepsilon$, where ε is positive and infinitesimal, and $y=v$, where v is an infinite natural number, then

3.3.11 $$(\exists z)\left[z\in N \wedge z>v \wedge |s_z - s|<\varepsilon\right]$$

holds in $*R$. This shows that $|s_n - s|<\varepsilon$ holds in $*R$ for some infinite n, and proves one half of the following theorem.

3.3.12 THEOREM. Let $\{s_n\}$ be a standard infinite sequence. Then the set of limit points of $\{s_n\}$ in R is given by $S'=\{^0s_n\}$ where n varies over the *infinite* natural numbers for which s_n is *finite* (and hence, possesses a standard part, ^0s_n).

PROOF.   We have just shown that if s is a limit point of $\{s_n\}$ then $s_n \simeq s$ for some infinite n and so $s = {}^0s_n$, $s\in S'$. Conversely, if $s = {}^0s_n$, where n is infinite, then $|s_n - s|$ is infinitesimal. Let $\varepsilon>0$, ε standard and let $v\in N$. Then there exists a natural number z (i.e., the number $z=n$) which is greater than v and such that $|s_z - s|<\varepsilon$. Thus, 3.3.11 is true in $*R$ and hence, belongs to K and is true also in R. This shows that s is a limit point of $\{s_n\}$ in R and completes the proof of 3.3.12.

In particular, if the standard sequence $\{s_n\}$ is bounded in R, then all s_n are finite in $*R$, by 3.3.3. It follows that the standard parts ^0s_n exist for all infinite n and hence, that the set S' is not empty. This proves

3.3.13 THEOREM (Bolzano–Weierstrass). A bounded infinite sequence possesses at least one limit point.

Let $\{s_{nm}\}$, $n=0,1,2,\ldots$, $m=0,1,2,\ldots$, be a double sequence in R. $\{s_{nm}\}$ may be regarded as a function defined on $N\times N$ and taking values in R. It is given by a three place relation, F, such that (n,m,s) satisfies F if and only if $s=s_{nm}$. The corresponding relation $*F$ in $*R$ defines a function from $*N\times *N$ into $*R$, which coincides with $\{s_{nm}\}$ on $N\times N$. Accordingly, we denote it also by $\{s_{nm}\}$.

By definition, a standard number s is the (double) limit of $\{s_{nm}\}$ in R, in symbols

$$\lim_{\substack{n\to\infty \\ m\to\infty}} s_{nm} = s$$

if for any standard $\varepsilon>0$ there exists a finite natural number v such that $|s_{nm}-s|<\varepsilon$ for all finite natural n and m greater than v.

The method used in the proof of 3.3.7 leads to

3.3.14 THEOREM. Let $\{s_{nm}\}$ be a standard double sequence and let s be a standard real number. Then

$$\lim_{\substack{n \to \infty \\ m \to \infty}} s_{nm} = s$$

in R if and only if $s_{nm} \simeq s$ for all infinite n and m.

In particular, we note

3.3.15 COROLLARY. Let $\{s_{nm}\}$ be standard. Then

$$\lim_{\substack{n \to \infty \\ m \to \infty}} s_{nm} = 0$$

in R if and only if s_{nm} is infinitesimal for all infinite n and m.

Cauchy's necessary and sufficient condition for convergence in the classical case (i.e., in R) states

3.3.16 THEOREM. A sequence $\{s_n\}$ converges if and only if

$$\lim_{\substack{n \to \infty \\ m \to \infty}} (s_n - s_m) = 0 .$$

By 3.3.15, this is equivalent to

3.3.17 THEOREM. A standard sequence $\{s_n\}$ converges (possesses a limit) in R if and only if $s_n \simeq s_m$ for all infinite n and m.

PROOF. of 3.3.17. Suppose that the standard sequence $\{s_n\}$ converges to the standard real number s in R. Then, for any infinite n and m, we have, by 3.3.7, $s_n \simeq s$ and $s_m \simeq s$. This shows that $s_n \simeq s_m$, as asserted, and proves that the condition of the theorem is necessary.

Suppose now that $s_n \simeq s_m$ for all infinite n and m, and let ω be any infinite natural number. We claim that s_ω is finite.

Suppose on the contrary that s_ω is infinite. Define a set of natural numbers A by

$$A = \{n \mid |s_\omega - s_n| < 1\} .$$

A is an internal set. For it is true in R that 'for every natural number x there exists a set of natural numbers y such that a natural number z is contained in y if and only if $|s_x - s_z| < 1$'. The statement just given in quotation marks can be formulated as a sentence of K and, accordingly, is true also in $*R$. Specializing the statement by taking ω for x, we obtain A for y.

A contains all infinite natural numbers for if n is infinite, $|s_\omega - s_n|$ is even infinitesimal. On the other hand, $|s_\omega| = |(s_\omega - s_n) + s_n| \leq |s_\omega - s_n| + |s_n|$ so that, for $s_n \in A$, $|s_\omega| \leq 1 + |s_n|$. But s_n is standard and hence finite for finite n while s_ω is infinite. This shows that A does not contain any finite N, A coincides with the set of infinite natural numbers. But this set is an external set (see 3.1.7) a contradiction which shows that s_ω is finite. This being the case, s_ω possesses a finite part $s = {}^0 s_\omega$. Then $s \simeq s_\omega$ and more generally $s \simeq s_n$ for any infinite n since $s_\omega \simeq s_n$. This shows that s is the limit of $\{s_n\}$, by 3.3.7 and completes the proof of 3.3.17.

Observe that the proof of the last theorem is independent of the accepted ('standard') definition of the limit of a sequence. We may if we wish *define* $\lim_{n \to \infty} s_n = s$ by the condition that $s_n \simeq s$ for all infinite n and we may then proceed directly to the development of the theory of convergence of infinite sequences.

Arguments similar to some of those used in the proof of 3.3.17 lead to the following results which apply to internal (and not only to standard) infinite sequences.

3.3.18 THEOREM.   Let $\{s_n\}$ be an internal sequence of real numbers such that $|s_n| \leq m$ for all finite n, where m is some fixed number in $*R$. Then there exists an infinite natural number v such that $|s_n| \leq m$ for all $n < v$.

PROOF.   If $|s_n| \leq m$ for *all* n, finite or infinite, then we may take for v any infinite natural number. If there is an n such that $|s_n| > m$, define a set of natural numbers A by

$$A = \{n \mid |s_n| > m\}.$$

Then A is a non-empty internal set of natural numbers. It therefore possesses a first element, which may serve as the v of the theorem.

3.3.19 THEOREM.   Let $\{s_n\}$ be an internal sequence of real numbers such that $|s_n| \leq m$ for all infinite n. Then there exists a finite natural number ν such that $|s_n| \leq m$ for all $n > \nu$.

Proof omitted.

3.3.20 THEOREM. Let $\{s_n\}$ be an internal sequence such that s_n is infinitesimal for all finite n. Then there exists an infinite natural number ν such that s_n is infinitesimal for all $n < \nu$.

PROOF. Consider the internal sequence $\{t_n\}$ where $t_n = n s_n$. For all finite n, $|t_n| = n|s_n|$ is infinitesimal and so $|t_n| \leq 1$. Hence, by 3.3.18, there exists an infinite ν such that $|t_n| \leq 1$ for all infinite $n < \nu$. For such n, $|s_n| \leq 1/n$, so that s_n is infinitesimal also for all infinite $n < \nu$. This proves the theorem.

3.3.21 THEOREM. Let $\{s_n\}$ be an internal infinite sequence which is finite for all infinite n. Then there exists a finite natural number ν such that s_n is finite for all $n > \nu$.

PROOF. It follows from the assumption of the theorem that the sequence $\{t_n\}$ which is given by $t_n = (1/n)s_n$, $n \geq 1$, satisfies $|t_n| \leq 1$ for all infinite n. It follows that $|t_n| \leq 1$ also for all finite $n > \nu$ where ν is a natural number, which exists by 3.3.19. For such n, $|s_n| \leq n$, s_n is finite.

3.4 Continuity and differentiation. Let $f(x)$ be a real-valued function of a real variable x in R which is defined in an open interval (a,b), a and b standard, $a < b$. On passing to $*R$, $f(x)$ is extended to a function which is defined for all numbers x in $*R$ such that $a < x < b$. By agreement we denote this function again by $f(x)$. (This is in fact the function which is given by a binary relation in $*R$ that is denoted by the same constant in K as the binary relation in R that defined the original $f(x)$.)

3.4.1 THEOREM. In order that the standard real number c be the limit of $f(x)$ in R as x approaches b (from below), $\lim_{x \to b} f(x) = c$, it is necessary and sufficient that $f(x) \simeq c$ for all $x \simeq b$, $a < x < b$.

PROOF. Suppose that $\lim_{x \to b} f(x) = c$ in the ordinary sense. Let ε be any standard positive number. Then there exists a standard positive h such that $|f(x) - c| < \varepsilon$ for $|x - b| < h$, $a < x < b$. Now, if $x \simeq b$ then $|x - b|$ is actually infinitesimal, and so $|x-b|<h$ and $|f(x)-c|<\varepsilon$. Since ε is arbitrary positive in R this shows that $f(x)\simeq c$, the condition of the theorem is necessary.

The condition is also sufficient. For suppose that it is satisfied and choose any positive infinitesimal h. Then the following semi-formalized statement is true in $*R$, for arbitrary standard positive ε.

3.4.2 $(\forall x)\left[\left[|x-b|<h \wedge a<x<b\right]\supset|f(x)-c|<\varepsilon\right].$

Thus, in $*R$, it is true that

3.4.3 $(\exists y)\left[y>0 \wedge \left[(\forall x)\left[\left[|x-b|<y \wedge a<x<b\right]\supset|f(x)-c|<\varepsilon\right]\right]\right].$

But 3.4.3 is admissible in R and, accordingly, must belong to K and must hold also in R. Since ε was standard positive but otherwise arbitrary we conclude directly from the interpretation of 3.4.3 in R that $\lim_{n\to b}f(x)=c$. This completes the proof of 3.4.1.

Corresponding results hold for the limit of $f(x)$ as x tends to a (from above) and for the limit of $f(x)$ as x tends to an interior point of the interval (a,b). We state our conclusion for the latter case as a theorem.

3.4.4 THEOREM. In order that the standard real number c be the limit of $f(x)$ in R as x approaches x_0, x_0 a standard point (number), $a<x_0<b$, it is necessary and sufficient that $f(x)\simeq f(x_0)$ for all $x \neq x_0$ such that $x\simeq x_0$.

The theorem still holds if $f(x)$ is defined in the interval (a,b) *except for* $x=x_0$.

With $f(x)$ as before, the classical definition of continuity at a point taken in conjunction with 3.4.4 leads immediately to

3.4.5 THEOREM. The function $f(x)$ is continuous at a standard point x_0, $a<x_0<b$, if and only if $f(x)\simeq f(x_0)$ for all $x\simeq x_0$.

This is equivalent to saying that $f(x)$ maps the monad of x_0 into the monad of $f(x_0)$.

Suppose now that $f(x)$ is defined also at b. Then 3.4.1 shows that $f(x)$ is continuous at b if and only if $f(x)\simeq f(b)$ for all $x\simeq x_0$, $x<b$, and a corresponding statement holds for a. Thus, in these cases also, the part of the monad of $x_0=b$ (or $x_0=a$) that belongs to the interval of definition of $f(x)$ is mapped into the monad of $f(b)$ (or of $f(a)$).

We are now in a position to prove the intermediate value theorem for continuous functions (Bolzano-Weierstrass).

3.4.6 THEOREM (Standard).   Let $f(x)$ be a function which is defined and continuous in the closed interval $a \le x \le b$, and such that $f(a) < 0$, $f(b) > 0$. Then there exists a point c, $a < c < b$ such that $f(c) = 0$.

The word 'standard' is added above in order to indicate that we are dealing with the standard situation, i.e., with the ordinary real numbers. We shall make the same insertion also on other occasions when the circumstances are not specified elsewhere in the text.

PROOF of 3.4.6   Let ω be an infinite natural number. Then the sequence $\{x_0 = a, x_1, x_2, \ldots, x_\omega = b\}$ which is defined by $x_j = a + (j/\omega)(b-a)$ is an internal sequence in $*R$. Indeed, since it is true in R that for every natural number n, there exists the sequence of numbers $\{x_j\}$, $j = 0, \ldots, n$, $x_j = a + (j/\omega)(b-a)$, the same must be true in $*R$. We note that $\{x_0, \ldots, x_\omega\}$ is a finite sequence in the sense of $*R$ although it contains an infinite number of elements in the absolute sense. Now $f(x_0) < 0$ and $f(x_\omega) > 0$. It follows that the set A of natural numbers j for which $f(x_j) \ge 0$ is not empty but excludes $j = 0$. A is an internal set which possesses a first element $j = \mu$. The difference $x_\mu - x_{\mu-1} = (1/\omega)(b-a)$ is infinitesimal and so $^0x_\mu = {}^0x_{\mu-1}$. We write $^0x_\mu = {}^0x_{\mu-1} = c$. If $a < c < b$ then $f(^0x_\mu) \simeq f(c)$ and $f(^0x_{\mu-1}) \simeq f(c)$ by 3.4.5. If $c = a$ or $c = b$ then these relations still hold by the remarks following 3.4.5. But $f(^0x_{\mu-1}) < 0$; since $f(c)$ is infinitely close to $f(^0x_{\mu-1})$ this implies that $f(c) \le 0$. Similarly, $f(^0x_\mu) \ge 0$ implies that $f(c) \ge 0$. Hence, $f(c) = 0$. This shows that $c \ne a$, $c \ne b$, c is the required argument value for which $f(c) = 0$. This completes the proof of 3.5.6.

Let the standard function $f(x)$ be defined in the open interval (a,b), a and b standard, $a < b$, and let x_0 be a standard point in the interior of (a,b). We say that $f(x)$ is bounded at x_0 in R if for some standard $h > 0$ and $m > 0$, $|f(x)| < m$ for all x such that $|x - x_0| < h$.

3.4.7 THEOREM. $f(x)$ is bounded at x_0 if and only if $f(x)$ is finite in the monad of x in $*R$.

PROOF.   If $f(x)$ is bounded at x_0 then '$|f(x)| < m$ for all x such that $|x - x_0| < h$' is true in R, for some standard $h > 0$, $m > 0$. The statement in quotation marks can be formulated within K and hence, holds also in $*R$. Now, in $*R$, $|x-x_0|<h$ for all x in the monad of x_0 since, for such x, $|x-x_0|$ is actually infinitesimal. Hence $|f(x)|<m$ for all x in the monad of x_0, the condition of the theorem is necessary.

Again, if the condition is satisfied, choose h infinitesimal and m positive infinite. Then it is true in $*R$, that $|f(x)|<m$ for all x for which $|x-x_0|<h$ since all such x are in the monad of x_0. Thus the statement 'there exist $y>0$ and $z>0$ such that $|f(x)|<z$ for all x such that $|x-x_0|<y$' holds in $*R$. But this sentence can be formulated within Λ, in the vocabulary of K and, accordingly, belongs to K and holds also in R. This shows that the condition of the theorem is also sufficient and completes its proof.

3.4.8   THEOREM.   With $f(x)$ and x_0 as before, if $f(x)$ is continuous at x_0 then it is bounded at x_0 (in R).

PROOF.   By 3.4.5, $f(x) \simeq f(x_0)$ for all x in the monad of x_0. Hence $f(x)$ is finite in the monad of x_0, $f(x)$ is bounded at x_0, by 3.4.7.

With $f(x)$ defined as before and $a<x_0<b$, where x_0 is standard, consider the function

$$k(x) = \frac{f(x) - f(x_0)}{x - x_0},$$

$k(x)$ is defined for all values of x in (a,b), both in R and in $*R$, except for $x=x_0$. The derivative of $f(x)$ at x_0, $f'(x_0)$ is given, classically, by $\lim_{x \to x_0} k(x)$ whenever that limit exists. Hence, by 3.4.4,

3.4.9   THEOREM.   In order that the standard number c be the derivative of $f(x)$ at x_0 it is necessary and sufficient that

$$\frac{f(x) - f(x_0)}{x - x_0} \simeq c$$

for all $x \neq x_0$ in the monad of x_0.

3.4.10   THEOREM.   If $f(x)$ is differentiable (possesses a derivative) of x_0 in R then $f(x)$ is continuous at x_0 in R.

PROOF.   If $f(x)$ is differentiable at x_0 then for all $x \neq x_0$ in the monad of x_0,

$f(x)-f(x_0)\simeq c(x-x_0)$ where c is the derivative of $f(x)$ at x_0. This shows that for such x, $f(x)-f(x_0)$ is infinitesimal, $f(x)$ is continuous at x_0, by 3.4.4.

The derivation of the addition rule for differentiation from 3.4.9 is trivial. Let us consider the product rule.

Let the standard function $g(x)$, like $f(x)$, be defined in the interval (a,b) and let $f(x)$ and $g(x)$ be differentiable at the standard point x_0, $a<x_0<b$, with derivatives $f'(x_0)$ and $g'(x_0)$, respectively. Putting $h(x)=f(x)g(x)$, we then have

$$h(x)-h(x_0)=f(x)g(x)-f(x_0)g(x_0)$$
$$=(f(x)-f(x_0))g(x)+f(x_0)(g(x)-g(x_0)).$$

Dividing by $x-x_0$, we obtain

$$\frac{h(x)-h(x_0)}{x-x_0}=\frac{f(x)-f(x_0)}{x-x_0}g(x)+f(x_0)\frac{g(x)-g(x_0)}{x-x_0}.$$

Now take $x\neq x_0$ in the monad of x_0. Then $g(x)$ belongs to M_0 (the ring of finite numbers in $*R$), by 3.4.7, 3.4.8, 3.4.10, and so does $f(x_0)$. Also,

$$\frac{f(x)-f(x_0)}{x-x_0}\simeq f'(x_0),\qquad \frac{g(x)-g(x_0)}{x-x_0}\simeq g'(x_0),\text{ by 3.4.9}.$$

Hence,

$$\frac{f(x)-f(x_0)}{x-x_0}g(x)+f(x_0)\frac{g(x)-g(x_0)}{x-x_0}\simeq f'(x_0)g(x)+f(x_0)g'(x_0)$$

and

$$\frac{h(x)-h(x_0)}{x-x_0}\simeq f'(x_0)g(x)+f(x_0)g'(x_0)\simeq f'(x_0)g(x_0)+f(x_0)g'(x_0),$$

since $g(x)\simeq g(x_0)$, by 3.4.5 and 3.4.10. Referring to 3.4.9, we now see that the derivative of $h(x)$ at x_0 exists and is equal to

$$h'(x_0)=f'(x_0)g(x_0)+f(x_0)g'(x_0).$$

With $f(x)$ defined in an interval (a,b), and $a<x_0<b$, $f(x)$, a,b, and x_0 standard as before, we say that $f(x)$ has a maximum at x_0 (in R, and hence in $*R$) if there exists a standard positive number h such that $f(x)\leq f(x_0)$ for all x for which $|x-x_0|<h$. One proves by methods used repeatedly already,

3.4.11 THEOREM. The function $f(x)$ possesses a maximum at x_0 if and only if $f(x) \le f(x_0)$ for all $x \simeq x_0$.

There is a corresponding result for minima.

3.4.12 THEOREM. Suppose $f(x)$ has a maximum at the standard point x_0, $a < x_0 < b$, and is differentiable at that point. Then $f'(x_0) = 0$.

PROOF. Choose x_1 and x_2 in the monad of x_0 such that $x_1 < x_0 < x_2$. Then $f(x_1) - f(x_0) \le 0$ and $f(x_2) - f(x_0) \le 0$, by assumption and so

$$\frac{f(x_1) - f(x_0)}{x_1 - x_0} \le 0 \quad \text{and} \quad \frac{f(x_2) - f(x_0)}{x_2 - x_0} \ge 0.$$

But

$$f'(x_0) \simeq \frac{f(x_1) - f(x_0)}{x_1 - x_0} \simeq \frac{f(x_2) - f(x_0)}{x_2 - x_0}$$

and the only standard number which has in its monad both non-positive (≤ 0) and non-negative (≥ 0) numbers is 0. Hence $f'(x_0) = 0$ as asserted. The same conclusion applies if $f(x)$ has a minimum at $x = x_0$.

3.4.13 THEOREM. Let the standard function $f(x)$ be defined and continuous in the closed interval $a \le x \le b$. Then $f(x)$ attains its absolute maximum at some standard point c, i.e., $f(c) \ge f(x)$ for all x in the closed interval (in R and hence, in $*R$).

PROOF. Define the sequence $\{x_0 = a, x_1, x_2, ..., x_\omega = b\}$ by $a_j = a + (j/\omega)$ $(b-a)$ as in the proof of 3.4.6. Consider the sequence of real numbers of $*R$, $\{f(x_0) = f(a), f(x_1), f(x_2), ..., f(x_\omega) = f(b)\}$. It is not difficult to see that this is an internal sequence. Now it is a true statement about R that for every natural number n and for every sequence of real numbers $\{r_0, r_1, ..., r_n\}$ there is a natural number j such that $r_j \ge r_i$, $i = 0, 1, ..., n$. This is a statement which can be formulated within K and, accordingly, holds also in $*R$. It follows that for some natural number j in $*R$, $f(x_j) \ge f(x_i)$, $i = 0, 1, ..., \omega$. Put $c = {}^0x_j$, then we claim that $f(c) \ge f(x)$ for all standard x in the closed interval $a \le x \le b$.

Indeed, the length of the sub-interval defined by the sequence $\{x_0, x_1, ..., x_\omega\}$ is infinitesimal since it is equal to $(b-a)/\omega$. It follows immediately that every standard number in the closed interval from a to b contains in its monad at least one (in fact, infinitely many) x_i. Now suppose that $f(c) < f(x)$ for some standard x in the interval. Choose an x_i such that $x = {}^0x_i$. Then $f(c) \simeq f(x_j)$ and $f(x) \simeq f(x_i)$. But $f(c)$ and $f(x)$ are standard numbers and so it will be seen that $f(c) < f(x)$ implies $f(c) + \eta_1 < f(x) - \eta_2$ for arbitrary infinitesimal η_1 and η_2. Hence, $f(x_j) < f(x_i)$. This contradicts the definition of x_j and proves the theorem.

Since the fact that $f(c) \geq f(x)$ for all x in the interval can be stated as a sentence of K it is true also in $*R$. Similarly, there is a standard number at which $f(x)$ attains its minimum in the interval.

3.4.14 THEOREM (Standard, Rolle). Let the function $f(x)$ be defined and continuous in the closed interval $a \leq x \leq b$ and differentiable in the open interval $a < x < b$, $a < b$ and such that $f(a) = f(b) = 0$. Then $f'(c) = 0$ for some c in the interior of the interval.

The proof is standard. If $f(x) = 0$ throughout the interval then $f'(c) = 0$ throughout the interval, so the conclusion is true, trivially. If $f(x) > 0$ somewhere in the interval then $f(x)$ has a positive absolute maximum in the interval, at a point c, say, and since $f(c) > 0$ it follows that $a < c < b$. Then $f(x)$ has a maximum at c in the sense of 3.4.12, and so $f'(c) = 0$. If $f(x) < 0$ somewhere in the interval we obtain the same conclusion by considering the absolute minimum. This proves Rolle's theorem.

We may now prove the mean value theorem by the usual linear transformation.

3.5 Integration. In this section, we shall consider the Cauchy integral. Let $f(x)$ be a standard function which is continuous in the interval $a \leq x \leq b$, where a and b are standard and $a < b$. By a *fine partition* of the interval (a,b) we mean a sequence $\{x_0, x_1, \ldots, x_\omega\}$, where ω is a natural number in $*N$, such that $x_0 = a < x_1 < x_2 < \ldots < x_\omega = b$ and $x_i - x_{i-1}$ is infinitesimal for $i = 1, 2, \ldots, \omega$. There are fine partitions which are internal, e.g., the sequence used in the proof of 3.4.6, which was defined by $x_j = a + (j/\omega)(b-a)$ where ω is an infinite natural number.

Let $\pi = \{x_0, x_1, \ldots, x_\omega\}$ be an internal fine partition and let $\xi = \{\xi_1, \xi_2, \ldots, \xi_\omega\}$ be an internal sequence such that $x_j \leq \xi_{j+1} \leq x_{j+1}$, $j = 0, 1, \ldots, \omega - 1$. There are such internal sequences, for example the sequence which is given by $\xi_j = x_{j-1}$, $j = 1, 2, \ldots, \omega$ and also the sequence given by $\xi_j = x_j$, $j = 1, 2, \ldots, \omega$. We define

3.5.1
$$S(\pi, \xi) = \sum_{j=1}^{\omega} f(\xi_j)(x_j - x_{j-1}).$$

$S(\pi, \xi)$ has a meaning in the sense that it is given by a functional which is the extension in $*R$ of the sum $\sum_{j=1}^{n} f(\xi_j)(x_j - x_{j-1})$ for sequences $\{x_0, x_1, \ldots, x_n\}$ and $\{\xi_1, \xi_2, \ldots, \xi_n\}$ satisfying the appropriate conditions, i.e., $x_0 < x_1 < \ldots < x_n$ and $x_j \leq x_{j+1}$, $_i = 0, 1, \ldots, n-1$ in R.

3.5.2 THEOREM. For any internal fine partition, and for any ξ satisfying the appropriate conditions as detailed,

$$\int_a^b f(x)\, dx = {}^0 S(\pi, \xi).$$

PROOF. We have to show that $\int_a^b f(x)\, dx \simeq S(\pi, \xi)$, i.e., that for any standard $\varepsilon > 0$

$$\left| \int_a^b f(x)\, dx - S(\pi, \xi) \right| < \varepsilon.$$

But the following statement is known to be true in R.
 'For every $\varepsilon > 0$ there exists a $\delta > 0$ such that

$$\left| \int_a^b f(x)\, dx - \sum_{j=1}^{n} f(\xi_j)(x_j - x_{j-1}) \right| < \varepsilon \quad \text{for any} \quad \{x_0, x_1, \ldots, x_n\},$$

$\{\xi_1, \xi_2, \ldots, \xi_n\}$, such that $x_0 = a < x_1 < \ldots < x_n = b$, $x_j \leq \xi_{j+1} \leq x_{j+1}$, $j = 0, 1, \ldots, n-1$, *and such that* $|x_j - x_{j-1}| < \delta$ *for* $j = 1, 2, \ldots, n$'.
  Now this statement can be formulated in the vocabulary of K and accordingly, belongs to K and holds also in $*R$. But the condition $|x_j - x_{j-1}| < \delta$, $_i = 1, 2, \ldots, \omega$ is certainly satisfied by any fine partition since in this case the differences $x_j - x_{j-1}$ are actually infinitesimal. Hence

$$\left| \int_a^b f(x)\, dx - \sum_{j=1}^{\omega} f(\xi_j)(x_j - x_{j-1}) \right| < \varepsilon.$$

This proves the theorem.

The above argument relies on the standard theory of Cauchy integration. This is necessary if we wish to *identify* $^0S(\pi,\xi)$ with the familiar integral. Alternatively, we may *define* the integral in question as the standard part of 3.5.1. In order to justify this definition, we have to show first that the standard part in question exists, i.e., that $S(\pi,\xi)$ is finite and then, that all the $S(\pi,\xi)$ in question are infinitely close to one another.

In order carry out the first step, we observe that $f(x)$ is finite throughout the interval $a \le x \le b$ in $*R$, by 3.4.7. (The proof of 3.4.7 is formulated for points x_0 such that $a < x_0 < b$ but can be adapted to $x_0 = a$, $x_0 = b$ by trivial changes.) Thus, for every infinite positive real number r, $|f(x)| < r$ for all x in the interval. Hence

$$|S(\pi,\xi)| = \left| \sum_{j=1}^{\omega} f(\xi_j)(x_j - x_{j-1}) \right| \le \sum_{j=1}^{\omega} |f(\xi_j)|(x_j - x_{j-1})$$

$$< \sum_{j=1}^{\omega} r(x_j - x_{j-1}) = r(b-a).$$

Now let ω be any infinite positive number. Then $r = \omega/(b-a)$ is positive infinite and so $|S(\pi,\xi)| < \omega(b-a)^{-1}(b-a) = \omega$. Let A be the set of positive numbers p such that $|S(\pi,\xi)| < p$. As we have just shown, this set includes all infinite positive numbers. Accordingly, we conclude that $|S(\pi,\xi)|$ is finite and possesses a standard part. The second step in the argument depends on the fact that a refinement of a given fine partition, i.e., the inclusion of additional x_j, does not change the standard part of $S(\pi,\xi)$. We omit the details.

With $f(x)$ as before, define $F(x) = \int_a^x f(x)\,dx, a \le x \le b$. $F(x)$ is, in the first instance, defined for x in R but, as usual, this definition is extended on passing to $*R$. Now for any standard x_0, $a < x_0 < b$, and for any positive standard h such that the number $x_0 + h$ still belongs to the closed interval $a \le x \le b$, we have

$$F(x_0 + h) - F(x_0) = \int_{x_0}^{x_0+h} f(x)\,dx.$$

We define a fine partition of the interval $x_0 \le x \le x_0 + h$ by $x_j = x_0 + (j/\omega)h$, $j = 0,1,\ldots,\omega$ where ω is some infinite natural number, and we define a corresponding sequence $\xi=\{\xi_1,...,\xi_\omega\}$ by $\xi_j=x_{j-1}$, $j=1,...,\omega$. Then, by 3.5.1 and 3.5.2,

3.5.3 $$F(x_0+h)-F(x_0)={}^0\left(\sum_{j=1}^{\omega} f(x_{j-1})(x_j-x_{j-1})\right)$$

$$={}^0\left(\frac{h}{\omega}\sum_{j=1}^{\omega} f(x_{j-1})\right)=h\,{}^0\left(\frac{1}{\omega}\sum_{j=1}^{\omega} f(x_{j-1})\right).$$

Now since $f(x)$ is continuous in the closed interval $x_0\leq x\leq x_0+h$ it attains its maximum and minimum for that interval at certain points ξ', ξ''. Then

$$f(\xi')\leq f(x_j)\leq f(\xi''), j=1,...,\omega, \qquad \text{and so}$$

$$f(\xi')\leq\frac{1}{\omega}\sum_{j=1}^{\omega} f(x_{j-1})\leq f(\xi''), \qquad \text{and hence}$$

$$f(\xi')\leq{}^0\left(\frac{1}{\omega}\sum_{j=1}^{\omega} f(x_{j-1})\right)\leq f(\xi''), \qquad \text{and hence, by 3.5.3}$$

3.5.4 $$f(\xi')\leq\frac{F(x_0+h)-F(x_0)}{h}$$

and

3.5.5 $$\frac{F(x_0+h)-F(x_0)}{h}\leq f(\xi'').$$

Now 3.5.4 and 3.5.5 can be formulated within K and, accordingly, hold also for infinitesimal $h\neq 0$. In this case, both ξ' and ξ'' are infinitely close to x_0 and so, by the continuity of $f(x)$,

$$f(\xi')\simeq f(\xi'')\simeq f(x_0).$$

Hence from 3.5.4 and 3.5.5,

3.5.6 $$\frac{F(x_0+h)-F(x_0)}{h}\simeq f(x_0)$$

for all positive infinitesimal h. A similar argument shows that 3.5.6 is true also for negative infinitesimal h. Hence, by 3.4.9,

3.5.7 THEOREM (Standard).   Let $f(x)$ be continuous for $a \le x \le b$. Then $F(x) = \int_a^x f(x)\,dx$ exists and $F'(x) = f(x)$ for all $a < x < b$.

This is the fundamental theorem of the Integral Calculus for continuous functions. For right and left derivatives respectively, the argument can be adapted so as to include $x = a$ and $x = b$ in the result.

Next, let us consider an example of an improper integral. Let the standard function $f(x)$ be defined and continuous for all $x \ge a$, a standard. Then the improper integral $\int_a^\infty f(x)\,dx$ is defined classically as $\lim_{b \to \infty} \int_a^b f(x)\,dx$, whenever that limit exists. Adapting the non-standard version of Cauchy's condition (Theorem 3.3.17) we prove without difficulty,

3.5.8 THEOREM.   In order that the improper integral $\int_a^\infty f(x)\,dx$ exist it is necessary and sufficient that $\int_\eta^\zeta f(x)\,dx \simeq 0$ for all positive infinite η and ζ.

Observe that $\int_\eta^\zeta f(x)\,dx$ exists, in $*R$, for all infinite η and ζ since $\int_c^d f(x)\,dx$ exists in R for all $c \ge a$, $d \ge a$. The integral $\int_\eta^\zeta$ possesses even for non-standard η and ζ the usual properties of an integral to the extent to which they can be formulated within our language. For instance,

$$\int_\zeta^\eta f\,dx = -\int_\eta^\zeta f\,dx \quad \text{and} \quad \int_\eta^\zeta f\,dx + \int_\zeta^\lambda f\,dx = \int_\eta^\lambda f\,dx.$$

It is natural to ask whether the integral $\int_\eta^\zeta f(x)\,dx$ can be approximated by a sum such as 3.5.1, even for infinite η and ζ. To answer this question, we shall suppose $\eta < \zeta$ and we shall confine ourselves to partitions $\pi = \{x_0 = \eta, x_1, \ldots, x_\omega = \zeta\}$ which are given by $x_j = \eta + (j/\omega)(\zeta - \eta)$, $j = 0, 1, \ldots, \omega$, ω an infinite integer, while $\zeta_j = x_{j-1}$, $j = 1, \ldots, \omega$. Then the right hand side of 3.5.1 becomes

$$\sum_{j=1}^\omega f(x_{j-1})\frac{\zeta - \eta}{\omega} = \frac{\zeta - \eta}{\omega}\sum_{j=0}^{\omega-1} f(x_j).$$

Notice that since $\zeta - \eta$ may now be infinite, $(\zeta - \eta)/\omega$ may be finite even for infinite ω. We shall include this possibility in our considerations to begin with.

3.5.9 THEOREM. With $f(x)$ defined as above, let ω be an arbitrary but fixed infinite natural number. Then there exists an *infinite* positive real number b such that for all η, ζ which satisfy $a \leq \eta < \zeta \leq b$,

3.5.10
$$\int_{\eta}^{\zeta} f(x)\,dx \simeq \frac{\zeta - \eta}{\omega} \sum_{j=0}^{\omega-1} f(x_j)$$

where the x_j are defined as above.

PROOF. Let n be any finite natural number. Then, for the given infinite ω and for any c and d such that $a \leq c \leq d \leq n$, the difference

$$\int_{c}^{d} f(x)\,dx - \frac{d-c}{\omega} \sum_{j=0}^{\omega-1} f(x_j)$$

is infinitesimal and so by 3.5.2, it is true in $^{*}R$ that

3.5.11
$$n \left| \int_{c}^{d} f(x)\,dx - \frac{d-c}{\omega} \sum_{j=0}^{\omega-1} f(x_j) \right| < 1, \qquad x_j = c + \frac{j}{\omega}(d-c).$$

If in 3.5.11, we replace ω by a finite natural number m then we obtain an inequality which may or may not be true for a particular n. However, if we express this inequality by a sentence $X(n,m)$ within our formal language then it is true in R that 'for every natural number m, if there exists any natural number y such that $\neg X(y,m)$ holds then there exists a smallest y of this kind'.

The usual argument now shows that the statement just given in quotation marks holds also in $^{*}R$. Accordingly, there must be a smallest number $y = n_0$ in $^{*}N$ which satisfies $\neg X(y,\omega)$ if any y satisfies $\neg X(y,\omega)$ at all. If such an n_0 exists then it must be infinite since $X(n,\omega)$ is true for all finite n. In that case, we put $b = n_0 - 1$. If $\neg X(y,\omega)$ is not true for any y, i.e., if $X(y, \omega)$ holds for all natural numbers y then we set b equal to an arbitrary positive infinite real number. In either case,

$$\left| \int_c^d f(x)\,dx - \frac{d-c}{\omega} \sum_{j=0}^{\omega-1} f(x_j) \right| < \frac{1}{n}$$

for all $a \le c < d \le n \le b$, which shows, for $n = b$, that

$$\int_c^d f(x)\,dx \simeq \frac{d-c}{\omega} \sum_{j=0}^{\omega-1} f(x_j) \quad \text{for} \quad a \le c < d \le b.$$

This proves the theorem. If we want to make sure that the sum on the right hand side involves a *fine* partition, we only have to take b infinite but smaller than $\min (n_0 - 1, \sqrt{\omega})$ if n_0 as above exists and as $\sqrt{\omega}$ otherwise. For, with this choice,

$$\frac{\zeta - \eta}{\omega} \le \frac{|\zeta| + |\eta|}{\omega} \le \frac{2}{\sqrt{\omega}},$$

so that the length of the intervals (x_j, x_{j+1}) is infinitesimal.

3.5.12 THEOREM. With $f(x)$ defined as above, the integral $\int_a^\infty f(x)\,dx$ exists if and only if for some infinite natural number ω and for some infinite positive real number b

$$\frac{\zeta - \eta}{\omega} \sum_{j=0}^{\omega-1} f(x_j) \simeq 0$$

for all infinite η, ζ such that $\eta < \zeta \le b$.

PROOF. Suppose the integral exists. Then $\int_\eta^\zeta f(x)\,dx \simeq 0$ for all positive infinite η and ζ, by 3.5.8. At the same time, for arbitrary infinite natural ω and for a number b which satisfies the conclusions of 3.5.11,

$$\frac{\zeta - \eta}{\omega} \sum_{j=0}^{\omega-1} f(x_j) \simeq \int_\eta^\zeta f(x)\,dx \quad \text{for} \quad a \le \eta < \zeta \le b.$$

Hence, for infinite η and ζ which satisfy $\eta < \zeta \le b$,

$$\frac{\zeta - \eta}{\omega} \sum_{j=0}^{\omega - 1} f(x_j) \simeq 0.$$

This shows that the condition of the theorem is necessary.

Now suppose that the condition of the theorem is satisfied. Thus, for some infinite natural ω and positive infinite b,

$$\frac{\zeta - \eta}{\omega} \sum_{j=0}^{\omega - 1} f(x_j)$$

is infinitesimal for all positive infinite η and ζ such that $\eta < \zeta \le b$. Now we know that, for the given ω and for sufficiently small positive infinite b'

$$\frac{\zeta - \eta}{\omega} \sum_{j=0}^{\omega - 1} f(x_j) \simeq \int_{\eta}^{\zeta} f(x)\,dx \qquad \text{for} \qquad a \le \eta < \zeta \le b'.$$

Hence, for positive infinite η and ζ such that $\eta < \zeta \le \min(b, b')$, $\int_{\eta}^{\zeta} f(x)\,dx \simeq 0$. Now let ε be any standard positive number, then $|\int_{\eta}^{\zeta} f(x)\,dx| < \varepsilon$ for $\eta < \zeta \le \min(b, b')$. Let A be the set of natural numbers n, such that $|\int_{\eta}^{\zeta} f(x)\,dx| < \varepsilon$ for all ζ and η such that $n \le \eta < \zeta \le \min(b, b')$. As we have just seen, A includes all infinite natural numbers smaller than $\min(b, b')$. (A includes all infinite natural numbers $\ge \min(b, b')$, trivially.) Also, A is an internal set, so it possesses a smallest element, which must therefore be finite. Accordingly, there exists a standard natural number n such that for all η and ζ for which $n \le \eta < \zeta \le \min(b, b')$, $|\int_{\eta}^{\zeta} f(x)\,dx| < \varepsilon$. Hence, for any standard η and ζ such that $n \le \eta < \zeta$, we have $|\int_{\eta}^{\zeta} f(x)\,dx| < \varepsilon$, in $^*R$ and hence also in R. But this shows that $\int_{a}^{\infty} f(x)\,dx$ satisfies the necessary and sufficient condition for the convergence of an improper integral in the classical sense and completes the proof of 3.5.12.

The last part of the proof makes use of an interesting phenomenon which appears in various forms. Thus, let $\{s_n\}$ be any standard infinite sequence. We know that a necessary and sufficient condition for $\{s_n\}$ to converge to 0 is that $s_n \simeq 0$ for all infinite n. However, an apparently weaker condition is already sufficient for convergence.

3.5.13 THEOREM.   Let $\{s_n\}$ be a standard infinite sequence. Suppose that there exists an infinite natural number v such that s_n is infinitesimal for all infinite $n < v$. Then $\lim_{n \to \infty} s_n = 0$ in R.

PROOF.   Let ε be a standard positive number. Then $|s_n| < \varepsilon$ for all infinite n such that $n < v$ since s_n is actually infinitesimal for such n. Thus, the set A which consists of all natural numbers μ such that $|s_n| < \varepsilon$ for all n which satisfy $\mu \leq n < v$ is not empty, and includes all infinite natural μ. Accordingly, the first μ which satisfies this condition is finite, $\mu = \mu_0$, say. Then $|s_n| < \varepsilon$ for all finite n such that $n \geq n_0$, both in $*R$ and in R. This shows that $\lim_{n \to \infty} s_n = 0$ and proves 3.5.13.

3.6 Differentials.   Let $\{a_1,...,a_n\}$ be a set of real numbers in $*R$, where n is a finite positive natural number. The set of all linear combinations $r_1 a_1 + \cdots + r_n a_n$ where $r_1,...,r_n$ are finite (i.e., belong to M_0) will be denoted by $O(a_1,...,a_n)$. Similarly, the set of all $r_1 a_1 + \cdots + r_n a_n$ such that $r_1,...,r_n$ are infinitesimal (belong to M_1) will be denoted by $o(a_1,...,a_n)$. In particular, $M_0 = O(1)$ and $M_1 = o(1)$.

For any given $a_1,...,a_n$, put $a = \max(|a_1|,...,|a_n|)$. Then it is easy to see that $O(a_1,...,a_n) = O(a)$, $o(a_1,...,a_n) = o(a)$. Both $O(a)$ and $o(a)$ are additive groups which admit M_0 as domain of (multiplicative) operators. M_0 is isomorphic to $O(a)$ as an additive group under the isomorphism under which $m \in M_0$ is mapped on $ma \in O(a)$. Under this isomorphism, M_1 is mapped on $o(a)$. It follows that the difference group $O(a)/o(a)$ is isomorphic to the additive group of R.

Let $y = f(x)$ be an internal function which is defined in an open interval $(a,b), a < b$, a and b standard. We choose an infinitesimal positive real number, calling it dx. For x in (a,b), we then put $dy = df = f(x+dx) - f(x)$. Thus, d may be regarded as an operator which maps any function in $*R$ on another function which is defined in a slightly smaller interval (i.e., $(a,b-dx)$). Applying d to dy, we obtain

$$d^2 y = d(dy) = f(x+2dx) - 2f(x+dx) + f(x),$$

which is defined for $a < x < b - 2dx$. More generally, putting $d^0 y = y$,

$d^n y = d(d^{n-1} y)$, $n = 1, 2, \ldots$, we obtain functions which are defined for $a < x < x - n\,dx$, respectively.

For standard $f(x)$ and standard x, $a < x < b$, it follows immediately from 3.4.9 that if the function $f(x)$ is differentiable at x then

$$\frac{dy}{dx} \simeq f'(x).$$

More generally, we propose to show that, under certain conditions which will be discussed presently,

3.6.1
$$\frac{d^n y}{dx^n} \simeq f^{(n)}(x), \ n = 1, 2, \ldots,$$

where dx^n stands for $(dx)^n$. Since 3.6.1 indicates equivalence modulo $M_1 = o(1)$, 3.6.1 implies that $(d^n y / dx^n) - f^{(n)}(x) \in o(1)$. Hence,

3.6.2
$$d^n y - f^{(n)}(x)\,dx^n \in o(dx^n).$$

We shall write $\alpha \simeq \beta \bmod o(a_1, \ldots, a_n)$ or $\alpha \simeq \beta \bmod O(a_1, \ldots, a_n)$ in order to indicate equivalence modulo $o(a_1, \ldots, a_n)$ or $O(a_1, \ldots, a_n)$ respectively. With this notation, 3.6.2 becomes

3.6.3
$$d^n y \simeq f^{(n)}(x)\,dx^n \bmod o(dx^n).$$

3.6.4 THEOREM. Let $f(x)$ be a standard function which is defined in the open interval $a < x < b$ and possesses continuous derivatives up to the nth order in that interval, for standard a, b, and n, then 3.6.1 and 3.6.3 hold for any finite real x such that $a < {}^0 x < b$.

PROOF. If $a < x < x + nh < b$ in R then, by a classical generalization of the mean value theorem, there exists a (standard) number θ, $0 < \theta < 1$ such that

3.6.5
$$\frac{\Delta^n f(x)}{h^n} = f^{(n)}(x + n\theta h),$$

where $\Delta^n f(x)$ is the nth formal difference for the interval length h. Thus $\Delta f(x) = f(x+h) - f(x)$, $\Delta^2 f(x) = f(x+2h) - 2f(x+h) + f(x)$, etc. Since this result is true in R it must be true also in $*R$. But, for $h = dx$, $\Delta^n f(x)$ becomes $d^n y$, and so

3.6.6
$$\frac{d^n y}{dx^n} = f^{(n)}(x + n\theta dx), \qquad 0 < \theta < 1.$$

By assumption, $a < {}^0x < b$ and $f^{(n)}(x)$ is continuous at ${}^0x$. Hence, $f^{(n)}({}^0x) \simeq f^{(n)}(x) \simeq f^{(n)}(x + n\theta dx)$ and so

$$\frac{d^n y}{dx^n} \simeq f^{(n)}(x).$$

This proves 3.6.4.

3.7   Total differentials.   Let $f(x_1, \ldots, x_n)$ be a function of n variables in R and let $(\xi_1, \ldots, \xi_n)$ be a point within the domain of definition of $f(x_1, \ldots, x_n)$ such that the function exists in the interior of a sphere S which is given by $(x_1 - \xi_1)^2 + (x_2 - \xi_2)^2 + \ldots + (x_n - \xi_n)^2 = r^2$, $r > 0$. The function extends to $*R$ in the usual way. Various results established previously on continuity and differentiation of functions apply also to functions of several variables. Continuity will be considered more systematically in the next chapter, within the general framework of topological and metric spaces. For the present we shall require only the following result.

3.7.1   THEOREM.   Suppose $f(x_1, \ldots, x_n)$ as introduced above is continuous at the point $\xi_1, \ldots, \xi_n$. Then $f(x'_1, \ldots, x'_n) \simeq f(\xi_1, \ldots, \xi_n)$ for all $(x'_1, \ldots, x'_n)$ such that the distance $\sqrt{\{(x'_1 - \xi_1)^2 + \cdots + (x'_n - \xi_n)^2\}}$ is infinitesimal.

PROOF.   For every standard positive ε there exists a standard positive δ such that $|f(x_1, \ldots, x_n) - f(\xi_1, \ldots, \xi_n)| < \varepsilon$ provided $\sqrt{\{(x_1 - \xi_1)^2 + \cdots + (x_n - \xi_n)^2\}} < \delta$. But $\sqrt{\{(x'_1 - \xi_1)^2 + \cdots + (x'_n - \xi_n)^2\}}$ is smaller than δ since it is even infinitesimal. Hence, $|f(x'_1, \ldots, x'_n) - f(\xi_1, \ldots, \xi_n)| < \varepsilon$. Since this is true for arbitrary standard positive ε, $|f(x'_1, \ldots, x'_n) - f(\xi_1, \ldots, \xi_n)|$ is infinitesimal. This proves the theorem.

Now suppose that the standard function $f(x_1, \ldots, x_n)$ exists and possesses continuous first derivatives within a sphere S as above. Taking $(x'_1, \ldots, x'_n)$ and $(h_1, \ldots, h_n)$ as two n-tuples of standard real numbers such that $(x'_1, \ldots, x'_n)$ and $(x'_1 + h_1, \ldots, x'_n + h_n)$ belong to the interior of S, we put $g(t) = f(x'_1 + h_1 t, \ldots, x'_n + h_n t)$.

By a classical result (which can be established also by means of Non-standard Analysis),

$$g'(t) = \frac{\partial f}{\partial x_1}\frac{dx_1}{dt} + \cdots + \frac{\partial f}{\partial x_n}\frac{dx_n}{dt} = \frac{\partial f}{\partial x_1}h_1 + \cdots + \frac{\partial f}{\partial x_n}h_n,$$

where the partial derivatives are taken for the appropriate $x_i = x_i' + h_i t$. Hence, by the mean value theorem

3.7.2 $g(1) - g(0) = f(x_1' + h_1, \ldots, x_n' + h_n) - f(x_1', \ldots, x_n')$

$$= \left(\frac{\partial f}{\partial x_1}\right)_\theta h_1 + \cdots + \left(\frac{\partial f}{\partial x_n}\right)_\theta h_n,$$

where the partial derivatives are taken at some point $(x_1' + \theta h_1, \ldots, x_n' + \theta h_n)$, $0 < \theta < 1$.

Passing to $*R$ we choose $(x_1', \ldots, x_n')$ such that $^0x_i' = \xi_i$, $i = 1, \ldots, n$ and we make the $h_i = dx_i$ infinitesimal but not all equal to zero. We *define* the *total differential* df by

$$df = f(x_1' + dx_1, \ldots, x_n' + dx_n) - f(x_1', \ldots, x_n').$$

Then, from 3.7.2,

3.7.3 $df = \left(\frac{\partial f}{\partial x_1}\right)_0 dx_1 + \cdots + \left(\frac{\partial f}{\partial x_n}\right)_0 dx_n + r_1 dx_1 + \cdots + r_n dx_n,$

where the subscript 0 indicates that the partial derivatives are taken at $(x_1', \ldots, x_n')$ and where

$$r_i = \left(\frac{\partial f}{\partial x_i}\right)_\theta - \left(\frac{\partial f}{\partial x_i}\right)_0, \qquad i = 1, \ldots, n.$$

We claim that the r_i are infinitesimal. Indeed, denoting by $(\partial f / \partial x_i)_\xi$ the partial derivatives of f at $(\xi_1, \ldots, \xi_n)$ and applying 3.7.1 twice (with the partial derivatives of f in place of f in the formulation of that theorem) we find that

$$\left(\frac{\partial f}{\partial x_i}\right)_0 \simeq \left(\frac{\partial f}{\partial x_i}\right)_\xi, \quad \left(\frac{\partial f}{\partial x_i}\right)_\theta \simeq \left(\frac{\partial f}{\partial x_i}\right)_\xi$$

and so

$$\left(\frac{\partial f}{\partial x_i}\right)_\theta \simeq \left(\frac{\partial f}{\partial x_i}\right)_0,$$

the r_i are infinitesimal, as asserted. It follows that $r_1 dx_1 + \cdots + r_n dx_n$ $\in o(dx_1, \ldots, dx_n)$. Referring to 3.7.3 we see that we have proved the following theorem.

3.7.4 THEOREM. Let $f(x_1, \ldots, x_n)$ be a standard function which is defined and possesses continuous first derivatives in the interior of a sphere with center $(\xi_1, \ldots, \xi_n)$ and radius $r > 0$ where $\xi_1, \ldots, \xi_n$, and r are standard. If $^0x'_1 = \xi_1, \ldots, ^0x'_n = \xi_n$ and $dx_1, \ldots, dx_n$ are infinitesimal, then

3.7.5 $$df \simeq \left(\frac{\partial f}{\partial x_1}\right)_0 dx_1 + \cdots + \left(\frac{\partial f}{\partial x_n}\right)_0 dx_n \bmod o(dx_1, \ldots, dx_n),$$

where $df = f(x'_1 + dx_1, \ldots, x'_n + dx_n) - f(x'_1, \ldots, x'_n)$ and where the subscripts 0 indicate that the partial derivatives are taken at $(x'_1, \ldots, x'_n)$.

3.8 Elementary Differential Geometry.

Quite early in the history of the Differential Calculus, infinitesimals were employed in the development of the theory of curves, and long after the classical ε, δ-method had displaced the naive use of infinitesimals in Analysis they survived in Differential Geometry and in several branches of Theoretical Physics. Even now there are many classical results in Differential Geometry which have never been established in any other way, the assumption being that somehow the rigorous but less intuitive, ε, δ-method would lead to the same result. So far as one can see without a complete check this assumption is usually correct. However, although no striking results will be obtained in this domain, it seems appropriate to investigate some simple examples from elementary Differential Geometry by means of the infinitesimals of Nonstandard Analysis.

Let the curve C in three-dimensional (x, y, z)-space over the real number R be given by

3.8.1 $$x = x(t), \quad y = y(t), \quad z = z(t), \qquad 0 \le t \le 1,$$

where $x(t)$, $y(t)$, $z(t)$ have continuous derivatives in their interval of definition.

Passing to $^*R$, we choose a positive infinitesimal dt and we put, correspondingly,

$$dx = x(t+dt) - x(t),$$
$$dy = y(t+dt) - y(t),$$
$$dz = z(t+dt) - z(t).$$

Evidently, dx, dy, dz may depend on t as well as on dt. Putting $P = (x(t), y(t), z(t))$ and $P' = (x(t+dt), y(t+dt), z(t+dt))$ we see that the length of the chord PP' is given by

3.8.2 $ds = \sqrt{\{dx^2 + dy^2 + dz^2\}},$

where it is understood that the positive value of the square root is taken on the right hand side. A point $P = (x, y, z)$ will be called *standard* if its coordinates, x, y, and z are standard real numbers. Let A and B be two standard points on the curve C, corresponding to $t = a$ and $t = b$, a and b standard, $0 < a < b < 1$, respectively. Consider the polygonal line $P_0 P_1 \ldots P_\omega$, $P_0 = A$, $P_\omega = B$, which is given by

3.8.3 $P_j = (x(t_j), y(t_j), z(t_j)),$ $t_j = a + j\,dt,$ $dt = \dfrac{b-a}{\omega},$ $j = 0,1,\ldots,\omega,$

where ω is an arbitrary but fixed infinite natural number. Putting

$$dx_j = x(t_{j+1}) - x(t_j), \quad dy_j = y(t_{j+1}) - y(t_j), \quad dz_j = z(t_{j+1}) - z(t_j),$$
$$j = 0,\ldots,\omega - 1,$$

we obtain for the length of the chord $P_j P_{j+1}$,

$$ds_j = \sqrt{\{dx_j^2 + dy_j^2 + dz_j^2\}}.$$

Then the total length of the polygonal line $P_0 P_1 \ldots P_\omega$ is

3.8.4 $l = \displaystyle\sum_{j=0}^{\omega-1} ds_j = \sum_{j=0}^{\omega-1} \sqrt{\{dx_j^2 + dy_j^2 + dz_j^2\}}.$

The length of the arc of the curve C from A to B is now given by ^0l and this may be regarded either as a definition or as an assertion, presuming the classical definition. In the former case we have to show that the definition is legitimate, more particularly that ^0l exists and that it is independent of the particular choice of ω. In any case

$$ds_j = \sqrt{\{dx_j^2 + dy_j^2 + dz_j^2\}} = \sqrt{\left\{\left(\frac{dx_j}{dt}\right)^2 + \left(\frac{dy_j}{dt}\right)^2 + \left(\frac{dz_j}{dt}\right)^2\right\}}\,dt.$$

We wish to show that dx_j/dt, dy_j/dt, dz_j/dt are, up to infinitesimals, equal to $x'(t_j)$, $y'(t_j)$, $z'(t_j)$, respectively. For standard t_j, this is confirmed by 3.6.4, for $n=1$. More generally, by the mean value theorem,

$$dx_j = x(t_{j+1}) - x(t_j) = x'(t_j + \theta \, dt) \, dt, \quad 0 < \theta < 1.$$

But $x'(t)$ is continuous, by assumption, and so $x'(t_j + \theta \, dt) \simeq x'(t_j)$. Hence $dx_j/dt \simeq x'(t_j)$, and similarly $dy_j/dt \simeq y'(t_j)$, $dz_j/dt \simeq z'(t_j)$ for $j = 0, 1, \ldots, \omega - 1$, and so

3.8.5
$$ds_j = \sqrt{\{(x'(t_j) + \xi_j)^2 + (y'(t_j) + \eta_j)^2 + (z'(t_j) + \zeta_j)^2\}} \, dt$$
$$= \sqrt{\{(x'(t_j))^2 + (y'(t_j))^2 + (z'(t_j))^2 + \varrho_j\}} \, dt$$

where $\xi_j, \eta_j, \zeta_j, \varrho_j$, are infinitesimal. Now it is easy to verify that if α and β are two non-negative real numbers and $\alpha \simeq \beta$ then, taking positive square roots, $\sqrt{\alpha} \simeq \sqrt{\beta}$. Hence, for $\alpha = (x'(t_j))^2 + (y'(t_j))^2 + (z'(t_j))^2$, $\beta = \alpha + \varrho_j$, 3.8.5 yields

$$ds_j = (\sqrt{\{(x'(t_j))^2 + (y'(t_j))^2 + (z'(t_j))^2\}} + \sigma_j) \, dt, \quad \sigma_j \simeq 0, \quad j = 0, 1, \ldots, \omega - 1.$$

Let $\sigma = \max_{0 \le j \le \omega - 1} |\sigma_j|$. σ exists, by transfer from R to $^*R$, since every sequence of n numbers in R possesses a maximum. Moreover, σ is infinitesimal since it coincides with one of the $|\sigma_j|$. Let

$$w(t) = (x'(t))^2 + (y'(t))^2 + (z'(t))^2, \quad 0 < t < 1,$$

then $w(t)$ is continuous. Furthermore, $\sqrt{w(t)}$ is continuous, for since $w(t) \simeq w(t')$ for t standard, $0 < t < 1$, and $^0t' = t$, it follows that $\sqrt{w(t)} \simeq \sqrt{w(t')}$. Now,

$$l = \sum_{j=0}^{\omega - 1} ds_j = \sum_{j=0}^{\omega - 1} \sqrt{w(t_j)} \, dt + \sum_{j=0}^{\omega - 1} \sigma_j \, dt$$

and

$$\left| \sum_{j=0}^{\omega - 1} \sigma_j \, dt \right| \le \sum_{j=0}^{\omega - 1} |\sigma| \, dt = \sigma(b - a) = \tau$$

where τ is infinitesimal. Hence

$$l = \sum_{j=0}^{\omega - 1} \sqrt{w(t_j)} \, dt + \lambda, \quad \lambda \simeq 0.$$

But $\sqrt{w(t)}$ is continuous in the closed interval $a \leq t \leq b$, so, by 3.5.2,

$$^0\left(\sum_{j=0}^{\omega-1} \sqrt{w(t_j)}\,dt\right) = \int_a^b \sqrt{w(t)}\,dt$$

and hence

3.8.6 $^0 l = \int_a^b \sqrt{\{(x'(t))^2 + (y'(t))^2 + (z'(t))^2\}}\,dt.$

3.8.6 shows that $^0 l$ exists and is independent of the particular choice of ω. It shows, moreover, that our definition of the length of an arc coincides with the classical definition.

Next, we consider the determination of the tangent. We assume that, for all t in the interval of definition,

3.8.7 $w(t) = (x'(t))^2 + (y'(t))^2 + (z'(t))^2 \neq 0.$

Let $P = (x_0, y_0, z_0)$ where $x_0 = x(t_0)$, $y_0 = y(t_0)$, $z_0 = z(t_0)$, $0 < t_0 < 1$, t_0 standard. Let dt be a positive infinitesimal number and put $x_1 = x(t_0 + dt)$, $y_1 = y(t_0 + dt)$, $z_1 = z(t_0 + dt)$, $dx = x_1 - x_0$, $dy = y_1 - y_0$, $dz = z_1 - z_0$, $Q = (x_1, y_1, z_1)$ so that Q also is on the curve. Then the distance between P and Q is given by

3.8.8 $PQ = \Delta = (dx^2 + dy^2 + dz^2)^{\frac{1}{2}} = [(x'(t_0 + \theta_1\,dt))^2 +$
$(y'(t_0 + \theta_2\,dt))^2 + (z'(t_0 + \theta_3\,dt))^2]^{\frac{1}{2}}\,dt, \qquad 0 < \theta_1 < 1,$
$0 < \theta_2 < 1, \quad 0 < \theta_3 < 1,$

where we have used the mean value theorem, Δ/dt is infinitely close to $\sqrt{w(t_0)}$, since $x'(t), y'(t), z'(t)$ are supposed continuous. It follows that Δ/dt is not infinitesimal.

The equation of the straight line through P and Q is, with τ as parameter,

3.8.9 $x = x_0 + \dfrac{dx}{\Delta}\tau, \qquad y = y_0 + \dfrac{dy}{\Delta}\tau, \qquad z = z_0 + \dfrac{dz}{\Delta}\tau.$

Since Δ/dt is not infinitesimal, we find without difficulty that

$$\frac{dx}{\Delta} \simeq \frac{x'(t_0)}{\sqrt{w(t_0)}}, \qquad \frac{dy}{\Delta} \simeq \frac{y'(t_0)}{\sqrt{w(t_0)}}, \qquad \frac{dz}{\Delta} \simeq \frac{z'(t_0)}{\sqrt{w(t_0)}}.$$

It follows that for any standard value of τ, the point given by 3.8.9 is infinitely close to

3.8.10 $x = x_0 + \dfrac{x'(t_0)}{\sqrt{w(t_0)}}\,\tau, \qquad y = y_0 + \dfrac{y'(t_0)}{\sqrt{w(t_0)}}\,\tau, \qquad z = z_0 + \dfrac{z'(t_0)}{\sqrt{w(t_0)}}\,\tau\,.$

More generally, for any finite τ the point given by 3.8.9 is infinitely close to the point obtained from 3.8.9 if we replace τ by $^0\tau$.

On the other hand, let

3.8.11 $\qquad x = \xi + \alpha\tau', \qquad y = \eta + \beta\tau', \qquad z = \zeta + \gamma\tau'\,,$

where $\xi, \eta, \zeta, \alpha, \beta, \gamma$ are standard, α, β, γ are direction cosines and τ' is a parameter such that for any finite value of τ in 3.8.9, the corresponding point is infinitely close to some standard point of 3.8.11. If $\tau = 0$ in 3.8.9 corresponds to $\tau' = \tau'_0$ then

$$x_0 = \xi + \alpha\tau'_0, \qquad y_0 = \eta + \beta\tau'_0, \qquad z_0 = \zeta + \gamma\tau'_0$$

and so 3.8.11 may also be written as

3.8.12 $\qquad x = x_0 + \alpha(\tau' - \tau'_0), \qquad y = y_0 + \beta(\tau' - \tau'_0), \qquad z = z_0 + \gamma(\tau' - \tau'_0)\,.$

Next, suppose that 3.8.9, for $\tau = 1$ is infinitely close to 3.8.11 for $\tau' = \tau'_1$. This yields

$$^0\!\left(\frac{dx}{\varDelta}\right) = \alpha(\tau'_1 - \tau'_0), \qquad ^0\!\left(\frac{dy}{\varDelta}\right) = \beta(\tau'_1 - \tau'_0), \qquad ^0\!\left(\frac{dz}{\varDelta}\right) = \gamma(\tau'_1 - \tau'_0)\,.$$

Squaring and adding, we obtain $(\tau'_1 - \tau'_0)^2 = 1$ and so

$$\alpha = \pm\,^0\!\left(\frac{dx}{\varDelta}\right), \qquad \beta = \pm\,^0\!\left(\frac{dy}{\varDelta}\right), \qquad \gamma = \pm\,^0\!\left(\frac{dz}{\varDelta}\right).$$

The sign is the same in all three of these formulae and can be adjusted to $+$ by adjusting the sign of the parameter. Then $^0(dx/\varDelta) = x'(t_0)/\sqrt{w(t_0)}$, etc., which shows that 3.8.11 coincides with 3.8.10.

The points of 3.8.9 which correspond to finite values of τ are precisely the *finite* points of that straight line, i.e., the points all of whose coordinates are finite. Thus, the tangent to C of P may be obtained (or defined) as the unique straight line l in R such that for every finite point p on the straight line through P and Q, Q a point on C which is infinitely close to P, there is a point on l which is infinitely close to p.

Instead of taking the straight line through the standard point P and a point Q which is infinitely close to P on C, we might equally well have taken the straight line through two distinct points P_1 and P_2 on C both of which are infinitely close to C.

Suppose now that the functions $x(t)$, $y(t)$, $z(t)$ have continuous second derivatives. In order to obtain the osculating plane at P, we may then consider the plane PQ_1Q_2 through points $P = (x_0, y_0, z_0)$, $Q_1 = (x_1, y_1, z_1)$, $Q_2 = (x_2, y_2, z_2)$, where

$$
\begin{aligned}
x_0 &= x(t_0), & y_0 &= y(t_0), & z_0 &= z(t_0), \\
x_1 &= x(t_0 - dt), & y_1 &= y(t_0 - dt), & z_1 &= z(t_0 - dt), \\
x_2 &= x(t_0 + dt), & y_2 &= y(t_0 + dt), & z_2 &= z(t_0 + dt),
\end{aligned}
$$

where dt is a positive infinitesimal number. The osculating plane to the curve at P is then obtained as the unique plane p in R such that for every finite point Q on PQ_1Q_2 there is a standard point on p which is infinitely close to Q.

The three (or more) dimensional space over $*R$ provides an example of a non-archimedean geometry which might be investigated for its own sake but this line will not be pursued here.

3.9 Remarks and references. The existence of non-standard models of Arithmetic was discovered by T. Skolem (SKOLEM [1934]). Skolem's method foreshadows the ultrapower construction and is of interest also in its own right. Among subsequent papers which are concerned wholly or in part with Non-standard Arithmetic we mention HENKIN [1949] (where the order type of non-standard models of Arithmetic is given), ROBINSON [1952, 1961], GILMORE and ROBINSON [1955], KEMENY [1958], MACDOWELL and SPECKER [1961], SCOTT [1961], RABIN [1961], MENDELSON [1961], MÜLLER [1961].

The first paper on Non-standard Analysis is ROBINSON [1961a] which uses the language of the Lower Predicate Calculus. The elements of the Differential and Integral Calculus are expounded in ROBINSON [1963] in this framework. LUXEMBURG [1962] develops the theory with special emphasis on the ultrapower construction. The use of higher order languages in Non-standard Analysis was initiated in ROBINSON [1962].

CHAPTER IV

GENERAL TOPOLOGY

4.1 Topological spaces. A topological space consists of a non-empty set A of individuals, called *points,* and a set Ω of subsets of A, said to be *the open sets* of the space such that (i) the empty set (of individuals) and the entire set of points A are included in Ω, (ii) the intersection of any two elements of Ω is in Ω and (iii) the union of any (finite or infinite) number of elements of Ω is in Ω. The reader is expected to be familiar with the elements of the theory of topological spaces.

We shall identify a given topological space with the full structure T which is based on the set of points of the space, A, and in which the set of open sets, Ω, is a specified entity of type $((0))$ which satisfies conditions (i), (ii), and (iii) above.

Now let $*T$ be an arbitrary but fixed enlargement of T (see 2.10.5 above), and let $*A$ be the set of individuals of $*T$ also to be called points, where $A \subset *A$. In agreement with our general convention the points of A will be called *standard points.* The *set of open sets* in $*T$, $*\Omega$, is a standard set. It is the set which has the same name as Ω in the vocabulary of K. The elements of $*\Omega$ are internal sets in $*T$. Moreover, some of them are standard sets, having names in K which are the names of corresponding sets in T. Thus, if U is an open set in T, and $*U$ is the corresponding set in $*T$ then $*U$ is by definition a standard set. Moreover, $U \subset *U$; for if $p \in U$ then this can be stated by a sentence of K and the interpretation of that sentence in $*T$ is that $p \in *U$.

To the extent to which they can be formulated within K, the properties of Ω are (where necessary, with the appropriate reinterpretation) possessed also by $*\Omega$. Thus, the empty set and $*A$ are contained in $*\Omega$ and the intersection of any two elements of $*\Omega$ is in $*\Omega$. The union of a subset Φ of $*\Omega$ also is in $*\Omega$, *provided Φ is an internal set,* for the reinterpretation of (iii) above in $*T$ applies only to this case.

Let p be any standard point, and let Ω_p be the set of all open sets which contain p (open neighborhoods of p). Then we define the *monad* of p, $\mu(p)$ by

4.1.1
$$\mu(p) = \bigcap_{U_\nu \in \Omega_p} {}^*U_\nu .$$

Thus, $\mu(p)$ is the intersection of all standard sets in $^*T$ which are open neighborhoods of p. Evidently, $p \in \mu(p)$. $\mu(p)$ may well be an external set. We shall see in due course that the present definition of a monad is compatible with that given in Chapter III.

Let $B = \{B_\nu\}$ be a base for the topology of T. Then it is easy to see that $\mu(p)$ is also the intersection of all $^*B_\nu$ which contain p.

4.1.2 THEOREM.   Let p be any standard point. There exists an open set V in $^*T$ (i.e., V is internal and an element of $^*\Omega$) such that V contains p and is contained in $\mu(p)$.

PROOF.   For a given standard point p, the following is a concurrent relation of type $((0),(0))$ in T.

'x is an open neighborhood of p (an open set which includes p) and y is an open neighborhood of p which is included in x'.

The x-domain of this relation coincides with the set of open neighborhoods of p. The relation is concurrent for if $U_1,...,U_n$ are open neighborhoods of p then $V = U_1 \cap ... \cap U_n$ is an open neighborhood of p which is included in $U_1,...,U_n$. Thus, there exists in $^*T$ an open set V, $V \in {}^*\Omega$, such that $p \in V$ and $V \subset {}^*U_\nu$ for all standard open sets $^*U_\nu$. But if so then $V \subset \cap {}^*U_\nu$, i.e., $V \subset \mu(p)$. This proves the theorem.

4.1.3 THEOREM.   Let p and q be two standard points such that $q \in \mu(p)$. Then $\mu(q) \subset \mu(p)$.

This is obvious.

Let S be any set of points in T. Any point p in T belongs to S if and only if it belongs to $^*S$.

4.1.4 THEOREM.   A set of points S in T is open if and only if for every point p in S, $\mu(p)$ belongs to $^*S$.

PROOF. The condition is necessary. For since $\mu(p)$ is the intersection of *all* standard open sets that include p it must be a subset of $*S$.

The condition is also sufficient. For let $p \in S$. Then $\mu(p) \subset *S$ by assumption. Hence, by 4.1.2, it is true in $*T$ that 'there exists an open neighborhood V of p, such that $V \subset *S$'. Reinterpreting the sentence in quotation marks in T, we find that there exists an open neighborhood V_p of p which is included in S. Hence

$$S = \bigcup_{p \in S} V_p$$

which shows that S is open.

Using the fact that a closed set is, by definition, the complement of an open set, we obtain.

4.1.5 THEOREM. A set of points S in T is closed if and only if for any standard point p which does not belong to S, the monad $\mu(p)$ has no point in common with $*S$.

For any S in T, the closure operation $S \rightarrow \bar{S}$, where $\bar{S}$ is the intersection of all closed sets that contain S, defines a mapping from the set of subsets of A into the set of closed subsets of A. On passing from T to $*T$, we find that the corresponding operation – still to be indicated by a bar – maps the set of internal subsets of $*A$ into the set of closed internal subsets of $*A$. Moreover, for any subset S of A, $\overline{*S} = *(\bar{S})$. For if s denotes the set S in K, and $\bar{s}$ denotes the closure of S, and c denotes the binary relation which holds between a set and its closure, then $\Phi_{((0),(0))}(c,s,\bar{s})$ belongs to K. Interpreting this sentence in $*T$, where s denotes $*S$ and $\bar{s}$ denotes $*(\bar{S})$, we obtain precisely $\overline{*S} = *(\bar{S})$.

4.1.6 THEOREM. A point p in T belongs to the closure $\bar{S}$ of a set S in T if and only if $\mu(p)$ has a point in common with $*S$.

PROOF. If $\mu(p)$ has a point in common with $*S$ then it has a point in common with every set that includes $*S$, in particular with $\overline{*S} = *(\bar{S})$. Hence, by 4.1.5, $p \in \bar{S}$, the condition is sufficient.

The condition is also necessary. For suppose that $\mu(p)$ and $*S$ are disjoint. By 4.1.2, there exists a set $V \in *\Omega$, $p \in V$ such that $V \subset \mu(p)$ and hence, such that V and $*S$ are disjoint. Let $W = *A - V$, then W is an internal set in *T such that W is closed, and $p \notin W$. Furthermore, $\overline{*S} \subset W$ since *$S \subset W$ and so p does not belong to $\overline{*S} = *(\bar{S})$. The corresponding fact in T is that p does not belong to $\bar{S}$. This proves the theorem.

The boundary of a set of points S in T is, by definition, the intersection of $\bar{S}$ and $\overline{A - S}$. The following theorem is an immediate consequence of 4.1.6.

4.1.7 THEOREM. A point p belongs to the boundary of a set S in T if and only if $\mu(p)$ has points in common both with *S and with $*(A - S) = *A - *S$.

Next we consider separation properties.

4.1.8 THEOREM. T is a Hausdorff (or $T_2 -$) space if and only if for any two distinct points p and q in T, the monads $\mu(p)$ and $\mu(q)$ are disjoint.

The condition is necessary. For if T is a $T_2 -$space then the given p and q have disjoint open neighborhoods U and V. Then $U \cap V = \emptyset$ and so *$U \cap *V = \emptyset$ and $\mu(p) \subset *U$, $\mu(q) \subset *V$, and hence $\mu(p) \cap \mu(q) = \emptyset$. This shows that the condition is necessary. Conversely, if $\mu(p) \cap \mu(q) = \emptyset$ then, 'there exist internal open sets U and V (in *T) such that $p \in U$, $q \in V$, $U \cap V = \emptyset$' as we see by choosing $U \subset \mu(p)$, $V \subset \mu(q)$, $p \in U$, $q \in V$. Such sets exist, by 4.1.2. Since the sentence in quotation marks holds also in *T with the appropriate verbal alterations, the theorem follows.

4.1.8 shows that for a Hausdorff space a point p in *T is in the monad of not more than one standard point q. We write $q = {}^0p$ and we call q the standard part of p.

We prove by the same method.

4.1.9. T is a $T_0 -$space if and only if for any two points p and q in T either $p \notin \mu(q)$ or $q \notin \mu(p)$.

Let S be any set in T, and let $\{U_v\}$ be the set of open sets in T which contain S. We then define the set $\mu(S)$ as the intersection $\bigcap_v *U_v$. Evidently, this generalizes the notion of a monad introduced previously. The method used in the proof of 4.1.2 leads to

4.1.10 THEOREM. Let S be any set in T. Then there exists an internal open set V in *T such that *$S \subset V \subset \mu(S)$.

The proof of the following theorem relies on 4.1.2 and 4.1.10, just as the proof of 4.1.8 relies on 4.1.2.

4.1.11 THEOREM. In order that the topological space T be regular it is necessary and sufficient that for every set of points S in T and for every point p in $T-S$, $\mu(S) \cap \mu(p) = \emptyset$. Similarly,

4.1.12 THEOREM. In order that the topological space T be normal it is necessary and sufficient that for any two disjoint open sets in T, S_1 and S_2, the intersection $\mu(S_1) \cap \mu(S_2)$ is empty.

The following characterization of *compactness* has important applications apart from its own intrinsic interest. A point p in *T will be called *near-standard* if there exists a standard point q such that $p \in \mu(q)$.

4.1.13 THEOREM. A topological space T is compact if and only if all points of *T are near-standard.

PROOF. Suppose that there exists a point p in *T which is not near-standard. Then p is not contained in the monad of any (standard) point in T. Thus, every point q in T possesses open neighborhoods U such that $p \notin$ *U. Let $\{U_v\}$ be the set of all open sets in T such that $p \notin U_v$. As we have just seen, $\bigcup_v \{U_v\} = A$, so that $\{U_v\}$ constitutes an open covering of T. If T is compact then this covering contains a finite subcovering $\{U_1, ..., U_k\}$, say, $k \geq 1$, such that

4.1.14 $$U_1 \cup U_2 \cup \cdots \cup U_k = A .$$

Now this equation can be formulated as a sentence of K which, interpreted in *T, yields,
$$^*U_1 \cup {}^*U_2 \cup \cdots \cup {}^*U_k = {}^*A .$$

But this equation entails that $p \in$ *U_j for some j, $1 \leq j \leq k$, a contradiction which shows that the condition of the theorem is necessary.

The condition is also sufficient. Suppose that T is not compact. Then there exists a set of open sets in T, to be called Ψ, such that the union $\bigcup \Psi$ is equal to A, but such that no finite subset of Ψ is a covering of T. Consider the relation $R(x,y)$ in T which holds between x and y if x is an element of Ψ and y is a point in T which does not belong to x, briefly

$x \in \Psi \wedge \neg y \in x$. Then R is concurrent. It follows that there exists a point p in $*T$ such that $p \notin *U$ for all $U \in \Psi$. Now let q be any point in T. Then $q \in V$ for some $V \in \Psi$, but $p \notin V$. It follows that $p \notin \mu(q)$. This completes the proof of 4.1.13.

By relativizing 4.1.13 to a set of points in T we obtain

4.1.15 COROLLARY. A set of points B in a topological space T is compact if and only if for every point $q \in *B$ there exists a standard point $p \in B$ such that $q \in \mu(p)$.

If we wish to consider several topological spaces simultaneously, we may think of their respective sets of points A_j as subsets of a single set of individuals A. Then the full structure M with set of individuals A contains full structures T_j which are based on the respective A_j. Correspondingly, if $*T$ is an enlargement of T then it contains enlargements $*T_j$ of the respective T_j (see above). We identify the topological spaces in question with the respective T_j. Going further, if we are given a set of topological spaces $\{T_v\}$ which are indexed in an index set $I = \{v\}$, we may assume that both the sets of points A_v and the index set I are contained in a full structure M, and we may proceed from there to an enlargement $*M$ of M which contains enlargements $*T_v$, $*I$ of the spaces T_v *and of the index set I*. It is irrelevant whether the several spaces and the index set overlap in M or are disjoint.

Within this setting, let $\{T_v\}$ be a set of topological spaces indexed in a set $I = \{v\}$. The Cartesian product $\Pi = \times_v \{T_v\}$ is the set of all functions $f(x)$ on I such that $f(v) \in A_v$ for all $v \in I$, where A is the set of points in T_v. Thus, Π is an entity within the structure M introduced above. We endow Π with the usual product topology and we shall from now on use the symbol Π in order to indicate the space *with* its topology. We recall that a base of the product topology is given by the sets

4.1.16 $U = \{f(x) | f(v_i) \in U_i\}, \qquad i = 1, \dots, k$

where $\{v_1, \dots, v_k\}$ is any finite sequence of elements of I, and U_i is an open set in T_{v_i}, $i = 1, \dots, k$.

4.1.17 THEOREM. Let $p = f(x)$ and $q = g(x)$ be two points in $*\Pi$ such that p is standard. Then $q \in \mu(p)$ if and only if $g(v) \in \mu(f(v))$ for all standard v.

PROOF.   If for some v_0, $g(v_0) \notin \mu(f(v_0))$, then there exists a standard open neighborhood of $f(v_0)$, $*U$, such that $g(v_0) \notin *U$. But then $g(x)$ does not belong to the standard open set $*V$ in $*\Pi$ which corresponds to

$$V = \{h(x) \mid h(v_0) \in U\}$$

although V contains $f(v)$. We conclude that the condition is necessary.

The condition is also sufficient. For if U is an element of the base of Π as given by 4.1.15, then $g(v_i) \in \mu(f(v_i))$, $i = 1, \ldots, k$ and so $g(v_i) \in *U_i$. Hence, $g(x) \in *U$, is required. This completes the proof of 4.1.17.

Observe that the condition of 4.1.17 makes no mention of the non-standard elements of $*I$.

4.1.18   THEOREM (Standard).   Suppose that the space T_v are Hausdorff spaces. Then Π is a Hausdorff space.

PROOF.   Let $p = f(x)$ and $q = g(x)$ be two distinct points in Π. Then $f(v_0) \neq g(v_0)$ for some $v_0 \in I$ and $\mu(f(v_0)) \cap \mu(g(v_0)) = \emptyset$. But then the monads $\mu(p)$ and $\mu(q)$ in $*\Pi$ cannot have any point in common, by 4.1.17.

4.1.19   THEOREM (Standard. Tychonoff).   Suppose that the spaces T_v are compact. Then Π is compact.

PROOF.   By 4.1.13 we only have to show that every point $p = f(x)$ in $*\Pi$ is near-standard. Now $f(v)$ is near-standard for all $v \in I$, by assumption. Hence, using the axiom of choice, we may determine a point in Π, $q = g(x)$ such that $f(v) \in \mu(g(v))$ for all $v \in I$. But then $p \in \mu(q)$, by 4.1.17. This completes the proof of 4.1.19.

We had to make use of the axiom of choice in order to determine $g(x)$. However, if the T_v are Hausdorff spaces every $f(v)$ determines uniquely the standard point to whose monad it belongs so that, in this case, $g(v)$ as a function is determined uniquely, and the axiom of choice is not employed.

4.2   Sequences, nets, mappings.   The discussion of infinite sequences of points in a topological space T involves both T and the natural numbers, N. Similarly as in 4.1, we shall suppose that both T and N are embedded in a full structure M. On passing to an enlargement $*M$ of M we then obtain within $*M$ enlargements $*T$ and $*N$ of T and N respectively, as explained. To an infinite sequence $\{p_n\}$ in M, where n ranges over N and the p_n are points in T there corresponds in $*M$ an infinite sequence, still to be denoted by $\{p_n\}$, which is defined for all n in $*N$, while the p_n now belong to $*T$. In accordance with our general terminology this $\{p_n\}$ is to be regarded as a standard sequence in $*M$.

Let T be a Hausdorff space. The point p is the *limit* of the sequence $\{p_n\}$ in T if for any given open neighborhood U of p there exists a natural number v such that $p_n \in U$ for $n > v$. In that case $p_n \in *U$ is certainly true for all infinite n (since for such n, $n > v$). This (compare the proof of 3.3.7) establishes the necessity of the condition in

4.2.1 THEOREM. Let $\{p_n\}$ be an infinite sequence in the Hausdorff space T. In order that p be the limit of $\{p_n\}$ in T it is necessary and sufficient that, in $*T$, $p_n \in \mu(p)$ for all infinite n.

To prove sufficiency, let U be an open neighborhood of p in T. Then it is true, in $*T$, that 'there exists a natural number v (i.e., any infinite natural number) such that $p_n \in *U$ for all $n > v$ (since in that case $p_n \in \mu(p) \subset *U$, by assumption)'. Excepting the words in parentheses, we may express the statement in quotation marks within K and reinterpret in M. This completes the proof of 4.2.1.

Since monads of distinct standard points in a Hausdorff space are disjoint there cannot be more than one limit to a given sequence.

For arbitrary T, p is a *limit point* of $\{p_n\}$ if for every open neighborhood U of p and every natural number v, $p_n \in U$ for some $n > v$.

4.2.2 THEOREM. Let $\{p_n\}$ and p be standard. Then p is a limit point of $\{p_n\}$ if and only if $p_n \in \mu(p)$ for some infinite natural number n.

Let U be an open neighborhood of p in $*T$ such that $U \subset \mu(p)$. Such a U exists by 4.1.2. Let v be any infinite natural number. If p is a limit point of $\{p_n\}$ then by the definition of a limit point which applies both in M and in $*M$, $p_n \in U$ for some $n > v$. But then $p_n \in \mu(p)$, proving necessity. We omit the proof of sufficiency.

4.2.3 THEOREM (Standard). In a compact space T, every sequence $\{p_n\}$ possesses a limit point p.

PROOF. Choose an arbitrary infinite n. Then p_n is near-standard, by 4.1.13, $p_n \in \mu(p)$ for some standard point p. p is a limit point of $\{p_n\}$ by 4.2.2.

According to the classical definition a *directed set* is a non-empty set J in which a transitive and reflexive relation $\prec$ is defined such that any two elements of J have a joint *upper bound*, i.e., such that for any a and b in J there exists a c in J for which $a \prec c$, $b \prec c$. This condition shows that the relation $x \prec y$ is concurrent. It follows that if $*J$ is an enlargement of J then there exist elements $b \in *J$ such that $a \prec b$ in $*J$ for all a in J. Such elements of $*J$ will be called *infinite*.

A *net* $f(x)$ is a mapping from a directed set J into a topological space T. $f(x)$ is said to *converge* to a point p in T if for every open neighborhood U of p there exists an $\alpha \in J$ such that $f(b) \in U$ for all b for which $\alpha \prec b$. (*Moore-Smith convergence*).

Similarly as before, we shall suppose that the given topological space and the directed set or sets under consideration are embedded in a single full structure M, so that we may deal with their enlargements within $*M$. In future corresponding provisions will be taken for granted without mention.

4.2.4 THEOREM. A net $f(x)$ converges to a point p in T if and only if $f(b) \in \mu(p)$ for all infinite b in J.

The proof is parallel to that of Theorems 3.3.7 and 4.2.1 and will be omitted.

A point p is a *cluster point (limit point)* of a net $f(x)$ if and only if for any open neighborhood U of p and for any $v \in J$, where J is the directed set which is the domain of $f(x)$, there exists an $n \in J$, such that $v \prec n$ and $f(n) \in U$.

4.2.5 THEOREM. In order that the point p be a cluster point of the net $f(x)$ in T it is necessary and sufficient that $f(n) \in \mu(p)$ for some infinite $n \in J$.

Proof omitted.

4.2.6 THEOREM (Standard). In order that a point p belong to the

closure of a set S in a topological space T it is necessary and sufficient
that there exist a net $f(x)$ whose values are in S and whose limit is p.

PROOF. Suppose p is the limit of a net $f(x)$ on the directed set J whose
values are in S. We have to show that $\mu(p)$ has a point in common with
*S (see 4.1.6). Now $f(n)\in A$ for all $n\in J$ and so $f(n)\in$*S for $n\in$*J, and in
particular for all infinite n in *J. But, for such n, $f(n)\in\mu(p)$ by 4.2.4, and so
*$S\cap\mu(p)\neq\emptyset$, which shows that the condition of the theorem is sufficient.

The condition is also necessary. Supposing $p\in S$, we define the directed
set J as the set of all open neighborhoods of p, with the relation of in-
clusion $\supset$ for $\prec$. Let $U\in J$, then we choose for $f(U)$ any element of
$U\cap S$ (which is a non-empty set). Then $f(x)$ converges to p. For if U is
any infinite element of *J then $U\subset\mu(p)$, since U is a subset of all standard
open neighborhoods of p in *T. Hence $f(U)\subset\mu(p)$, showing that p is the
limit of $f(x)$.

Next we consider mappings from a topological space T into a topological space S. As before, we suppose that T and S are embedded in
some full structure M and we consider their enlargements *T and *S
within an enlargement *M of M.

Let $f(x)$ be a mapping from (all of) T into S. By definition, or as an
immediate consequence of the definition, $f(x)$ is classically continuous at
a point p in T if for each open neighborhood U of $f(p)$ there is an open
neighborhood V of p such that the image of V under f, $f(V)$ is contained
in U, $f(V)\subset U$.

4.2.7 THEOREM. In order that the standard mapping $f(x)$ be continuous, at the standard point p it is necessary and sufficient that
$f(\mu(p))\subset\mu(f(p))$.

PROOF. The condition is necessary. For suppose $f(x)$ is continuous at p.
Let U be any standard neighborhood of $f(p)$. Then we have to show that
$f(\mu(p))\subset U$. Now, since $f(x)$ is continuous at p, $f(V)\subset U$ for some
standard open neighborhood V of p. But then $f(\mu(p))\subset f(V)\subset U$.

The condition is also sufficient. Suppose it is satisfied and let U be a
standard open neighborhood of $f(p)$. Let V be an open neighborhood of
p in *T such that $V\subset\mu(p)$. Such a set exists by 4.1.2. Then $f(V)\subset f(\mu(p))$
$\subset\mu(f(p))\subset U$. Hence, it is true in *$M$ that 'there exists an open neigh-

borhood Y of p such that $f(Y) \subset U'$. The statement in quotation marks must be true also in M. This proves the theorem.

4.2.8 THEOREM.   Let $f(x)$ be a homeomorphism from T onto S. Then, for every point p in $T, f(\mu(p)) = \mu(f(p))$.

This theorem is an immediate consequence of 4.2.7. It shows that the notion of a monad is invariant under standard topological transformations.

Let $*T$ be an enlargement of the topological space T. The internal open sets of $*T$ do not, in general, constitute a topology of $*T$, for it may be that the union of a set of internal open sets is not an internal set at all. However, it is true that $*T$ as well as the empty set of points are internal open sets. Moreover, since the intersection of any two open sets in T is an open set in T we find, on passing to $*T$, that the intersection of any two internal open sets is an internal open set. It follows that the internal open sets in $*T$ constitute a base for a topology in $*T$. We shall call it the *Q-topology*. The corresponding topological notions will be indicated by prefixing the letter Q, e.g., Q-open set, Q-interior, Q-boundary.

Observe that, in addition to the Q-notions, we also have the notions which are induced in $*T$ by the notions of T. For example, in T there exists the relation $R(x,y)$ of type $((0),(0))$ which assigns to every set of points x its boundary, y. The extension of this relation to $*T$ assigns to every *internal* set in $*T$ another internal set, its *boundary* (without quali-fication). Thus, we obtain, at first sight, two kinds of boundaries. How-ever, the following considerations show that the situation is less compli-cated then might be expected.

4.2.9 THEOREM.   Let B be an internal set of points in $*T$. Then B is Q-open if and only if it is open.

PROOF.   If B is an internal open set then it belongs to the specified base of the Q-topology. Hence, B is Q-open. If B is internal and Q-open, con-sider the set Γ of all internal open sets which are included in B. Then Γ is an internal set. Moreover, the union of an internal set of internal open sets is open, for this assertion is obtained by the reinterpretation in $*T$ of a formal sentence of K which states, for T, that the union of a set of open sets is open. But the union of Γ is precisely B. This proves 4.2.9.

Given an internal set B, we may characterize the Q-closure of B, D say, as the complement of the union of all internal open sets which are disjoint with B. But this characteristic shows that D is also the closure of B, $D = \bar{B}$. Hence

4.2.10 THEOREM. The closure and the Q-closure of an internal set coincide. The interior and the boundary of an internal set coincide with its Q-interior and Q-boundary, respectively.

4.3 Metric spaces. Let T be a metric space with distance function $\varrho(x,y)$. The definition and discussion of T involve T and the real numbers, R, so we assume as usual that T and R are embedded, simultaneously, in a full structure M. On passing to an enlargement $*M$ of M we can develop the non-standard theory of the given metric space.

The topology of T is defined, as usual, by specifying as base the set of all *open balls* B, where $B = \{q \mid \varrho(p, q) < r\}$ for a point p in T and a positive real r.

For *any* point p in $*T$ we define the monad of p, $\mu(p)$ as the set of all points q such that $\varrho(p,q)$ is infinitesimal. Distinct monads are disjoint. For a topological space, the notion of a monad $\mu(p)$ was defined only for standard p. For this case, the notion of a monad defined previously coincides with that given just now. Thus, the results given previously for topological spaces and involving monads of standard points apply also here. We write $p \simeq q$ if $\varrho(p,q) \simeq 0$ i.e., if $q \in \mu(p)$.

A point p in $*T$ will be called *finite* if there exists a standard point q such that $\varrho(p,q)$ is *finite*. Compare this with the definition of a near-standard point. p is near-standard if p is in the monad of a standard point q, i.e., in metric terminology if $\varrho(p,q)$ is *infinitesimal*. Thus, a near-standard point p is finite. It possesses a unique standard part, ^0p.

A metric space is *bounded* if there exists a real number m such that $\varrho(p,q) \le m$ for all points p and q in the space.

4.3.1 THEOREM. A metric space T is bounded if and only if all points of $*T$ are finite.

PROOF. The condition is necessary. For if p is an arbitrary point in $*T$ and q is any standard point then $\varrho(p,q) \le m$ for some finite m (as in T).

Hence $\varrho(p,q)$ is finite. Conversely, if T is not bounded, let q be a fixed point of T. Then it is true in T that 'for every real number in m there exists a point p such that $\varrho(p,q) > m$'. (For if not then T would be bounded, with bound $2m$). The statement in quotation marks can be formulated in K and hence, holds also in $*M$. Choosing m positive infinite in $*R$, we find that $\varrho(p,q)$ is infinite. But if so then $\varrho(p,q')$ cannot be finite for any standard point q' for $\varrho(q',q)$ is standard, hence finite, and $\varrho(p,q) \leq \varrho(p,q') + \varrho(q',q)$. This completes the proof of 4.3.1.

4.3.2 THEOREM. (Standard).   A compact metric space is bounded.

PROOF.   For any p in $*T$, $\varrho(p,q)$ is infinitesimal for some standard q, by 4.1.13. Hence, p is finite. This proves the theorem.

The condition that $\varrho(p,q)$ be finite, defines an equivalence relation in $*T$. Indeed, $\varrho(p,p)$ is finite, $\varrho(p,q)$ is finite if $\varrho(q,p)$ is finite and if $\varrho(p,q)$ and $\varrho(q,r)$ are finite then $\varrho(p,r)$ is finite. This equivalence relation divides $*T$ into a number of disjoint subsets which will be called the *galaxies* of $*T$. The finite points of $*T$ constitute its *principal galaxy*.

Let $\{p_n\}$ be a standard infinite sequence, and let p be a standard point. Then it is an immediate consequence of the corresponding theorems for topological spaces (4.2.1 and 4.2.2) or, alternatively, of the corresponding results for real numbers (3.3.7 and 3.3.12) that p is the limit of $\{p_n\}$ if $p_n \simeq p$ for *all* infinite n and that p is a limit point of $\{p_n\}$ if $p_n \simeq p$ for *some* infinite n.

Let B be any (internal or external) set of points in $*T$. Then the set of points ^0B in T is defined by

$$^0B = \{p | (\exists q)[q \in B \wedge \varrho(p,q) \simeq 0]\}$$

or, which is the same, by

$$^0B = \{p | B \cap \mu(p) \neq \emptyset\}.$$

4.3.3 THEOREM.   If B is an internal set of $*T$ then ^0B is closed.

PROOF.   Let p be a point in T which belongs to the closure of ^0B. Then the open ball of radius $\frac{1}{2n}$ about p, n a finite positive integer, contains a point $p' \in {}^0B$. The open ball of radius $\frac{1}{2n}$ about p' in turn contains a point

$q \in B$. Then $\varrho(p,q) < 1/n$. Now let D be the set of natural numbers such that the intersection of B and of the open ball of radius $1/n$ about p is not empty. We have just seen that D includes all finite positive integers. The usual argument shows that D contains also an infinite positive integer. Thus, for some infinite n, there exists a point $q_0 \in B$ such that $\varrho(p,q_0) < 1/n$. But then $q_0 \in \mu(p)$, $p \in {}^0 B$. This proves the theorem.

4.3.4 THEOREM. If B is a standard set then ${}^0(*B) = \bar{B}$ where $\bar{B}$ is the closure of B in T.

PROOF. If B is a standard set then $B \subset *B$ and so $B \subset {}^0(*B)$. Moreover, ${}^0(*B)$ is closed, by 4.3.3, so $\bar{B} \subset {}^0(*B)$. On the other hand if $p \notin \bar{B}$ then $\mu(p) \cap *B = \emptyset$, so $p \notin {}^0(*B)$. This proves $\bar{B} = {}^0(*B)$.

4.3.5 THEOREM. If $\mu(p)$ is an internal set then p is isolated.

PROOF. Suppose $\mu(p)$ is an internal set. Then the set of natural numbers D which is defined by

$$ D = \left\{ n \,\middle|\, (\forall q) \left[\left[\varrho(p,q) < \frac{1}{n} \right] \supset \left[q \in \mu(p) \right] \right] \right\} $$

is internal. If p is not isolated, then D cannot contain any finite natural number. For in that case, if n is a finite natural number, there exists a standard point $q \neq p$ such that $\varrho(p,q) < 1/n$. But then $\varrho(p,q)$ is a standard positive number and so $q \notin \mu(p)$. This shows that n does not satisfy the defining condition for D and hence, does not belong to D. On the other hand, it is evident that D includes all infinite natural numbers. But the set of infinite natural numbers is external (see 3.1.7). This contradiction proves the theorem. Observe that if p is isolated then $\mu(p) = \{p\}$.

A Cauchy sequence in T is a sequence of points $\{p_n\}$ in T such that

$$ \lim_{\substack{n \to \infty \\ m \to \infty}} \varrho(p_n, p_m) = 0. $$

As an immediate consequence of 3.3.15, we obtain

4.3.6 THEOREM. $\{p_n\}$ is a Cauchy sequence if and only if all p_n belong to the same monad for infinite n.

T is *complete* if every Cauchy sequence in T has a limit.

4.3.7 THEOREM. Suppose T is complete and let p be a point of $*T$ which is not near-standard. Then there exists a standard positive ε such that $\varrho(p,q) \geq \varepsilon$ for all standard points q.

PROOF. Given T and p which satisfy the assumptions of the theorem suppose that no ε as asserted exists. For every finite positive integer n, we may then determine a standard point q_n such that $\varrho(p,q_n) < 1/n$. Then $\{q_n\}$ is a Cauchy sequence, since

$$\varrho(q_n,q_m) \leq \varrho(q_n,p) + \varrho(p,q_m) < \frac{1}{n} + \frac{1}{m}\,.$$

But T is complete so there exists a standard point q such that $\lim_{n \to \infty} q_n = q$, i.e., such that $q_n \in \mu(q)$ for all infinite n. On the other hand, since $\varrho(p,q_n) < 1/n$ for all finite n, the usual argument shows that this inequality is true also for sufficiently small infinite n. For such n, $\varrho(p,q_n)$ is infinitesimal and, at the same time $\varrho(q_n,q)$ is infinitesimal as q is the standard point which is the limit of $\{q_n\}$. But $\varrho(p,q) \leq \varrho(p,q_n) + \varrho(q_n,q)$ so that $\varrho(p,q)$ is infinitesimal. Thus, $p \in \mu(q)$, p is near-standard, a contradiction which proves the theorem.

4.3.8 THEOREM. Let $\{p_n\}$ be an internal sequence, and let ε be a standard positive number such that $\varrho(p_j,p_k) \geq \varepsilon$ for all finite j and k. Then there exists a natural number n for which p_n is not near-standard.

PROOF. Suppose on the contrary that p_n is near-standard for all natural numbers n. Define a standard sequence $\{q_n\}$ by $q_n = {}^0 p_n$ for all finite n. q_n is then defined automatically also for infinite n.

$\varrho(p_n,q_n)$ is infinitesimal for all finite n. Refering to 3.3.20, we conclude that there exists an infinite natural number ν such that $\varrho(p_n,q_n)$ is infinitesimal for all $n < \nu$. Choosing m infinite and smaller than ν, and putting $q = {}^0 p_m$, we then have $q_m \in \mu(q)$, so that q is a limit point of $\{q_n\}$, by 4.2.2. This implies that there exists an increasing sequence of standard natural numbers $n_1 < n_2 < n_3 < \dots$, such that $\varrho(q,q_{n_j}) < \tfrac{1}{4}\varepsilon$. Hence, for $j \neq k$, $n_j \neq n_k$ and

$$\varrho(q_{n_j},q_{n_k}) \leq \varrho(q,q_{n_j}) + \varrho(q,q_{n_k}) < \tfrac{1}{2}\varepsilon\,.$$

But

$$\varrho(p_{n_j},q_{n_j})\simeq 0,\qquad \varrho(p_{n_k},q_{n_k})\simeq 0$$

and, using the triangle inequality,

$$|\varrho(p_{n_j},p_{n_k})-\varrho(q_{n_j},q_{n_k})|\leq\varrho(p_{n_j},q_{n_j})+\varrho(p_{n_k},q_{n_k}).$$

Hence,

$$\varrho(p_{n_j},p_{n_k})\simeq\varrho(q_{n_j},q_{n_k}),$$

and thus,

$$\varrho(p_{n_j},p_{n_k})<\tfrac{1}{2}\varepsilon.$$

This contradicts our assumption on $\{p_n\}$ and proves 4.3.8.

4.3.9 THEOREM. The space T is complete if and only if for every standard Cauchy sequence $\{p_n\}$, there exists at least one infinite natural number ν such that p_ν is near standard.

The proof is straightforward and will be omitted.

4.3.10 THEOREM. Let $\{p_n\}$ be an internal sequence such that the points p_n belong to the same monad μ_0 for all finite n. Then there exists an infinite natural number ν such that $p_n\in\mu_0$ for all $n<\nu$.

For the proof, choose $p_0\in\mu_0$ and apply 3.3.10 for $s_n=\varrho(p_0,p_n)$.

4.3.11 THEOREM (Standard). Suppose every infinite sequence in a metric space T has a limit point. Then T is compact.

This is the converse of 4.2.3, for metric spaces. To prove it, we shall make use of the axiom of choice. Suppose that T is not compact. It then follows from that there exists a point p in $*T$ which is not near-standard. We distinguish two possibilities.

(i) There exists a standard $\varepsilon>0$ such that $\varrho(p,q)\geq\varepsilon$ for all q in T. In that case, we define an infinite sequence $\{q_n\}$ of points in T as follows. We choose an arbitrary point of T as q_0. Next we choose a point q_1 in T such that $\varrho(q_0,q_1)\geq\varepsilon$. Such a point q_1 exists. For the statement 'there exists a point x such that $\varrho(q_0,x)\geq\varepsilon$' is true in $*M$ (where we may choose $x=p$) and hence, in M. Next we choose a point q_2 in T such that

$\varrho(q_0,q_2)\geq\varepsilon$ and $\varrho(q_1,q_2)\geq\varepsilon$. Such a point q_2 exists. For the statement 'there exists a point x such that $\varrho(q_0,x)\geq\varepsilon$ and $\varrho(q_1,x)\geq\varepsilon$' is true in $^*M$ (where we may choose $x=p$) and hence, in M. Having chosen $q_0,...,q_n$, $n\geq2$, we next choose q_{n+1} in T such that $\varrho(q_0,q_{n+1})\geq\varepsilon$, $\varrho(q_q,q_{n+1})\geq\varepsilon,...$, $\varrho(q_n,q_{n+1})\geq\varepsilon$, where the consideration of p in $^*T$ again shows that such a point q_{n+1} exists in T. In this way we determine an infinite sequence $\{q_n\}$ in T such that $\varrho(q_j,q_m)\geq\varepsilon$ for $j\neq m$. Evidently, $\{q_n\}$ does not have a limit point.

(ii) For every standard $\varepsilon>0$, there exist a point q in T such that $\varrho(p,q)<\varepsilon$. In that case we may select a sequence of points in T, $\{q_n\}$, such that $\varrho(p,q_n)<(n+1)^{-1}$ for all finite natural numbers n. Now suppose that $\{q_n\}$ has a limit point q in T. Since p is not near-standard, there exists a finite positive integer m such that $\varrho(p,q)\geq1/m$. Also, since q is a limit point of $\{q_n\}$ there exists a finite $n>m$ such that $\varrho(q_n,q)<\frac{1}{2m}$. At the same time, $\varrho(p,q_n)<(n+1)^{-1}<(2m+1)^{-1}$. Hence,

$$\varrho(p,q)<\varrho(p,q_n)+\varrho(q_n,q)<\frac{1}{2m+1}+\frac{1}{2m}<\frac{1}{m}.$$

This contradiction shows that $\{q_n\}$ cannot have a limit point and completes the proof of the theorem.

Referring back to 4.3.3 we observe that the set ^0B can be defined by $^0B=\{p|B\cap\mu(p)\neq\emptyset\}$ for any internal or external set of points in a general topological space T. Under certain conditions we are then still able to obtain the conclusion of 4.3.3 as shown by the following theorem.

4.3.12 THEOREM. If B is an internal set of points in $^*T$, where T is a topological space which satisfies the first axiom of countability then ^0B is closed.

PROOF. Let p be a point in T which belongs to the closure of ^0B. By assumption, there exists a sequence of open neighborhoods of p, $\{U_n\}$ such that every neighborhood of p contains some U_n. Replacing $\{U_n\}$ by the sequence $\{V_n\}$, where $V_n=U_0\cap U_1\cap\cdots\cap U_n$, if necessary, we see that we may as well suppose from the outset that $U_0\supset U_1\supset U_2\supset\cdots$. Then $U_\omega\subset\mu(p)$ for all infinite ω. For if U is any open set which includes p, then $U\supset U_n$ for some finite n, and hence, certainly $U\supset U_\omega$. Also, since p belongs to the closure of ^0B, any $^*U_n$, n finite, contains a point of $^0B,q$, and hence contains $\mu(q)$. But $\mu(q) \cap B \neq \emptyset$ and so $^*U_n \cap B \neq \emptyset$ for all finite n. We conclude that $U_\omega \cap B \neq \emptyset$ also for some infinite ω (otherwise N would be an internal set, contrary to what has been proved). But $U_\omega \subset \mu(p)$, and so $\mu(p) \cap B \neq \emptyset$, $p \in {}^0B$, ^0B is closed. This proves the theorem.

4.4 Topologies in $^*T$. Let D be an internal open set in $^*T$. For every point $p \in D$, there exists an open ball B with radius $r > 0$ and center p such that $B \subset D$, for this is a property of open sets in T which can be formulated within K and hence, is true also in $^*T$. Now let Γ be the set of all internal open balls which are subsets of D. Γ is an internal set and so its union is an internal open set. But this is precisely D. Thus, we may take the set of all open balls in $^*T$ as a basis for the Q-topology of $^*T$. Recall that Theorems 4.2.9 and 4.2.10 are then applicable.

The Q-topology turns $^*T$ into a Hausdorff space.

A set of points B in $^*T$ will be called an S-ball if there is a point p in $^*T$ and a standard positive number r such that B consists of all points q in $^*T$ such that $^0(\varrho(p,q)) < r$. We denote this set by $S(p,r)$.

Consider the intersection of two S-balls, $S(p,r)$ and $S(p',r')$, and suppose that it is not empty. Then for any point q in the intersection,

$$\varrho(p,q) + \varrho(q,p') \geq \varrho(p,p')$$

and hence

$$^0(\varrho(p,q)) + {}^0(\varrho(q,p')) \geq {}^0(\varrho(p,p')),$$

so that

$$^0(\varrho(p,p')) < r + r'.$$

With the same assumption on q let $t = \varrho(p,q)$, $t' = \varrho(p',q)$, so that $^0t < r$, $^0t' < r'$. Put $\eta = \min(r - {}^0t, r' - {}^0t')$ so that η is standard positive. We claim that

$$S(q,\eta) \subset S(p,r), \qquad S(q,\eta) \subset S(p',r').$$

Indeed, for any $q' \in S(q,\eta)$, we have

$$\varrho(p,q') \leq \varrho(p,q) + \varrho(q,q')$$
$$^0(\varrho(p,q')) \leq {}^0(\varrho(p,q)) + {}^0(\varrho(q,q')) < {}^0t + r - {}^0t = r,$$

and so $q' \in S(p,r)$. The second half of the assertion is proved in the same way.

We have shown that every point in the intersection of two S-balls is the center of an S-ball which is included in that intersection. It follows that the S-balls may serve as a basis for a topology in *T. We call it the *S-topology*. It is not difficult to see that every set which is open in the S-topology is open also in the Q-topology. Thus, the Q-topology is finer than the S-topology. We shall indicate the notions of the S-topology by prefixing the letter S to the appropriate term, e.g., S-open set, S-interior, S-boundary.

4.4.1 THEOREM. A monad is the union of the open balls contained in it and is thus a Q-open set. A monad is also the intersection of all S-balls that contain it. Hence, it is the intersection of all S-open sets that contain it.

The first half is obvious. As for the second half, let μ_0 be any monad, so that $\mu_0 = \mu(p)$ for some point p in *T. For any $q \in \mu(p)$ and for any S-ball $S(p,r)$ we then have $q \in S(p,r)$ since ${}^0(\varrho(p,q)) = 0$. On the other hand, if $q \notin \mu(p)$ then ${}^0(\varrho(p,q)) = r_0 > 0$. In this case, q is not contained in the S-ball $S(p,r_0)$. This proves our assertion.

4.4.2 THEOREM. Let B be any internal set in *T. Then the point p belongs to the S-interior of B if and only if $\mu(p) \subset B$.

PROOF. If p belongs to the S-interior of B then $S(p,r) \subset B$ for some $r > 0$. But $\mu(p) \subset S(p,r)$ so the condition is necessary.

The condition is also sufficient, for suppose it is satisfied and let D be the internal set which is defined by

$$D = \{n \mid (\forall q) [\varrho(p,q) < \frac{1}{n} \supset q \in B]\},$$

D contains all infinite natural numbers. It follows that D contains also a finite positive natural number, v. Since the points q of $S(p,1/v)$ all satisfy $\varrho(p,q) < 1/v$, we conclude that $S(p,1/v) \subset B$, p is in the S-interior of B.

4.4.3 THEOREM. Let B be any internal set in *T. Then the point p belongs to the S-closure of B if and only if $\mu(p)$ has a point in common with B. And p belongs to the S-boundary of B if and only if $\mu(p)$ has points in common both with B and with the complement of B.

We define a relation of equivalence, $\simeq$, in $*T$ by the condition that $p \simeq q$ if p and q belong to the same monad, i.e., if $\varrho(p,q) \simeq 0$. (This is compatible with the relation introduced previously for the real numbers.) The relation $\simeq$ gives rise to a sort of quotient structure T_μ, as follows. The *points* of T_μ are the *monads* of $*T$. Let ψ be the function which maps every $p \in *T$ on the point $\xi = \mu(p)$ in T_μ. Then the open sets of T_μ shall be the images of the S-open sets of $*T$ under ψ. The space T, regarded as a subspace of $*T$ is homeomorphic, under ψ, to the subspace T_s of T_μ that consists of the monads which contain standard points.

Consider the distance function $\varrho(x,y)$ in $*T$. It is not difficult to see that if $p \simeq p'$, $q \simeq q'$ then $\varrho(p,q) \simeq \varrho(p',q')$. Hence $^0(\varrho(p,q)) = {}^0(\varrho(p',q'))$ provided one, and hence the other of these standard parts exist, i.e., provided $\varrho(p,q)$ is finite. For points ξ and η in T_μ such that $\varrho(p,q)$ is finite for $p \in \xi$, $q \in \eta$, i.e., such that p and q belong to the same galaxy we now *define* $\varrho(\xi,\eta)$ by $\varrho(\xi,\eta) = {}^0(\varrho(p,q))$. Moreover, in that case we shall also say that ξ and η belong to the same *galaxy*. Thus, the galaxies of T_μ are the images of the galaxies in $*T$ under ψ. The reader will verify without difficulty that the function $\varrho(\xi,\eta)$ provides a metric for each galaxy, and that the topology provided by this metric coincides, for each galaxy G, with the topology induced in G by the specified topology of T_μ.

4.4.4 THEOREM. *Every galaxy in $*T$ is both S-open and S-closed. Every galaxy in T_μ is both open and closed.*

Indeed, let G be a galaxy in $*T$. Then

$$G = \bigcup_{p \in G} S(p,1).$$

Hence, G is open. The complement of G is the union of all the other galaxies, which are also open. Hence, G is closed. The second part of the theorem now follows from the fact that ψ maps S-open sets onto open sets, by the definition of the topology of T_μ.

The subspace T_s of T_μ was introduced above as the homeomorphic image of T under ψ. It is not difficult to see that all points of T_s belong to the same galaxy and that ψ is an isometry on T.

4.4.5 THEOREM. Suppose T is complete. Then T_s is closed in T_μ.

PROOF. Let ξ be a point which is in the complement of T_s and let $p \in \xi$. Then p is not near-standard. Let ε be a standard positive number such as exists according to the conclusion of 4.3.7. Let B be the open ball of radius $\frac{1}{2}\varepsilon$ about ξ. Then B does not include any point of T_s. Thus, the union of all open balls which are disjoint with T_s coincides with the complement of T_s. This shows that T_s is closed.

The next theorem provides a closely related result for the case that T is not necessarily complete.

4.4.6 THEOREM. The closure of T_s in T_μ is isometric to the completion of T.

PROOF. Let C be the completion of T. C consists of equivalence sets of Cauchy sequences of T taken with respect to the equivalence relation $\pi \sim \chi$ if $\lim_{n \to \infty} \varrho(p_n, q_n) = 0$, where $\pi = \{p_n\}$, $\chi = \{q_n\}$. In order to determine the required isometric mapping from C onto $\bar{T}_s$ (the closure of T_s) proceed as follows. Let c be an element of C and let $\pi = \{p_n\}$ be a Cauchy sequence which belongs to c. Then the p_n all belong to the same monad μ_c for infinite n. Moreover, the condition $\lim_{n \to \infty} \varrho(p_n, q_n) = 0$ shows that if $\chi = \{q_n\}$ also belongs to c then $\varrho(p_n, q_n)$ is infinitesimal for infinite n and so q_n also belongs to μ_0, for such n. Thus, the relation $\mu_c = \Psi(c)$ defines a mapping from C into T_μ. For any given c, μ_c belongs to $\bar{T}_s$ since, by the definition of distance in T_μ, $\varrho(\mu_c, \mu(p_n))$ tends to 0 as n tends to infinity, and $\mu(p_n) \in T_s$. At the same time, we see without difficulty that different points of C are mapped into different points of $\bar{T}_s$.

Now let μ_0 be any point of $\bar{T}_s$ and let $p \in \mu_0$. For any standard positive integer n, there exists a point q_n of T such that $\varrho(p, q_n) < 1/n$. We then see as in the proof of 4.3.7 that $\chi = \{q_n\}$ is a Cauchy sequence. Moreover, $\varrho(p, q_n)$ is infinitesimal for infinite n so if χ is an element of the point c in C then $\mu_0 = \Psi(c)$. This shows that the mapping is from C onto $\bar{T}_s$. We leave it to the reader to verify that it is an isometry, completing the proof of 4.4.6.

The following notion is related to the S-topology. Let $\{p_n\}$ be an (internal or external) infinite sequence in *T, i.e., a mapping from the natural numbers of *N into *T. We shall say that the point p in *T is an F-limit for $\{p_n\}$ if for every *standard* positive ε there exists a *finite* natural

number v such that $\varrho(p,p_n)<\varepsilon$ for all *finite* $n>v$. Notice that if p is an
F-limit of $\{p_n\}$ then any other point in $\mu(p)$ also is an F-limit of $\{p_n\}$, and
that conversely, any two F-limits of p_n must belong to the same monad.
Notice also that the question whether or not p is an F-limit for $\{p_n\}$
depends only on the terms of the sequence which have finite subscripts.
However, if $\{p_n\}$ is an internal sequence then the fact that p is an F-limit
of $\{p_n\}$ affects the behavior of the sequence also up to some infinite subscript as shown by the following theorem.

4.4.7 THEOREM. Let $\{p_n\}$ be an internal sequence and let p be an
F-limit of $\{p_n\}$. Then there exists an infinite natural number λ such that
$p_n \simeq p$ for all infinite $n<\lambda$.

PROOF. We define an internal sequence of non-negative real numbers,
$\{s_n\}$, by $s_n=\varrho(p_n,p)$. Then the assumption of the theorem implies that
for every finite positive integer k there exists a finite positive integer
v_k such that $s_n<1/k$ for finite $n>v_k$. Moreover, we may suppose that
$v_1<v_2<v_3<\cdots$. Next, we define a *standard* sequence $\{t_n\}$ by $t_n=0$ for
$n\leq v_1$ and by $t_n=1/k$ for $v_k<n\leq v_{k+1}$. This defines t_n, in the first instance,
for all finite n and then, by passing from M to $*M$, for all infinite n. Then
$\lim_{n\to\infty}t_n=0$ and so $t_n\simeq 0$ for all infinite n. Also, the set of natural numbers
$D=\{n|t_n\geq s_n\}$ includes all finite natural numbers greater than v_1. Since D
is an internal set it must therefore include also all infinite numbers less
than some infinite λ. Then $s_n\simeq 0$ for $n<\lambda$ and hence $p_n\simeq p$ for such n.
This completes the proof of 4.4.7.

4.5   Functions, limits, continuity in metric spaces.   Let $f(x)$ be a function
which is defined on a set of points B in a metric space T and which takes
values in a metric space S (i.e., a *mapping* from B into S.) When stating
that a function is defined on a set B we mean by this that B is *included in*
the domain of definition of $f(x)$. Let p be a point which belongs to the
closure of B. Classically, s is the limit of $f(x)$ as x tends to p, $\lim_{x\to p}f(x)=s$
(where B is ignored in the notation) if for every $\varepsilon>0$ there exists a $\delta>0$
such that $\varrho(f(q),s)<\varepsilon$ for all q in $B-\{p\}$ such that $\varrho(q,p)<\delta$. The same
formal definition is then satisfied by $f(x),B,p,s$ also in $*M$ (which is the
chosen enlargement of the full structure M that includes T and S). We
prove without difficulty (compare 3.4.1)

4.5.1 THEOREM. Let $f(x)$ be a standard function from the standard set D in $*T$ (so that $D = *B$) into $*S$ and let p be a standard point which belongs to the closure of D. In order that $\lim_{x \to p} f(x) = s$ in D, where s is a standard point in $*S$, it is necessary and sufficient that

$$f(\mu(p) \cap (D - \{p\})) \subset \mu(s).$$

Now let $f(x)$ be a standard function which maps the standard set $D = *B$ into $*S$. Then the classical metric definition of continuity at a point – $\lim_{q \to p} f(q) = f(p)$ – together with 4.5.1 yields the following theorem which may also be obtained by the relativization of 4.2.7.

4.5.2 THEOREM. In order that the standard function $f(x)$, which maps the standard set $D = *B$ into $*S$, be continuous at the standard point p it is necessary and sufficient that $f(\mu(p) \cap D) \subset \mu(f(p))$.

Classically, the mapping $f(x)$ is uniformly continuous on B if for any given $\varepsilon > 0$ there is a $\delta > 0$ such that $\varrho(f(p),f(q)) < \varepsilon$ for all p and q in B such that $\varrho(p,q) < \delta$. The corresponding non-standard condition is given by

4.5.3 THEOREM. In order that the standard mapping from the standard set $D = *B$ into $*S$ be uniformly continuous it is necessary and sufficient that $f(p) \simeq f(q)$ whenever $p \simeq q$, p and q in D.

The proof is straightforward and will be omitted.

From 4.5.2 and 4.5.3 it follows immediately that for compact B, continuity implies uniform continuity. For suppose B is compact. Then all points p and q in $*B$ are near-standard. Hence, if $p \simeq q$, $p,q \in D$, then $p \simeq p_0$, $q \simeq p_0$ for some standard $p_0 \in B$. Hence, if $f(x)$ is continuous on B, $f(p) \simeq f(p_0)$, $f(q) \simeq f(p_0)$. Hence, $f(p) \simeq f(q)$, $f(x)$ is uniformly continuous.

Next we consider functions which are not (necessarily) standard. It turns out that the S-topology provides results which have classical applications. Throughout, $f(x)$ is supposed to be defined on a set of points of $*T$ and to take values in $*S$. (Recall that the term S-topology was introduced in 4.4, and is unrelated to the name of the present range space, S.)

$f(x)$ is said to be S-bounded on a set D if $f(x)$ is defined on B and if there exists a point q in $*S$ and a standard number m such that $\varrho(f(p),q) \leq m$ for all p in D.

4.5.4 THEOREM.   In order that the internal function $f(x)$ be S-bounded on the internal set D it is necessary and sufficient that all points of $f(D)$ (the set of images of points of D) belong to the same galaxy in *S.

PROOF.   The condition is clearly necessary. To see that it is also sufficient suppose that $f(x)$ is not S-bounded on D. Let F be the set of natural numbers which is defined by

$$F = \{n|\text{ for every point } q \text{ in } {}^*S \text{ there is a } p \in D \text{ such that } \varrho(f(p),q)>n\}.$$

Then F is an internal set which contains all finite natural numbers. It follows that F contains also an infinite natural number, v. Choose an arbitrary $p_0 \in D$ and put $q=f(p_0)$. Then by the definition of v, $\varrho(f(p), f(p_0))>v$ for some p in D. But this implies that $f(p)$ and $f(p_0)$ belong to different galaxies and proves the theorem.

   Let $f(x)$ be an internal function whose domain of definition includes a monad μ_0. Let $p \in \mu_0$, so that $\mu_0 = \mu(p)$. Then there exists a standard $\varepsilon > 0$ such that $f(x)$ is defined for all q for which $\varrho(p,q) \le \varepsilon$. (For $f(x)$ is defined for all q such that $\varrho(p,q) \le 1/n$, n infinite, and so the set of n for which this is true must include also finite numbers.)

4.5.5 THEOREM.   Suppose the internal function $f(x)$ is defined on a monad $\mu_0 = \mu(p)$ such that $f(\mu_0)$ is a subset of the galaxy G. Then there exists a standard $\varepsilon > 0$ such that $f(B) \subset G$ where B is the closed ball $\{q|\varrho(p,q) \le \varepsilon\}$.

PROOF.   Consider the product $h(q)=\varrho(f(p),f(q))\varrho(p,q)$. For any $q \in \mu_0$, $h(q)$ is infinitesimal since $\varrho(f(p),f(q))$ is finite and $\varrho(p,q)$ is infinitesimal. Thus, '$h(q)<1$ for all q such that $\varrho(p,q) \le 1/n$', is true for all *infinite* n. But if so then it must be true also for some finite positive $n=v$. We put $\varepsilon = 1/v$. Then for any q which is *not* in μ_0, $\varrho(p,q)$ is not infinitesimal and so $1/\varrho(p,q)$ is finite. Hence, for q not in μ_0 which belong to the set $\{q|\varrho(p,q) \le \varepsilon\}$,

$$\varrho(f(p),f(q))=\frac{1}{\varrho(p,q)}h(q)<\frac{1}{\varrho(p,q)}$$

is finite. This shows that $f(q)$ belongs to the same galaxy as $f(p)$ and proves the theorem.

Let $f(x)$ be defined on the set D and let p be a point which belongs to the S-closure of D. We say that the point s in $*S$ is an S-limit of $f(x)$ as x tends to p in D if for every *standard* $\varepsilon > 0$ there exists a *standard* $\delta > 0$ such that $\varrho(f(q),s) < \varepsilon$ for all q in $D - \{p\}$ for which $\varrho(q,p) < \delta$.

4.5.6 THEOREM. Suppose D is an internal set, p a point of the S-closure of D, and $f(x)$ an internal function which is defined on D. Then the point s is an S-limit of $f(x)$ as x tends to p in D if and only if

$$f(\mu(p) \cap (D - \{p\})) \subset \mu(s).$$

PROOF. Suppose s is an S-limit of $f(x)$ as x tends to p in D. If $q \in \mu(p)$ then q satisfies the condition $\varrho(q,p) < \delta$ for all standard positive δ. Hence, if q is at the same time in $D - \{p\}$ then $(f(q),s) < \varepsilon$ for arbitrary standard positive ε. This shows that $f(q) \in \mu(s)$, from which it follows that the condition is necessary.

Suppose that the condition is satisfied, and let ε be any standard positive number. Then '$\varrho(f(q),s) < \varepsilon$ for all q in $D - \{p\}$ for which $\varrho(q,p) < 1/n$' where n is any infinite integer since for such n, $q \in \mu(p)$. But then the statement in quotation marks must be true also for some finite n. Setting $\delta = 1/n$ we see that the condition of the theorem is also sufficient.

Observe that if s is an S-limit of $f(x)$ as x tends to p, as above, then the same applies to any other point in $\mu(s)$. Moreover, all these are S-limits also for x tending to any other point of $\mu(p)$.

Let $f(x)$ be defined on D, then $f(x)$ is S-*continuous* at the point p in D if $f(p)$ is an S-limit of $f(x)$ as x tends to p in D. Equivalently, $f(x)$ is S-continuous at p in D if for every standard $\varepsilon > 0$ there exists a standard $\delta > 0$ such that $\varrho(f(p),f(q)) < \varepsilon$ for any q in D such that $\varrho(p,q) < \delta$. Observe that this yields the correct notion of continuity for the S-topology although the specified basis for that topology consisted of open balls $\{q \mid {}^0\varrho(p,q) < \varepsilon\}$ for any p and standard positive ε. However, the condition that for every standard $\varepsilon > 0$ there exists a standard $\delta > 0$ such that ${}^0\varrho(f(p),f(q)) < \varepsilon$ for any q for which ${}^0\varrho(p,q) < \delta$ is equivalent to that given above.

As an immediate consequence of 4.5.6, we obtain

4.5.7 THEOREM. Let $f(x)$ be an internal mapping from an internal set D

into $^*S$. Then $f(x)$ is S-continuous at a point p in D if and only $f(\mu(p) \cap D)$ $\subset \mu(f(p))$.

Reference to 4.5.2 shows that this is exactly the condition for continuity of a standard function on a standard set at a standard point. We may in fact see directly that in that case continuity and S-continuity coincide. This is not the case in general. For example, if T and S are both given by the system of real numbers regarded as a metric space then the function defined by

$$f(0)=0, \qquad f(x)=\alpha \sin \frac{1}{x} \text{ for } x \neq 0, \alpha \text{ infinitesimal}, \alpha \neq 0,$$

is internal. $f(x)$ is S-continuous but not continuous at the point $x=0$. On the other hand, the function $f(x)=x^2$ is continuous but not S-continuous at any point $x=\omega$ where ω is infinite.

Let $f(x)$ be defined on the set of points D in $^*T$. We shall say that $f(x)$ is *uniformly* S-continuous in D if for every standard $\varepsilon>0$ there exists a standard $\delta>0$ such that $\varrho(f(p),f(q))<\varepsilon$ for any two points p and q in D for which $\varrho(p,q)<\delta$.

4.5.8 THEOREM. Let $f(x)$ be an internal function which is defined and S-continuous in (i.e., at all points of) an internal set D. Then $f(x)$ is uniformly S-continuous on D.

The theorem states, briefly, that, for internal functions and sets, S-continuity implies uniform S-continuity. Suppose the assumptions of the theorem on $f(x)$ and D are satisfied. It then follows from 4.5.7 that for every pair of points p,q in D, $p \simeq q$ implies $f(p) \simeq f(q)$. Now let ε be any standard positive number. Let F be the set of natural numbers n which satisfy the condition that for all p and q in D for which $\varrho(p,q)<1/n$, $\varrho(f(p),f(q))<\varepsilon$. Then F is an internal set which includes all infinite natural numbers. Accordingly, it includes also some finite positive integer, ν. Putting $\delta=1/\nu$ we see that $f(x)$ satisfies the condition of uniform S-continuity on D. This proves the theorem.

4.5.9 THEOREM. Let $f(x)$ be a function which is defined and S-continuous at all points of a set D which is S-connected (i.e., connected in the S-topology). Then the points of $f(D)$ all belong to the same galaxy in $*S$.

PROOF.   Let $p \in D$. We have to show that for all $q \in D$, $\varrho(f(p), f(q))$ is finite.

Let F be the set of all $q \in D$ such that $\varrho(f(p), f(q))$ is finite. Then $p \in F$. For any $q \in F$, $f(x)$ is S-continuous at q in D. Hence, there exists a positive standard ε_q such that $^0\varrho(f(q), f(r)) < 1$ for all $r \in D$ for which $^0\varrho(q, r) < \varepsilon_q$. For such r,

$$\varrho(f(p), f(r)) \leq \varrho(f(p), f(q)) + \varrho(f(q), f(r)) \leq \varrho(f(p), f(q)) + 1.$$

Hence $r \in F$. Thus, F is a union of sets which are S-open in D and is itself S-open in D.

If $F = D$ then we have finished. If F is a proper subset of D let H be the set of points $r \in D$ for which there is a point $q \in D - F$ such that $\varrho(f(q), f(r))$ is finite. Then $H \supset D - F$. Repeating the above argument we see that H is S-open in D. Moreover, F and H are disjoint. For if $r \in F \cap H$, then $\varrho(f(p), f(r))$ is finite, and $\varrho(f(q), f(r))$ is finite for some $q \in D - F$, and hence $\varrho(f(p), f(q))$ is finite by the triangle inequality. But this implies $q \in F$, a contradiction which shows that $H = D - F$. Thus, D is the union of the two disjoint sets F and H which are S-open in D. This contradicts the assumption that D is S-connected and proves that $F = D$. Accordingly, $\varrho(f(p), f(q))$ is finite for all p and q in D. This completes the proof.

Let $f(x)$ be a function which is defined on a set of points D in $*T$. The set ^0D was defined previously as the set of all standard points p such that $\mu(p)$ has a point in common with D. We define a function $^0f(x)$, to be called the *standard part* of $f(x)$ on a certain subset of ^0D as follows. Given $p \in {}^0D$ suppose that there exists a standard point q (in S and $*S$) such that $f(p') \simeq q$ for all points p' in $\mu(p) \cap D$. We then *define* $^0f(p) = q$. Let $D' \subset {}^0D$ be the set of points at which $^0f(x)$ is defined in this way. Passing to $*M$, we get a definition of $^0f(x)$ on $*D'$.

Some care is required in handling the definition of $^0f(x)$. Within our framework, the domain of a function is implicit in its definition, and the above D is supposed to be a subset of that domain. It follows that by extending D to a set D_1 we may actually destroy $^0f(x)$ at some point p. This will happen if, for a given $p \in {}^0D$, $f(p') \simeq q$ for all points $p' \in \mu(p) \cap D$ but not for some $p' \in \mu(p) \cap (D_1 - D)$. It cannot occur if D coincides with the domain of $f(x)$. And cannot arise, for a given $p \in {}^0D$, if $\mu(p) \subset D$.

4.5.10 THEOREM. Let the internal function $f(x)$ be defined on an internal set D such that $f(x)$ takes only near-standard values on D and is S-continuous on (i.e., at all points of) D. Then the function ${}^{0}f(x)$ exists and is uniformly continuous on ${}^{0}D$.

PROOF. Let $p\in{}^{0}D$. Since $f(x)$ is S-continuous on D, 4.5.7 shows that for any two points p_1 and p_2 in D which belong to $\mu(p)$, $f(p_1)\simeq f(p_2)$. Moreover, $f(p_1)$ is near-standard, by assumption, so ${}^{0}f(p)={}^{0}(f(p_1))={}^{0}(f(p_2))$. This shows that ${}^{0}f(x)$ is defined on ${}^{0}D$.

Let ε be a standard positive number. Since $f(x)$ is S-continuous on D it is actually uniformly S-continuous on D, by 4.5.8. It follows that there exists a standard $\delta>0$ such that $\varrho(f(p_1),f(q_1))<\varepsilon$ for all p_1 and q_1 in D such that $\varrho(p_1,q_1)<\delta$. We claim that at the same time $\varrho({}^{0}f(p),{}^{0}f(q))<2\varepsilon$ for all p and q in ${}^{0}D$ such that $\varrho(p,q)<\delta$. Indeed, let p and q be points of ${}^{0}D$ which satisfy this condition. By the definition of ${}^{0}D$ there exist points $p_1\in\mu(p)$ and $q_1\in\mu(p)$ which belong to D. Then $p\simeq p_1$, $q\simeq q_1$ and hence, by the triangle inequality, $\varrho(p,q)\simeq\varrho(p_1,q_1)$. This shows that $\varrho(p_1,q_1)<\delta$. Hence, by the definition of δ, $\varrho(f(p_1),f(q_1))<\varepsilon$. But ${}^{0}f(p)\simeq f(p_1)$ and ${}^{0}f(q)\simeq f(q_1)$, and so, again by the triangle inequality $\varrho({}^{0}f(p),{}^{0}f(q))\simeq\varrho(f(p_1),f(q_1))$. We conclude that $\varrho({}^{0}f(p),{}^{0}f(q))<2\varepsilon$, as asserted. This shows that $f(x)$ is uniformly continuous on ${}^{0}D$ and completes the proof of the theorem.

With the assumptions of 4.5.9, let p be a point in ${}^{0}D$ and let q be a point in $\mu(p)$ which, at the same time, belongs to both D and $*({}^{0}D)$. Then ${}^{0}f(p)\simeq f(q)$ by the definition of ${}^{0}f(x)$ and ${}^{0}f(q)\simeq{}^{0}f(p)$ by the continuity of ${}^{0}f(x)$. Hence ${}^{0}f(q)\simeq f(q)$.

4.6 Sequences of functions. Compact mappings. Let $\{f_n(x)\}$ be a sequence of functions which are defined on a set of points B in the metric space T and whose range is in the metric space S. According to the classical definition $\{f_n(x)\}$ converges *uniformly* on B, to a function $f(x)$, if for every positive ε there exists a natural number $\nu=\nu(\varepsilon)$ such that $\varrho(f(p),f_n(p))<\varepsilon$ for all p in B and for all $n>\nu$.

4.6.1 THEOREM. The sequence of standard functions $\{f_n(x)\}$ converges to the standard function $f(x)$ uniformly on $B\subset T$ (and hence, on $*B$) if and only if $f(p)\simeq f_n(p)$ for all $p\in *B$ and for all infinite n.

PROOF. Suppose that $\{f_n(x)\}$ converges to $f(x)$ uniformly on B. The classical definition then implies that for any standard $\varepsilon>0$ there exists a finite natural number v such that the following statement holds in M. 'For every $p \in B$ and $n>v$, $\varrho(f(p),f_n(p))<\varepsilon$'. Replacing B by $*B$ we then obtain a statement which is true in $*M$. This shows that, for infinite n, $\varrho(f(p),f_n(p))$ is smaller than any standard positive number and, accordingly, is infinitesimal. Thus, the condition of the theorem is necessary.

Sufficiency is established by noting that, for given standard positive ε, the condition $\varrho(f(p),f_n(p))<\varepsilon$ is satisfied for all $n>v$ where v is an arbitrary infinite natural number. Transferring back from $*M$ to M, we complete the proof.

4.6.2 THEOREM (Standard).   If the functions of the sequence $\{f_n(x)\}$ are defined and continuous on a set B and converge to a function $f(x)$ uniformly on B then $f(x)$ is continuous on B.

PROOF. Let $p \in B$ and $q \in \mu(p) \cap *B$. Then we have to show that $\varrho(f(p),f(q))$ is infinitesimal, in other words that for any given standard $\varepsilon>0$, $\varrho(f(p), f(q))<\varepsilon$. Now the classical definition of uniform convergence (see above) when applied to $*M$ shows that there exists a finite natural number v, such that $\varrho(f(r),f_n(r))<\frac{1}{4}\varepsilon$ for $n>v$ and for all r in $*B$. Thus, putting $n=v+1$,

$$\varrho(f(p),f(q)) \leq \varrho(f(p),f_n(p))+\varrho(f_n(p),f_n(q))+\varrho(f_n(q),f(q))<\varepsilon,$$

since $\varrho(f_n(p),f_n(q))$ is infinitesimal. This completes the proof.

Observe that in proving 4.6.2 we have not made use of the non-standard condition of 4.6.1. However, that condition will be applied presently in connection with a classical result on equicontinuity.

Working with standard notions, let $\{f_n(x)\}$ be a sequence of functions which map a metric space T into a compact metric space S. Suppose the sequence is uniformly equicontinuous on T. That is to say, for every $\varepsilon>0$ there exists a $\delta>0$ such that $\varrho(f_n(p),f_n(q))<\varepsilon$ provided $\varrho(p,q)<\delta$, $n=0,1,2,3,....$. Passing from M to $*M$, we see that the condition of uniform equicontinuity affirms that for given standard $\varepsilon>0$ there exists a standard $\delta>0$ such that $\varrho(p,q)<\delta$ implies $\varrho(f_n(p),f_n(q))<\varepsilon$ for any points p,q in $*T$ and for any natural number n, finite or infinite. This shows that $f_n(x)$ is uniformly S-continuous on $*T$.

Now let ω be an arbitrary but fixed infinite natural number. Then $f_\omega(x)$ takes only near-standard values since *all* points of $*S$ are near-standard. It follows that $F(x) = {}^0 f_\omega(x)$ exists and is uniformly continuous on T, by 4.5.10.

Suppose that T is compact. Then the points of $*T$ are near-standard and so, by the remarks made at the end of section 4.5, $F(p) \simeq f_\omega(p)$ for all p in $*T$. This discussion leads to –

4.6.3   THEOREM (Standard. Ascoli). Let $\{f_n(x)\}$ be an equicontinuous sequence of functions which map a compact metric space T into a compact metric space S. Then there exists a sequence of natural numbers $\{n_j\}$, $n_0 < n_1 < n_2 < \cdots$ such that the sequence $\{f_{n_j}(x)\}$, $j = 0,1,2,\ldots$ converges uniformly on T to a function $F(x)$ which is uniformly continuous on T.

To prove this theorem, we introduce $F(x) = {}^0 f_\omega(x)$ as above. Then $F(x)$ is uniformly continuous on T. It remains to select a suitable sequence $\{n_j\}$. Since S is compact, $\varrho(F(p), f_\omega(p))$ is infinitesimal for all points p in $*T$. Hence, for any standard $\varepsilon > 0$ and any finite natural number ν, it is true in $*M$ that there exists a natural number n greater than ν (i.e., $n = \omega$) such that $\varrho(F(p), f_n(p)) < \varepsilon$ for all points p in $*T$. Passing from $*M$ to M, we conclude that there exists a finite natural number $n > \nu$ such that $\varrho(F(p), f_n(p)) < \varepsilon$ for all points p in T. We put $\varepsilon_j = (j+1)^{-1}$, $j = 0,1,2,\ldots$ and determine the smallest $n = n_0$ such that $\varrho(F(p), f_{n_0}(p)) < \varepsilon_0$ for all points p in T. Next, we determine the smallest $n = n_1 > n_0$ such that $\varrho(F(p), f_{n_1}(p)) < \varepsilon_1$ for all p in T and then the smallest $n = n_2 > n_1$ such that $\varrho(F(p), f_{n_2}(p)) < \varepsilon_2$ for all p in T. Continuing in this way we obtain, without using the axiom of choice, an infinite sequence $\{n_j\}$, $n_0 < n_1 < n_2 < \cdots$ such that $\varrho(F(p), f_n(p)) < \varepsilon_j$, $j = 0,1,2,\ldots$. This completes the proof of 4.6.5.

If the $f_n(x)$ are real-valued, Ascoli's theorem is proved on the assumption that the functions are uniformly and collectively bounded. The appropriate space S is then provided by some finite interval of real numbers.

To conclude this section we shall prove some theorems on compact mappings. Recall that a set of points B in a metric space T is *bounded* if there is a (standard) real number m such that $\varrho(p,q) \leq m$ for all p and q in B. Recall also that a point p in $*T$ is *finite* if $\varrho(p,q)$ is finite for some

standard point q. A mapping $f(x)$ from a metric space T into a metric
space S is called *compact* if every bounded set of points B in T is mapped
by f into a set of points $f(B)$ whose closure $\overline{f(B)}$ is compact. The notion
is best known for the case that T and S are normed linear spaces and f is
a linear operator (see Chapter VII below).

4.6.4 THEOREM. Let $f(x)$ be a mapping from a metric space T into a
metric space S. Then $f(x)$ is compact if and only if it maps every finite
point in $*T$ on a near-standard point in $*S$.

PROOF. Suppose $f(x)$ is compact and let p be a finite point in $*T$. Since
p is finite, $\varrho(p,q)\leq m$ for some standard q and some standard real number
m. Consider the set $B=\{r\,|\,\varrho(r,q)\leq m\}$ in T. This is a bounded set (with
bound $2m$) in T, so the closure of $f(B)$ is compact. Moreover, $q\in *B$, so
$f(q)\in f(*B)=*(f(B))$ and hence, $f(q)\in *(\overline{f(B)}) =*(\overline{f(B)})$. But $\overline{f(B)}$ is
compact, by assumption and so the points of $*(\overline{f(B)})$, including $f(q)$, are
near-standard. This proves that the condition of the theorem is necessary.

The condition is also sufficient. Let B be a bounded set in T. By applying 4.3.1 to B (i.e., by taking B as the metric space of that theorem)
we find that the points of $*B$ are all finite. If the condition of the theorem
is satisfied then it follows that the points of $f(*B)$ are near-standard. We
have to show that, moreover, all $q\in *(\overline{f(B)})=\overline{f(*B)}$ are near-standard.
Let η be any positive infinitesimal number. Since q belongs to the closure
of $f(*B)$, $\varrho(q,p)<\eta$ for some $p\in f(*B)$. Then $q\simeq p$ and p is near-standard,
and so q also is near-standard. This shows that $\overline{f(B)}$ is compact (since
$\overline{f(B)}$ is closed) and completes the proof of 4.6.4.

4.6.5 THEOREM (Standard). Let $\{f_n(x)\}$ be a sequence of compact
mappings from a metric space T into a complete metric space S. Suppose
that $\{f_n(x)\}$ converges to a function $f(x)$ uniformly on T. Then $f(x)$ is
compact.

PROOF. Let p be any finite point in $*T$. It follows from 4.6.4 that the
point $f_n(p)$ are near-standard for all finite n. We have to show that $f(p)$
is near-standard.

Suppose $f(p)$ is not near-standard. It then follows by 4.3.7 that there
exists a standard $\varepsilon>0$ such that $\varrho(f(p),q)\geq\varepsilon$ for all points q in S. But

$\{f_n(x)\}$ converges to $f(x)$ *uniformly*, so $\varrho(f(p),f_n(p)) < \frac{1}{2}\varepsilon$ for sufficiently large finite n. For such n, $\varrho(f(p),q) < \frac{1}{2}\varepsilon$ where $q = {}^0(f_n(p))$. This contradiction proves the theorem.

4.7   Euclidean space.   Let T be the n-dimensional Euclidean space. Thus, there is a function $p_j = k(p,j)$ which yields for every point p of T and for every natural number j, $1 \leq j \leq n$ a real number p_j which is the jth coordinate of p. Conversely, for every sequence $(p_1,...,p_n)$ of real numbers there is a unique point with these coordinates. As a result, we may if we wish refrain from introducing T separately from R, and we may regard T as the full structure based on the set of all n-tuples of real numbers. Looked at in this way, the set of points in T is A^n where A is the set of standard real numbers and the set of points in the enlargement, $*T$, is $*(A^n) = (*A)^n$, where $*A$ is the set of all real numbers in $*R$. The results obtained previously on metric spaces are applicable, where the distance function in T and $*T$ is given by $\varrho(p,q) = \sqrt{\{(x_1 - y_1)^2 + (x_2 - y_2)^2 + \cdots + (x_n - y_n)^2\}}$ where $p = (x_1, x_2, ..., x_n)$ and $q = (y_1, y_2, ..., y_n)$. In particular, the distance of a point p from the origin, or *norm of* p is given by

$$\|p\| = \varrho(p,o) = \sqrt{\{p_1^2 + p_2^2 + \cdots + p_n^2\}},$$

where $o = (0,0,...,0)$. The finite points of $*T$, i.e., the elements of its principal galaxy, are those points whose norm is finite. For $n = 1$, the principal galaxy consists of the real numbers which were called finite previously, so that our terminology is consistent.

   Theorem 4.1.13 provides a very simple proof of the theorem of Heine–Borel.

4.7.1   THEOREM (Standard. Heine–Borel).   Let B be a set of points in A^n which is closed and bounded. Then B is compact.

PROOF.   We have to show that for every $p \in *B$ there exists a $q \in B$ such that $\varrho(p,q) \simeq 0$. Let $p = (p_1, p_2, ..., p_n)$. Then p is finite, by 4.3.1, since B is bounded. Hence $p_1, p_2, ..., p_n$ are finite and $q = ({}^0p_1, {}^0p_2, ..., {}^0p_n)$ exists. Moreover,

$$\varrho(p,q) = \sqrt{\{(p_1 - {}^0p_1)^2 + (p_2 - {}^0p_2)^2 + \cdots + (p_n - {}^0p_n)^2\}} \simeq 0.$$

The point q belongs to B since $p \in \mu(q)$ and B is closed. This completes the proof.

Let D be the closed unit ball in A^n, $D = \{p \mid \|p\| \leq 1\}$. Then the interior of D, which is the same as its Q-interior, is given by $\|p\| < 1$ while the boundary of D is the unit sphere. At the same time, the S-interior of D is given by ${}^0\|p\| < 1$ and is thus a proper subset of the interior of D, and the S-boundary of D consists of the points p for which ${}^0\|p\| = 1$ and this contains the boundary of D as a proper subset. The S-interior of D is not an internal set. This is an immediate consequence of the following theorem.

4.7.2 THEOREM. The only internal sets of points in $*(A^n)$ which are S-open are the empty set and the entire space.

PROOF. Let D be an internal set in $*(A^n)$ such that $D \neq \emptyset$ and $D \neq *(A^n)$, and let p and q be points which belong to D and to the complement of D, respectively. Let F be the straight segment from p to q, i.e., the set of points $r(\lambda) = (1 - \lambda)p + \lambda q$, $0 \leq \lambda \leq 1$. Let G be the set of all numbers λ, $0 \leq \lambda \leq 1$, such that $r(\lambda') \in D$ for all $1 \leq \lambda' \leq \lambda$. Then G is an internal set. G is not empty since it contains $\lambda = 0$ and if $\lambda_1 \in G$ and $0 \leq \lambda_2 < \lambda_1$, then $\lambda_2 \in G$. Arguing from the corresponding situation in R, we conclude that there exists a number λ_0, $0 \leq \lambda_0 \leq 1$ such that $\lambda \in G$ for $0 \leq \lambda < \lambda_0$ and $\lambda \notin G$ for $\lambda_0 < \lambda \leq 1$.

Suppose that D is S-open. If $r(\lambda_0) \in D$ then $\lambda_0 < 1$ since $r(1) = q \notin D$. Also, there exists a standard positive ε such that $s \in D$ for $\varrho(s, r(\lambda_0)) < \varepsilon$. But this implies the existence of a $\lambda_1, \lambda_0 < \lambda_1 \leq 1$, such that $r(\lambda_1) \in D$ for all λ for which $\lambda_0 \leq \lambda < \lambda_1$. This contradicts the definition of λ_0 and shows that we have to assume $r(\lambda_0) \notin D$. In that case $\lambda_0 \neq 0$ since $r(0) = p \in D$. We may therefore select a λ_1 such that $0 < \lambda_1 < \lambda_0$ and $r(\lambda_1) \in D$ and

$$\varrho(r(\lambda_1), r(\lambda_0)) = \|r(\lambda_1) - r(\lambda_0)\| = |\lambda_1 - \lambda_0| \varrho(p, q)$$

is infinitesimal. But then there exists a standard $\varepsilon > 0$ such that $s \in D$ for all points s for which $\varrho(r(\lambda_1), s) < \varepsilon$. This applies in particular to $s = r(\lambda_0)$ and forces us to conclude that $r(\lambda_0) \in D$. Accordingly, we have arrived at a contradiction, which proves the theorem.

Any straight segment in $*(A^n)$ is connected, as we may see by transfer from A^n. On the other hand, a straight segment in $*(A^n)$ need not be S-connected. For example, in $*A^1 = *A$, the set D of points (numbers) given by $0 \leq x \leq a$ where a is positive infinite is not S-connected since it can be decomposed into two S-open sets consisting of the finite and of the infinite elements of D, respectively. However, it is not difficult to show that a straight segment in $*(A^n)$ which consists of finite points only is indeed S-connected.

4.8 Remarks and references. Theorem 4.1.13 was presented at the Berkeley Symposium on Model Theory 1963 (ROBINSON [1965]), with a proof of 9.1.7 as an application. Immediately afterwards, S. Kripke, in response to a question by M. O. Rabin, produced a corresponding proof for Tychonoff's theorem (4.1.19 above).

CHAPTER V

FUNCTIONS OF A REAL VARIABLE

5.1 Measure and integration. In the present chapter, we shall consider a number of selected topics in the Theory of Functions of a real variable. In the discussion, particular care will be necessary in order to avoid confusion between the different kinds of finiteness and infinity.

Let R be the full structure on the set of ordinary real numbers, A, and let *R be an enlargement of R with set of individuals *$A \supset A$, all to be called real numbers, as in previous chapters. We shall say that an internal sequence of entities in *R, $\{a_n\}$, is *finite* if n varies over a set of natural numbers which is absolutely finite, i.e., n varies over subscripts $n \leq \nu$, where ν is a finite natural number. We shall say that $\{a_n\}$ is *Q-finite* if there exists a natural number ν, which may be finite or infinite, such that for all subscripts n which occur in the sequence, $n \leq \nu$. Finally, we shall say that $\{a_n\}$ is *infinite* if n ranges over all the natural numbers of *N, or over a set which is in internal one-to-one correspondence with the numbers of *N, e.g., the set of numbers of *N greater than some natural ν.

Let $\{a_n\}$ be an infinite internal sequence of numbers in *R. Then the sum $\sum_{n=0}^{\infty} a_n$ is given by the classical condition (whenever that sum exists). That is to say, $\sum_{n=0}^{\infty} a_n = a$, if for every positive ε in *R there exists a natural number ν in *N such that $|\sum_{n=0}^{k} a_n - a| < \varepsilon$ for $k > \nu$. The expressions $\sum_{n=0}^{k} a_n$ all have a meaning, by transfer from R, as sums of Q-finite sequences. The sum of an infinite series in R may also be regarded as a function, σ, from infinite sequences into real numbers. As such, σ extends to certain infinite sequences in *R and, whenever it exists, satisfies the classical condition detailed above.

Let U be any open set in *R. We see, by transfer from R, that is the union of a Q-finite or infinite sequence of disjoint open intervals $\{J_n\}$. In either case, we may write formally $U = \bigcup_{n=0}^{\infty} \{J_n\}$, if necessary by setting $J_n = \emptyset$ for sufficiently high n. Denoting the length of an interval J by

$m(J)$ – whenever that length exists – we write $m(U) = \sum_{n=0}^{\infty} m(J_n)$, whenever the sum on the right hand side exists, and we call $m(U)$ the measure of U. We can also obtain $m(U)$ by transfer from R to $*R$ of the set function which defines the measure of (some, but not all) open sets in R. The measure of a closed set in $*R$, whenever it exists, is then obtained by complementation in the usual way. Notice that if U is measurable (possesses a measure) then for any $\varepsilon > 0$ in $*R$, $\sum_{n=k+1}^{\infty} m(J_n) < \varepsilon$ for sufficiently large k. Making ε infinitesimal, we may therefore write $U = U_1 \cup U_2$ where $U_1 = \bigcup_{n=0}^{k} \{J_n\}$ is the union of a Q-finite sequence of open intervals and $U_2 = \bigcup_{n=k+1}^{\infty} \{J_n\}$ has infinitesimal measure, $m(U_2) \simeq 0$.

Now let B be any standard linear set (i.e., set of points on the real line, set of real numbers. It will be understood from now on until the end of this chapter that a *set of points* is a linear set.) If B is Lebesgue measurable then we denote its Lebesgue measure by $m(B)$. On passing to the enlargement $*R$, the set function $m(x)$ extends to certain additional internal sets of points. Evidently, this notion of a measure is compatible with that introduced earlier for open or closed sets.

5.1.1 THEOREM. Let B be a standard set of points. B is Lebesgue measurable if and only if there exist internal sets of points, B_o and B_i, open and closed respectively, such that $B_o \supset *B \supset B_i$ and such that both B_o and B_i possess finite measures which satisfy $m(B_o) \simeq m(B_i)$. Moreover, in that case, $m(B) = {}^{\circ}(m(B_o)) = {}^{\circ}(m(B_i))$.

By a finite measure, we mean a measure which is a finite real number. Similarly, the measure of a set is *infinite* if it exists as an infinite real value in $*R$. In order to avoid confusion, we refrain from introducing ∞ ('infinity') as a value, as is frequently done in Lebesgue theory.

PROOF of 5.1.1. Suppose the standard set of points B is Lebesgue measurable. Then it is true, in R, that 'for every positive x there exists an open set of points y such that $B \subset y$ and $m(B) \leq m(y) < m(B) + x$'. Transferring the statement in quotation marks from R to $*R$ and taking $x = \eta$ where η is positive infinitesimal, we see that there exists an internal open set of points B_o such that $*B \subset B_o$ and $m(B) \leq m(B_o) < m(B) + \eta$. This shows that $m(B) = {}^{\circ}(m(B_o))$. Similarly, it is true, in R, that 'for every positive x there exists a closed set y such that $y \subset B$ and $m(B) - x < m(y) \leq m(B)$'. Transferring to $*R$ and taking $x = \eta$ again positive infinitesimal, we conclude

that there exists an internal closed set B_i such that $B_i \subset {}^*B$ and $m(B) - \eta$
$< m(B_i) \leq m(B)$, and hence ${}^0(m(B_i)) = m(B)$.

Conversely, suppose that there exist internal sets B_0 and B_i, open and
closed respectively, such that $B_0 \supset {}^*B \supset B_i$ and such that $m(B_0)$ and $m(B_i)$
exist and are finite and $m(B_0) \simeq m(B_i)$. Let ${}^0(m(B_0)) = {}^0(m(B_i)) = a$. Then
we are going to show that a is both the outer and the inner measure of B
and hence, is the measure of B. Indeed, let ε be any standard positive
number. Then it is true, in ${}^*R$, that 'there exists an open set (i.e., B_0) such
that ${}^*B \subset B_0$ and $m(B_0) < a + \varepsilon$'. Transferring to R, we conclude that
$m_0(B) \leq a$, where $m_0(B)$ is the outer measure of B. A similar argument
shows that $m_i(B) \geq a$ where $m_i(B)$ is the inner measure of B. But
$m_i(B) \leq m_0(B)$ and so we may conclude that $m_i(B) = m_0(B) = a$, a is the
measure of B. This completes the proof of the theorem.

Instead of taking the notion of Lebesgue measure for granted we may,
in view of 5.1.1, assume only that the notion of a measure has been introduced for open and closed sets, as above, and we may then *define* the
measure of a standard set B as ${}^0(m(B_0)) = {}^0(m(B_i))$, provided ${}^0(m(B_0))$
and ${}^0(m(B_i))$ exist and are finite for certain internal sets B_0 and B_i, which
are open and closed, respectively, and such that $B_0 \supset {}^*B \supset B_i$. It is not
difficult to show that this definition is unique whenever it applies.

As an immediate consequence of 5.1.1 we have

5.1.2 THEOREM.   In order that a standard set B be of measure 0 it is
necessary and sufficient that there exist an internal open set D such that
${}^*B \subset D$ and $m(D) \simeq 0$.

We also observe that if an internal set D has infinitesimal measure (i.e.,
$m(D) \simeq 0$) and if F is an internal subset of D then F is measurable and
$m(F) \simeq 0$.

Now suppose that the standard set B is measurable, and choose internal
sets B_0 and B_i as in 5.1.1. Then B_0 is open, and B_i is closed and so
$D = B_0 - B_i$ is open. Also, $m(B_0) \simeq m(B_i)$ and so $M(D) = m(B_0) - m(B_i)$
is infinitesimal. It follows that the measures of the sets $D_1 = B_0 - {}^*B$ and
$D_2 = {}^*B - B_i$, which are subsets of D, are infinitesimal as well. Given any
standard $\varepsilon > 0$ it is therefore true, in ${}^*R$, that for any measurable set ${}^*B$,
there exists an open set, B_0, and a set of measure $< \varepsilon$, D_1, such that ${}^*B \cup D_1 = $
B_0 and ${}^*B \cap D_1 = \emptyset$; and there exists a closed set B_i and a set of measure $< \varepsilon$,

D_2 such that $B_i \cup D_2 = {}^*B$ and $B_i \cap D_2 = \emptyset$. Transferring this result to R, we obtain a well known classical fact. Moreover, as shown previously, we may write B_0 as the union of two disjoint sets, B_1 and B_2 where B_1 is the union of a Q-finite sequence of intervals and B_2 is an open set of infinitesimal measures. Then $'B = (B_1 \cup B_2) - D_1$ where $B_1 \cap B_2 = \emptyset$, $D_1 \subset B_1 \cup B_2$, B_1 is the union of a Q-finite sequence of intervals and (for any given standard $\varepsilon > 0$), $m(B_2) < \varepsilon$, $m(D_1) < \varepsilon'$. Transferring to R, we obtain

5.1.3 THEOREM (Standard. Lebesgue). If the set B is measurable and ε is any positive real number then B can be represented in the form $B = (B_1 \cup B_2) - D_1$ where $B_1 \cap B_2 = \emptyset$, $D_1 \subset B_1 \cup B_2$, B_1 is the union of a finite number of intervals, and B_2 and D_1 are measurable and such that $m(B_2) < \varepsilon$, $m(D_1) < \varepsilon$.

Having defined the Lebesgue measure of a linear set in R, and by transfer in ${}^*R$, by either standard or non-standard methods, we may then discuss Lebesgue integration. Let $f(x)$ be a standard function of a real variable, defined on a measurable standard set B and bounded and measurable on that set. Thus, there exist standard real numbers α and β such that $\alpha \le f(x) \le \beta$ for all x in B; and for any real numbers y_1 and y_2 such that $\alpha \le y_1 \le y_2 \le \beta$, the set $\{x \mid y_1 \le f(x) \le y_2\}$ is measurable. Observe that there is no need to specify more particularly whether $f(x)$ is bounded (or measurable) in R or in ${}^*R$ for since $f(x)$ is a standard function it will be bounded (or measurable) in R if and only if it is bounded (or measurable) in ${}^*R$.

Now let Π be an internal fine partition of the interval $\alpha \le y \le \beta$. That is to say (see section 3.5) that Π is a Q-finite internal sequence $\{y_n\}$, $y_0 = \alpha < y_1 < y_2 < \cdots < y_k = \beta$ such that $y_{n+1} - y_n$ is infinitesimal for $n = 0$, $1, \ldots, k-1$. In particular, Π may be given by $y_n = \alpha + (n/k)(\beta - \alpha)$, $n = 0$, $1, \ldots, k$, where k is an infinite natural number. The measure of the set $\{x \mid y_n \le f(x) < y_{n+1}\}$, $n = 0, 1, \ldots, k-1$, will be denoted by m_n.

Consider the sums

5.1.4
$$s = \sum_{n=0}^{k-1} y_n m_n \quad \text{and} \quad S = \sum_{n=0}^{k-1} y_{n+1} m_n.$$

Then

$$\alpha \sum_{n=0}^{k-1} m_n = \alpha m(B) \le s \le S \le \beta \sum_{n=0}^{k-1} m_n = \beta m(B),$$

so that both s and S are finite. Moreover, since every finite sequence of real numbers in R possesses a maximum it follows, by transfer to $*R$, that the Q-finite sequence $\{y_{n+1}-y_n\}$, $n=0,1,...,k-1$ possesses a maximum, which will be denoted by η. η is non-negative and infinitesimal since it is one of the $y_{n+1}-y_n$. Then

5.1.5 $$|S-s|=S-s=\sum_{n=0}^{k-1}(y_{n+1}-y_n)m_n\leq\sum\eta m_n=\eta m(B).$$

Since $m(B)$ is a standard real number, this shows that $S\simeq s$, $^0S=\,^0s$. Moreover, if the internal partition Π_1 is obtained from Π by subdividing some of the intervals (y_n,y_{n+1}) of Π, and s_1 and S_1 are the corresponding sums 5.1.4 then it is not difficult to see that $s\leq s_1\leq S_1\leq S$ and so $^0s=\,^0s_1=\,^0S_1=\,^0S$. By combining any two internal fine partitions (as in the standard procedure) we may then show that $^0S=\,^0S_1$ for any two internal fine partitions Π and Π_1. Accordingly, we may *define* the Lebesgue integral $\int_B f(x)dx$ by

5.1.6 $$\int_B f(x)\,dx=\,^0S$$

and we might take this definition as the starting point of the theory of Lebesgue integration. Alternatively, we may take the standard definition of the Lebesgue integral for granted and we may then prove 5.1.6. The reader will have no difficulty in verifying this assertion.

For an unbounded standard function $f(x)$, which is defined and measurable on the measurable set B, we may define $\int f(x)dx$ as a limit by the following standard procedure. For any two standard numbers α and β, $\alpha<\beta$ we define the function $f_{\alpha\beta}(x)$ on B by

$$f_{\alpha\beta}(x)=f(x) \quad \text{for} \quad \alpha\leq f(x)<\beta,$$
$$f_{\alpha\beta}(x)=0 \quad \text{for} \quad f(x)<\alpha \quad \text{and for} \quad f(x)\geq\beta.$$

Then $f_{\alpha\beta}(x)$ is measurable and bounded on B. The integrals $\int_B f_{\alpha\beta}(x)dx$ therefore exist, and we may consider the expression

5.1.7 $$\lim_{\substack{\alpha\to-\infty\\\beta\to\infty}}\int_B f_{\alpha\beta}(x)\,dx.$$

If 5.1.7 exists then it is, by definition, the integral $\int_B f(x)dx$.

Let $\alpha < \beta < \gamma$. Then

$$\int_B f_{\alpha\beta}(x)\,dx + \int_B f_{\beta\gamma}(x)\,dx = \int_B f_{\alpha\gamma}(x)\,dx.$$

Formulating the Cauchy condition for the existence of the limit 5.1.7 and passing to *R, we obtain by the usual procedure the following test.

5.1.8 THEOREM.   The integral $\int_B f(x)\,dx$ exists if and only if the integrals $\int_B f_{\alpha\beta}(x)\,dx$ and $\int_B f_{-\beta,-\alpha}(x)\,dx$ are infinitesimal for all positive infinite α and $\beta > \alpha$.

Independently of the existence of the limit 5.1.7 we may define the integral $\int_B f_{\alpha\beta}(x)\,dx$ for non-standard as for standard α, β, $\alpha < \beta$, as follows. We introduce an internal fine partition Π of the interval $\alpha \le y \le \beta$ and consider the sums

5.1.9 $s(\alpha,\beta,\Pi) = \sum_{n=0}^{k-1} y_n m_n$ and $S(\alpha,\beta,\Pi) = \sum_{n=0}^{k-1} y_{n+1} m_n$.

Our previous argument (see 5.1.5) still shows that

5.1.10 $s(\alpha,\beta,\Pi) \simeq S(\alpha,\beta,\Pi)$.

On the other hand, we may deduce by transfer from R that

5.1.11 $s(\alpha,\beta,\Pi) \le \int_B f_{\alpha\beta}(x)\,dx \le S(\alpha,\beta,\Pi)$

and this shows, in conjunction with 5.1.10, that

5.1.12 $S(\alpha,\beta,\Pi) \simeq \int_B f_{\alpha\beta}(x)\,dx$

for all internal fine partitions Π of $\alpha \le y \le \beta$. Combining this result with 5.1.8, and choosing the particular type of fine partition mentioned earlier, we arrive at the following conclusion.

The integral $\int_B f(x)\,dx$ exists if and only if, for all α, β such that $\alpha < \beta$ and α and β are both infinite negative or both infinite positive, the sums $\sum_{n=0}^{k-1} y_{n+1} m_n$ are infinitesimal; where k is a positive integer such that $(1/k)(\beta-\alpha)$ is infinitesimal and where $y_n = \alpha + (n/k)(\beta - \alpha)$, $n = 0, 1, \ldots, k$. Moreover, if $\sum_{n=0}^{k-1} y_{n+1} m_n$ is infinitesimal for a particular α, β, and for some k as detailed, then this condition will be satisfied for all such k. Finally, if $\int_B f(x)\,\mathrm{d}x$ exists and if α is any negative infinite number and β is any positive infinite number, and $(1/k)(\beta-\alpha)$ is infinitesimal then the integral $\int_B f(x)\,\mathrm{d}x$ is given by

$$\int_B f(x)\,\mathrm{d}x \simeq {}^0\!\left(\sum_{n=0}^{k-1} y_{n+1} m_n\right),$$

where

$$y_n = \alpha + n/k\,(\beta-\alpha), \qquad n = 0,\dots,k\,,$$

and

$$m_n = m\left(\{x \,|\, y_n \leq f(x) < y_{n+1}\}\right), \qquad n = 0,\dots,k-1\,.$$

We may consider notions of measure directly in $*R$. This is analogous to a step taken previously in topology (section 4.4). We select for consideration the following definition.

Let D be a set of points in $*R$, which may be internal or external. Suppose that there exists a measurable open set B in R such that $D \subset *B$. We then define the *outer S-measure* of D, Som(D) as the greatest lower bound, in R, of the measures of all standard open sets that contain D. And we define the *inner S-measure* of D, Sim D, as the lowest upper bound, in R, of the measures of all standard closed sets that are contained in D. Thus, both the outer and the inner S-measure of a set D, whenever they are defined, are standard real numbers. Evidently, Sim$(D) \leq$ Som(D). If, moreover, Sim$(D) =$ Som(D) then we call this joint value the *S-measure* of D and we denote it by Sm(D).

For any sets of points B and F in R, $B \subset F$ if and only if $*B \subset *F$. It follows that for every open set of measure m, say, which includes B there exists a standard open set of measure m which includes $*B$. It follows that the outer S-measure of $*B$ coincides with the outer Lebesgue measure of B and of $*B$. Similarly, the inner S-measure of $*B$ will be seen to coincide with the inner Lebesgue measure of B and hence, of $*B$. We conclude that $*B$ is S-measurable (possesses an S-measure) if and only if B is measurable and that these measures coincide whenever they exist. However, a set may be S-measurable also in other cases. For example, let μ be a monad which includes a standard point, a, so that $\mu = \mu(a)$. Then μ is a subset of standard intervals of arbitrarily small length and so Sm$(\mu) =$ Som$(\mu) = 0$. Observe that our definition does not apply to monads that consist of infinite points for such monads are not contained in standard open sets of finite measure. However, it would be possible to extend the definition so as to include this case, e.g., by permitting translation (shift).

A standard procedure shows that the S-measure is countably additive. Thus (as can also be seen directly) a set which consists of the monads of a countable number of finite points is of S-measure zero.

An application of the notion of S-measure will be found in the next section.

5.2 Sequences of functions. Let $\{f_n(x)\}$ be a standard sequence of standard functions defined in a standard set $D = {}^*B$ (where B is a set of standard points) and let $f(x)$ be a standard function which is defined on B. We know (see section 4.5) that $\{f_n(x)\}$ converges to $f(x)$ uniformly on B if and only if $f_n(a) \simeq f(a)$ for all $a \in D$ and for all infinite n.

According to a classical definition, $\{f_n(x)\}$ (regarded as a sequence of functions in R) is uniformly convergent at a point a, a belonging to B or to the closure of B, if and only if the following condition is satisfied. For every $\varepsilon > 0$ there exist a $\delta > 0$ and a natural number ν such that, for all b in B, $|b - a| < \delta$ and $n > \nu$, $m > \nu$ imply $|f_n(b) - f_m(b)| < \varepsilon$. In this definition, $a, b, \varepsilon, \delta, n, m$ are all supposed to be standard. A simple non-standard characterization of points of uniform convergence is provided by the following theorem.

5.2.1   THEOREM.   Let a be a standard point which belongs to B or to the closure of B. Then a is a point of uniform convergence for the sequence $\{f_n(x)\}$ if and only if for all points $b \in D$ (where $D = {}^*B$), which belong to the monad of a, $f_n(b) \simeq f_m(b)$ for all infinite n and m.

PROOF.   Suppose that a is a point of uniform convergence for $\{f_n(x)\}$, let ε be any standard positive number, and let $\delta > 0$ and ν be suitable standard numbers such as exist according to the definition of a point of uniform convergence. Then for any point $b \in \mu(a) \cap D$, we have $|b - a| < \delta$, and for any infinite n and m, we have $n > \nu$, $m > \nu$. Hence, for such b, n, m, $|f_n(b) - f_m(b)| < \varepsilon$. Since ε is arbitrary, except for being positive standard, we conclude that $|f_n(b) - f_m(b)|$ is infinitesimal, $f_n(b) \simeq f_m(b)$, as asserted.

In order to prove that the condition is also sufficient we suppose that it is satisfied, and we specify a standard positive ε. Within ${}^*R$, the condition that $|f_n(b)-f_m(b)|<\varepsilon$ for $|b-a|<\delta$, $n>v$, $m>v$, is then satisfied by infinitesimal positive δ and infinite v. Transferring from $*R$ to R, we conclude that a is a point of uniform convergence.

Suppose now that $\{f_n(x)\}$ converges to $f(x)$ at all points of B and of $D=*B$ where the entities involved are standard, as before. In that case, we have

5.2.2 THEOREM. Let a be a standard point which belongs to B or to the closure of B. Then a is a point of uniform convergence for $\{f_n(x)\}$ if and only if for all $b\epsilon\mu(a)\cap D$, and for all infinite n, $f_n(b)\simeq f(b)$.

Indeed, if $f_n(b)\simeq f(b)$ for all infinite n then $f_m(b)\simeq f(b)\simeq f_n(b)$ for all infinite n and m. Referring to 5.2.1 we therefore conclude that the condition of the theorem is sufficient. The condition is also necessary, for if $b\epsilon D$ then $\lim_{n\to\infty}f_n(b)=f(b)$ holds in $*R$. It follows that if η is any positive infinitesimal number then there exists an infinite natural number v such that $|f_n(b)-f(b)|<\eta$ for $m>v$. This shows that $f_m(b)\simeq f(b)$. But if n is any other infinite natural number then $f_n(b)\simeq f_m(b)$ for any b in the monad of a, by 5.2.1. We conclude that $f_n(b)\simeq f(b)$ for all infinite n. This completes the proof of the theorem.

We shall call any point of D at which $f_n(b)$ is not infinitely close to $f(b)$ for some infinite n, a *point of intrinsic non-uniformity*. We have added the word *intrinsic* in order to distinguish such points from points which are not points of uniform convergence in the sense of the classical definition. Such points are, by definition, standard. On the contrary, since we have assumed that $\lim_{n\to\infty}f_n(x)=f(x)$ on D, the points of intrinsic non-uniformity are by necessity non-standard. The above argument does show that the intrinsic points of non-uniformity are precisely those points $b\epsilon D$ for which there exist infinite n and m such that $f_n(b)$ is not infinitely close to $f_m(b)$.

5.2.3 THEOREM. Let $\{f_n(x)\}$ be a standard sequence of standard functions $f_n(x)$ which are defined and measurable on a standard measurable set $D=*B$. Suppose that B is bounded (or, which is the same, that the points of D are all finite) and that $\{f_n(x)\}$ converges at all points of D. Then the set of intrinsic points of non-uniformity of $\{f_n(x)\}$ is of S-measure zero.

PROOF. We shall rely on the classical theorem of Egoroff, which affirms that, on the assumptions stated in 5.2.3 and for arbitrary standard $\eta > 0$, the convergence of $\{f_n(x)\}$ is uniform in a measurable standard subset D' of D such that $m(D') \geq m(D) - \eta$.

Let G be the set of intrinsic points of non-uniformity of $\{f_n(x)\}$ and, for given η, let $D_\eta = D - D'$. Then $m(D_\eta) \leq \eta$. Also, since $\lim_{n \to \infty} f_n(x) = f(x)$ uniformly on D', the set of intrinsic points of non-uniformity of $\{f_n(x)\}$ in D' is empty, i.e., $G \subset D_\eta$. D_η is a standard set of measure not exceeding η, so there exists a standard open set F_η of measure not exceeding 2η such that $D_\eta \subset F_\eta$ and hence $G \subset F_\eta$. This shows that G has an outer S-measure not greater than 2η. Since η is standard positive, otherwise arbitrary, we conclude that $\mathrm{Sm}(G) = 0$, as claimed.

Conversely, it is not difficult to deduce Egoroff's theorem from 5.2.3. However, no independent proof of 5.2.3 is known, nor does Non-standard Analysis appear to offer any advantages in the proof of Egoroff's theorem as such. Nevertheless the recasting of the classical notions and results of the theory of uniform and non-uniform convergence in the language of Non-standard Analysis may be said to have an interest of its own. Let us mention that it can be carried a good deal further still.

We shall now show how Egoroff's theorem, as applied to $*R$, provides a straightforward proof of Cantor's lemma in the theory of trigonometrical series.

5.2.4 THEOREM (Standard. Cantor). Suppose that $s_n(x) = a_n \cos nx + b_n \sin nx$ tends to zero as n tends to infinity almost everywhere in an interval $a \leq x \leq b$, $a < b$. Then a_n and b_n tend to zero as n tends to infinity.

PROOF. We have to show that $\lim_{n \to \infty} a_n^2 + b_n^2 = 0$, or, which is the same, that $a_\omega^2 + b_\omega^2 \simeq 0$ for all infinite natural numbers ω. To prove this, for given infinite ω, it is sufficient to consider the case that $a_\omega^2 + b_\omega^2 \neq 0$. We define a real number η by

$$\cos \eta = \frac{a_\omega}{\sqrt{\{a_\omega^2 + b_\omega^2\}}}, \qquad \sin \eta = \frac{b_\omega}{\sqrt{\{a_\omega^2 + b_\omega^2\}}}, \qquad 0 \leq \eta < 2\pi,$$

where the positive square root is chosen in the denominator. Then

$$s_n(x) = a_\omega \cos \omega x + b_\omega \sin \omega x = (a_\omega^2 + b_\omega^2)^{\frac{1}{2}} \cos (\omega x - \eta).$$

In every interval $\xi \leq x < \xi + 2\pi/\omega$, $\cos(\omega x - \eta) \geq \frac{1}{2}$ over one or two sub-intervals of total length $\frac{2}{3}\pi/\omega$. Also, the numbers $\xi_j = a + (2j\pi)/\omega$ divide $a \leq x \leq b$ into $[(b-a)\omega/2\pi]$ intervals of length $2\pi/\omega$ (where $[z]$ denotes the largest positive integer not greater than z, as usual), plus a fraction of such an interval. Hence, $\cos(\omega x - \eta) \geq \frac{1}{2}$ on an internal set of points D in the interval $a \leq x \leq b$ whose measure is $\frac{1}{3}(b-a) - \varepsilon$, ε infinitesimal, and hence, is greater than $\frac{1}{4}(b-a)$. On the other hand, by Egoroff's theorem, there exists a standard subset F of $a \leq x \leq b$, of measure $\frac{7}{8}(b-a)$ such that $s_n(x)$ converges uniformly on F. It follows that $s_\omega(\xi)$ is infinitesimal for all $\xi \in F$. But D and F cannot be disjoint since they are subsets of the interval $a \leq x \leq b$, and $m(D) + m(F) > \frac{1}{4}(b-a) + \frac{7}{8}(b-a) > b-a$. Let $\xi \in D \cap F$. Then $s_\omega(\xi) = \{a_\omega^2 + b_\omega^2\}^{\frac{1}{2}} \cos(\omega\xi - \eta) \simeq 0$ but $\cos(\omega\xi - \eta) \geq \frac{1}{2}$. Hence,

$$\sqrt{\{a_\omega^2 + b_\omega^2\}} \leq 2 s_\omega(\xi)$$

and further, $a_\omega^2 + b_\omega^2 \simeq 0$. This proves the theorem.

5.3 Distributions. Working in R, we denote by C^∞ the class of real-valued functions which are defined for all real numbers and possess derivatives of all orders everywhere. Passing to $*R$, we shall consider certain functions of the class $*C^\infty$ and we shall show that they provide adequate representations for the common type of Schwartz distributions.

Let F (*the test space*) be the subset of C^∞ which consists of the functions $g(x)$ with finite support, i.e., for which there exists a real number c (in R) such that $g(x) = 0$ for all $|x| \geq c$. Let Q_0 be the set of real-valued internal functions $f(x)$ in $*R$ which are defined for all real x and are square integrable over any interval of finite numbers (i.e., $\int_a^b (f(x))^2 \, dx$ exists, as a Lebesgue integral in $*R$, for any $a < b$ such that both a and b are finite real numbers) and such that the integrals $\int_{-\infty}^\infty f(x)g(x)\,dx$ are finite for all $g(x) \in F$. We shall use the inner product notation for such an integral,

$$(f,g) = \int\limits_{-\infty}^{\infty} f(x)g(x)\,dx.$$

The sum of two elements of Q_0 is an element of Q_0 and if $f(x) \in Q_0$ and λ is a finite number then $\lambda f(x) \in Q_0$.

Evidently, every standard function which belongs to $*C^\infty$, belongs also to Q_0. Moreover, if $g(x) \in F$ and $h(x) \in C^\infty$ then $h(x)g(x) \in F$. It follows that if $f(x) \in Q_0$ and $h(x) \in C^\infty$ then $\int_{-\infty}^{\infty} f(x) h(x) g(x) \, dx$ is finite for all $g(x)$ in F and so $f(x) h(x) \in Q_0$. (Notice that here as elsewhere, we are using the symbols $h(x)$ and $g(x)$, ambiguously, both for functions in R and for their extensions in $*R$).

Let Q_1 be the class of functions of Q_0 such that $(f,g) \simeq 0$ for all $g(x) \in F$. Then Q_1 also is additive, and if $f(x) \in Q_1$ and λ is a finite real number then $\lambda f(x) \in Q_1$; and if $f(x) \in Q_1$ and $h(x) \in C^\infty$ then $f(x) h(x) \in Q_1$.

For every $f(x) \in Q_0$, we define a functional Δ_f on the set F by

$$\Delta_f[g] = {}^0(f,g) \qquad \text{for all } g(x) \in F.$$

If $g_1(x)$ and $g_2(x)$ are elements of F and λ is a standard real number we then have

$$\Delta_f[g_1 + g_2] = {}^0(f, g_1 + g_2) = {}^0((f,g_1) + (f,g_2)) = {}^0(f,g_1) + {}^0(f,g_2)$$
$$= \Delta_f[g_1] + \Delta_f[g_2],$$

and similarly,

$$\Delta_f[\lambda g_1] = {}^0\lambda \Delta_f[g_1] = \lambda \Delta_f[g_1].$$

This shows that the functionals Δ_f are linear from the algebraic point of view. We denote the set of these functionals by Φ, and we denote the mapping $f \to \Delta_f$ from Q_0 onto Φ by ψ. Both Q_0 and Φ constitute Abelian groups with respect to addition, and the mapping ψ is a homomorphism from Q_0 onto Φ. The kernel of this homomorphism is Q_1 and the quotient group Q_0/Q_1 is isomorphic to Φ. If $f \sim g$ is the equivalence relation in Q_0 which is defined by $f - g \in Q_1$ then the elements of Q_0/Q_1 are the equivalence classes with respect to this relation. We shall call them *pre-distributions*. Thus $f(x)$ and $g(x)$ in Q_0 belong to the same pre-distribution if $f - g \in Q_1$.

We shall now show that Φ consists of all functionals on F which are linear from the algebraic point of view. More particularly, we shall show that for any given functional $\Delta[g]$ on F which is linear from the algebraic point of view there exists in $*R$ a polynomial $p_\Delta(x)$ (i.e., $p_\Delta(x) \in *\Pi$, where Π is the class of polynomials in R) such that $(p_\Delta, g) = \Delta[g]$ for all $g(x)$ in F. Since a polynomial is square-integrable on any interval (in R and hence, in $*R$) and since $(p_\Delta, g) = \Delta[g]$ is finite for all $g(x)$ in F, we conclude that $p_\Delta(x) \in Q_0$.

In order to prove the existence of such a polynomial $p_\Delta(x)$ for given

$\Delta[g]$ we first consider a finite set of linearly independent functions $g_j(x) \in F$, $j = 1,...,m$, $m \geq 1$. Working in R, we claim that for every set of real numbers $a_1,...,a_m$, there exists a polynomial $p(x)$ such that

5.3.1
$$(p,g_j) = \int_{-\infty}^{\infty} p(x)g_j(x)\,dx = a_j, \quad j = 1,...,m.$$

Since the functions $g_j(x)$ have finite support, there exists a positive real ξ such that $g_j(x) = 0$ for $|x| \geq \xi$, $j = 1,...,m$. And a simple transformation of the independent variable $(x' = x/2\xi)$ shows that we may, without loss of generality, suppose that $\xi < 1$. Then 5.3.1 becomes

5.3.2
$$\int_{-1}^{1} p(x)g_j(x)\,dx = a_j, \quad j = 1,...,m,$$

where the $g_j(x)$ vanish in the neighborhood of the end points of the intervals of integration.

We represent the functions $g_j(x)$ by their Legendre expansions

$$g_j(x) = a_0^j P_0(x) + a_1^j P_1(x) + a_2^j P_2(x) + \cdots, \quad j = 1,...,m,$$

where, in view of the stated restrictions on the $g_j(x)$, the series on the right hand side converges to the correct value on the closed interval $-1 \leq x \leq 1$, the convergence being uniform in any interval $-\theta \leq x \leq \theta$ for $\theta < 1$. Since the $g_j(x)$ are linearly independent, by assumption, the matrix

$$\begin{pmatrix} a_0^1 & a_1^1 & a_2^1 \cdots \\ \vdots & \vdots & \vdots \\ a_0^m & a_1^m & a_2^m \cdots \end{pmatrix}$$

must be of rank m. It follows that there are subscripts $0 \leq j_1 \leq j_2 \leq \cdots \leq j_m$ such that the matrix

$$\alpha = \begin{pmatrix} a_{j_1}^1 & a_{j_2}^1 \cdots a_{j_m}^1 \\ \vdots & \vdots & \vdots \\ a_{j_1}^m & a_{j_2}^m \cdots a_{j_m}^m \end{pmatrix}$$

is non-singular. Writing

$$\alpha^{-1} \begin{pmatrix} g_1(x) \\ \vdots \\ g_m(x) \end{pmatrix} = \begin{pmatrix} h_1(x) \\ \vdots \\ h_m(x) \end{pmatrix}$$

we then obtain

$$h_l(x) = P_{j_l}(x) + k_l(x), \quad l = 1,\dots,m,$$

where the Legendre expansions of the functions $k_l(x)$ do not contain terms involving $P_{j_1}(x),\dots,P_{j_m}(x)$. Furthermore, defining $b_1,\dots,b_m$ by

$$\alpha^{-1}\begin{pmatrix} a_1 \\ \vdots \\ a_m \end{pmatrix} = \begin{pmatrix} b_1 \\ \vdots \\ b_m \end{pmatrix},$$

we see that 5.3.2 is equivalent to

5.3.3 $$\int_{-1}^{1} p(x)\, h_l(x)\, dx = b_l, \qquad l = 1,\dots,m.$$

We now set

5.3.4 $$p(x) = c_1 P_{j_1}(x) + c_2 P_{j_2}(x) + \cdots + c_m P_{j_m}(x).$$

Then

$$\int_{-1}^{1} p(x)\, k_l(x)\, dx = 0, \qquad l = 1,\dots,m,$$

and so 5.3.3 becomes

$$\int_{-1}^{1} p(x)\, P_{j_l}(x)\, dx = b_l, \qquad l = 1,\dots,m.$$

But $\int_{-1}^{1}(P_n(x))^2\, dx = 2/(2n+1)$ and so 5.3.4 is satisfied by $c_l = \tfrac{1}{2}(2j_l+1)b_l$, $l = 1,\dots,m$. This shows that

5.3.5 $$p(x) = \tfrac{1}{2}(2j_1 + 1)\, b_1 P_{j_1}(x) + \tfrac{1}{2}(2j_2 + 1)\, b_2 P_{j_2}(x) + \cdots + \tfrac{1}{2}(2j_m+1)b_m P_{j_m}(x)$$

is a polynomial which satisfies 5.3.2, as required.

Now let $\Delta[g]$ be any functional on F which is linear in the algebraic sense, and let $g_1,\dots,g_k$ be an arbitrary finite set of elements of F. Then we claim that there exists a polynomial $p(x)$ such that

5.3.6 $$\int_{-\infty}^{\infty} p(x)\, g_j(x)\, dx = \Delta[g_j], \qquad j = 1,\dots,k.$$

Indeed, to satisfy 5.3.6, we only have to select a linearly independent subset of the g_j, e.g., $g_1,...,g_m$, and to satisfy 5.3.1 for $a_j = \Delta[g_j], j = 1,...,m$. This implies that the first m equations of 5.3.6 are satisfied directly while the remaining $k - m$ equations (if any) are satisfied by virtue of the linearity of $\Delta[g]$. (If $k = 1$ and $g_1(x) = 0$ the assertion is trivially true.)

We have just shown that, for a given functional Δ as above, the relation $T(g,p)$ which is defined as follows is concurrent in R.

'g is a function which belongs to the set F and p is a polynomial and $\int_{-\infty}^{\infty} p(x) g(x) \, dx = \Delta[g]$'.

This shows that a $p_\Delta(x)$ as required exists in $*R$.

Instead of selecting $p_\Delta(x)$ as a polynomial in x we might equally well have determined it as a trigonometrical polynomial (or in terms of certain other sequences of orthogonal polynomials). Again, since the behavior of $p_\Delta(x)$ for infinite values of x makes no difference to the value of (p_Δ, g) for $g(x) \in F$, we may modify $p_\Delta(x)$ in such a way that the function resulting from the modification still belongs to $*C^\infty$ but vanishes for $|x| \geq a$, a a positive infinite. This can be done, for example, by multiplying $p_\Delta(x)$ by a function $q(x) \in *C^\infty$ which has the property that $q(x) = 1$ for $|x| \leq \frac{1}{2}a$ and $q(x) = 0$ for $|x| \geq a$. The fact that such a function $q(x)$ exists can be established by transfer from R since it is known that there are functions in C^∞ with the stated property for standard a.

Our result shows that every pre-distribution (element of Q_0/Q_1) contains at least one function of the class $*C^\infty$. We write $P_0 = Q_0 \cap *C^\infty$, $P_1 = Q_1 \cap *C^\infty$, so that P_1 and P_0 are again additive groups and P_1/P_0 is isomorphic to Q_1/Q_0 and to Φ.

In the Schwartz theory of distributions, a linear functional $\Delta[g]$ on F is a distribution only if it satisfies a further requirement, of continuity, as follows. Let $\{g_n(x)\}$ be a sequence of elements of F such that $g_n(x) = 0$ for all n, and for $|x| > a$, for some standard real a and such that $g_n^{(k)}(x) \to 0$ uniformly in x as n tends to infinity for $k = 0,1,2,3,...$. Then the requirement is that $\lim_{n \to \infty} \Delta[g_n] = 0$. We shall say that the pre-distribution α is a Q-distribution, if the image of the elements of α in Φ satisfies the condition of continuity just mentioned, i.e., is a distribution in the sense of Schwartz's theory.

Thus, to every distribution in the sense of the theory of Schwartz there corresponds a Q-distribution, which is a particular class of non-standard functions. For example, to the Dirac Delta function as a distribution there corresponds the Q-distribution which contains, among others, the function $(a/\pi)^{\frac{1}{2}} \exp(-ax^2)$ where a is any positive infinite real number. On one hand, the standard theory has the advantage that in it, the Delta function is unique by its very definition, whereas in the non-standard theory uniqueness is achieved only by combining the candidates for the rule of Delta function, such as the various $(a/\pi)^{\frac{1}{2}} \exp(-ax^2)$ into a single equivalence class. On the other hand, the fact that the elements of the equivalence classes are proper functions helps to make the theory more intuitive. However, we should mention here that even in standard theory, distributions can be given a degree of concreteness through representing them by equivalence classes of sequences of functions.

5.3.7 THEOREM. Let α be a pre-distribution, and let $f(x)$ and $g(x)$ be two elements of α which possess continuous derivatives for all x. Then $f'(x)$ and $g'(x)$ belong to the same pre-distribution.

PROOF. On the assumptions of the theorem, the functions $f'(x)$ and $g'(x)$ are square integrable over any interval. Also, for every $h(x) \in F$, integration by parts yields $(f',h) = -(f,h')$. Since $h'(x)$ also is an element of F, (f,h') is finite. Hence, (f',h) is finite, and so $f'(x)$, and similarly, $g'(x)$, belong to Q_0. Now consider the difference $k(x) = f(x) - g(x)$. Since f and g belong to the same pre-distribution, (k,h) is infinitesimal for all $h(x) \in F$, and so $(k',h) = -(k,h')$ is infinitesimal for all $h(x) \in F$. This shows that $k'(x) \in Q_1$, and hence that $f'(x)$ and $g'(x)$ belong to the same pre-distribution.

5.3.8 COROLLARY. On the assumptions of 5.3.7, if $f(x)$ and $g(x)$ belong to the same Q-distribution then the pre-distribution to which $f'(x)$ and $g'(x)$ belong is a Q-distribution.

As we have seen, every pre-distribution α contains a continuously differentiable function (even a polynomial) $f(x)$, and 5.3.7 shows that the pre-distribution β to which $f'(x)$ belongs is then determined uniquely. We call β the derivative of α and we write $\beta = \alpha'$, or $\beta = d\alpha/dx$.

The following theorems provide some information on the structure of Q_0 and P_0.

5.3.9 THEOREM. The only standard function which belongs to P_1 is the zero function, $f(x) \equiv 0$.

PROOF. $f(x)$ is continuous, by assumption. If $f(x)$ does not vanish everywhere in $*R$ and hence, does not vanish everywhere in R, then $f(x_0) \neq 0$
for some standard x_0. A familiar procedure now shows how to construct
a function $g(x) \in F$ which is nowhere negative, and is positive only in some
sufficiently small neighborhood of x_0, such that $(f,g) = \int_{-\infty}^{\infty} f(x) g(x) \, dx$
$\neq 0$. And since (f,g) is a standard number, it follows that it cannot be
infinitesimal. In more detail, supposing, without loss of generality, that
$f(x_0) > 0$, let a be a standard positive number such that $f(x) > \frac{1}{2} f(x_0)$ for
$x_0 - a \leq x \leq x_0 + a$ and determine a function $h(x) \in F$ such that $h(x) \geq 0$
everywhere and

$$h(x) = 1 \quad \text{for} \quad |x| \leq \tfrac{1}{2}a, \qquad h(x) = 0 \quad \text{for} \quad |x| \geq a.$$

Putting $g(x) = h(x - x_0)$, we then have $g(x) \in F$ and

5.3.10 $$\int_{-\infty}^{\infty} f(x) g(x) \, dx = \int_{x_0-a}^{x_0+a} f(x) g(x) \, dx > \tfrac{1}{2} a f(x_0),$$

as required.

5.3.11   THEOREM.   In order that an internal function $f(x)$ which is
defined for all x and is square integrable over every interval of finite
numbers, belong to Q_0 it is sufficient that $\int_a^b (f(x))^2 \, dx$ be finite for all
finite a, b.

PROOF.   Let $g(x) \in F$, and let $g(x) = 0$ for $|x| \geq c > 0$, c standard. Then

5.3.12 $$\left(\int_{-\infty}^{\infty} f(x) g(x) \, dx \right)^2 = \left(\int_{-c}^{c} f(x) g(x) \, dx \right)^2$$

$$\leq \int_{-c}^{c} (f(x))^2 \, dx \int_{-c}^{c} (g(x))^2 \, dx,$$

by Schwarz's inequality. But $\int_{-c}^{c} (g(x))^2 \, dx$ is finite since it is a standard
number, and so $\int_{-\infty}^{\infty} f(x) g(x) \, dx = (f,g)$ is finite, $f \in Q_0$, as asserted.
  Another application of 5.3.12 proves

5.3.13 Theorem. Let $f(x) \in Q_0$. In order that $f(x) \in Q_1$ it is sufficient that $\int_a^b (f(x))^2 \, dx$ be infinitesimal for all finite a, b.

In general, a pre-distribution does not possess a numerical value at a point, since two functions may well belong to the same Q-distribution or pre-distribution yet differ, at a given point, by an amount which is not infinitesimal. For example, the functions $(a/\pi)^{\frac{1}{2}} \exp(-ax^2)$ belong to the (distribution for the) Dirac Delta function for all positive infinite a, yet for $x = 0$, the functions equal $\sqrt{(a/\pi)}$ and thus may take any positive infinite value whatever.

Let x_0 be any standard real number. We shall say that the pre-distribution α is *standard* at x_0 if there exist elements $f(x)$ of α such that $f(x)$ is S-continuous at x_0.

5.3.14 Theorem. Let α be a pre-distribution which is standard at the standard point x_0. Then there exists a standard y_0 such that $f(x_1) \simeq y_0$ for all $f(x)$ in α which are S-continuous at x_0 and for all x_1 in the monad of x_0 i.e. for $x_1 \simeq x_0$.

Proof. Let $f(x) \in \alpha$ be S-continuous at x_0. We show first that $f(x_0)$ is finite. Supposing that this is not true and adding, without essential loss of generality the assumption that $f(x_0)$ is positive (so that $f(x_0)$ is positive infinite) we then have $f(x_1) > \frac{1}{2} f(x_0)$ for all $x_1 \simeq x_0$ since $f(x_1) \simeq f(x_0)$ for such x_1. Defining the internal set of *positive* real numbers D by

$$D = \{b \mid b > 0 \wedge f(x_1) > \tfrac{1}{2} f(x_0) \text{ for } |x_1 - x_0| < b\}$$

we then see that D contains all positive infinitesimal numbers. But if the set of positive infinitesimal numbers were internal, then the entire monad of zero would be internal and we have already seen that this is not the case. It follows that D contains also numbers which are not infinitesimal and hence, contains also positive standard numbers. Denoting one of these by a, we then define the functions $h(x)$ and $g(x)$ as in the proof of 5.3.9, and obtain $\int_{-\infty}^{\infty} f(x) g(x) \, dx > \frac{1}{2} a f(x_0)$, where the right hand side is positive infinite. But this is impossible since $g(x) \in F$. We conclude that $f(x_0)$ is finite.

Now let $f_1(x_1)$ and $f_2(x)$ be two elements of α which are S-continuous at x_0. We propose to show that $f_1(x_0) \simeq f_2(x_0)$. Let $f(x) = f_1(x) - f_2(x)$

and suppose that $f(x_0)$ is not infinitesimal and (without loss of generality) that it is positive. Since $f(x)$ is the difference between two elements of a pre-distribution, it belongs to Q_1. It follows that (f,g) is infinitesimal for all $g \in F$. On the other hand, $f(x)$ is S-continuous at x_0 and so, as in the first part of the proof, there exists a positive standard a such that $f(x_1) > \frac{1}{2} f(x_0)$ for $|x_1 - x_0| < a$. Defining $g(x)$ and $h(x)$ as before, we obtain again $(f,g) > \frac{1}{2} a f(x_0)$, which contradicts the earlier conclusion that (f,g) is infinitesimal. Thus, $f_1(x_0) \simeq f_2(x_0)$ as asserted. Putting $y_0 = {}^0(f_1(x_0))$ for some $f_1(x) \in \alpha$ which is S-continuous at x_0, we then have $y_0 = {}^0(f(x_0))$ for any $f(x) \in \alpha$ which is S-continuous at x_0 and hence, by S-continuity, $y_0 = {}^0(f(x_1))$ also for any other x_1 in the monad of x_0. This completes the proof of 5.3.14.

If the pre-distribution α is standard at the standard point x_0, then we regard the y_0 which exists according to 5.3.14 as the value of α at x_0 and we write $y_0 = \alpha(x_0)$. In this sense $\alpha(x)$ is a real valued function (in R) which is defined for certain values of the argument. For example, if δ is the pre-distribution which contains $(a/\pi)^{\frac{1}{2}} \exp(-ax^2)$ for positive infinite a (i.e., the Q-distribution which corresponds to the Dirac Delta function) then $\delta(x)$ exists and equals zero for all standard $x \neq 0$.

If $f(x)$ is a standard function which belongs to a distribution α then $\alpha(x) = f(x)$ for all standard x. And if $\alpha(x_0)$ and $\beta(x_0)$ exist for certain pre-distributions α and β and for some standard x_0 then $\gamma = \alpha \pm \beta$ is standard at x_0 and $\gamma(x_0) = \alpha(x_0) \pm \beta(x_0)$.

5.3.15 THEOREM. Let α be a pre-distribution and let $f(x)$ be a standard continuous function (so that $f(x)$ belongs to Q_0). Suppose that for all standard x, α is standard and $\alpha(x) = f(x)$. Then $f(x) \in \alpha$.

PROOF. Let β be the pre-distribution which contains $f(x)$. Then β is standard for all standard x and $\beta(x) = f(x)$. Consider the pre-distribution $\gamma = \alpha - \beta$. γ is standard for all standard x and, for such x, $\gamma(x) = \alpha(x) - \beta(x) = f(x) - f(x) = 0$. We propose to show that γ is the zero pre-distribution, i.e., $\gamma = Q_1$. For this purpose, we only have to verify that if $k(x) \in \gamma$ and $g(x) \in F$ then (k,g) is infinitesimal.

Since $g(x)$ has finite support there exists a standard positive c such that $g(x) = 0$ for $|x| \geq c$. Let a be any standard real number such that $|a| \leq c$. We know that there exists a function $f_a(x) \in \gamma$ such that $f_a(x_1) \simeq 0$ for all

$x_1 \simeq a$. It follows that, for any specified standard $\varepsilon > 0$, the assertion
'$|f_a(x_1)| < \varepsilon$ provided $|x_1 - a| < \eta_a$' is true for all infinitesimal η_a. The usual
argument establishes that there even exists a positive standard η_a such
that $|f_a(x_1)| < \varepsilon$ for all x_1 such that $|x_1 - a| < \eta_a$. Having chosen a suitable
η_a for each a in the interval $-c \leq x \leq c$, for a pre-assigned ε, we denote
the open intervals $a - \eta_a < x < a + \eta_a$ by U_a. The theorem of Heine–Borel
now shows that there are among the U_a a finite number of open intervals,
$U_1, U_2, \ldots, U_m$, whose union includes the interval $-c \leq x \leq c$. To each U_j
there belongs a function $f_a(x)$ such that $|f_a(x)| < \varepsilon$ in that interval. It will
be renamed $f_j(x)$. Moreover, we may shorten the intervals U_j as necessary
(and perhaps omit some of them) so as to ensure that no point of
$-c \leq x \leq c$ is included in more than two of the U_j and moreover, so that
the total length of the overlap, together with the length of the intervals
by which $\cup_{j=1}^m U_j$ extends beyond $-c \leq x \leq c$ is as small as we please.
Thus, we may certainly assume that the total length of the intervals U_j
does not exceed $4c$. Let $\lambda = \max |g(x)|$. We shall make use of the fact that
$g(x)$ can be written as a sum of functions $g_j(x) \in F$

$$g(x) = g_1(x) + g_2(x) + \cdots + g_m(x)$$

such that $g_j(x) = 0$ outside U_j, and $|g_j(x)| \leq \lambda$ on U_j, $j = 1, \ldots, m$. Then

$$\int_{-\infty}^{\infty} k(x)g(x)\,dx = \int_{-c}^{c} k(x)g(x)\,dx = \sum_{j=1}^m \int_{U_j} k(x)g_j(x)\,dx$$

$$= \sum_{j=1}^m \int_{U_j} (k(x) - f_j(x))g_j(x)\,dx + \sum_{j=1}^m \int_{U_j} f_j(x)g_j(x)\,dx.$$

But $\int_{U_j}(k(x) - f_j(x))g_j(x)\,dx$ must be infinitesimal since $k(x)$ and $f_j(x)$
belong to the same pre-distribution, γ. At the same time

$$\left| \sum_{j=1}^m \int_{U_j} f_j(x)g_j(x)\,dx \right| \leq \sum_{j=1}^m \int_{U_j} \varepsilon\lambda\,dx \leq 4c\varepsilon\lambda.$$

We conclude that

$$\left| \int_{-\infty}^{\infty} k(x)g(x)\,dx \right| \leq \eta + 4c\varepsilon\lambda,$$

where η is positive infinitesimal. Since ε is standard positive, but otherwise arbitrary this inequality shows that (k,g) is infinitesimal. Hence, $\gamma = Q_1$, and hence, $\alpha = \beta$, $f(x) \in \alpha$. This completes the proof of 5.3.15.

The last theorem shows that to every standard function $f(x)$ which is of class C^0, i.e., which exists and is continuous for all x, there corresponds a unique pre-distribution α such that $\alpha(x) = f(x)$ for all standard x, and this is precisely the pre-distribution which contains $f(x)$. In general it will cause no confusion if we say that in this case α *is* a function. Moreover, if the pre-distribution α is a function then it is actually a Q-distribution. For if $\{g_n(x)\}$ is a sequence of functions from F such that $g_n(x) = 0$ for all n and $|x| \geq c$, c a standard number, and $\lim_{n \to \infty} g_n(x) = 0$ uniformly for $|x| < c$, then $(f, g_n) = {}^0(f, g_n)$ tends to zero as n tends to infinity, and this is sufficient to ensure that α is a Q-distribution.

Let $f(x) \in C^\infty$ be an element of the pre-distribution α (so that α is a function and a Q-distribution), and let β be any pre-distribution. Then the set of functions $f(x)g(x)$, $g(x) \in \beta$ is a subset of Q_0 and moreover, is a subset of a single pre-distribution γ. We shall say that γ is the *product* of α and β, $\gamma = \alpha\beta$. The reason why we cannot define the product $\alpha\beta$ for arbitrary pre-distributions (or Q-distributions) α and β is that, in general, we have no assurance that the products $f(x)g(x)$, where $f(x) \in \alpha$ and $g(x) \in \beta$, all belong to the same pre-distribution. This is the price paid for the transition from internal functions as such to equivalence classes of such functions.

5.3.16 THEOREM. Let $f(x) \in C^\infty$, be an element of the pre-distribution α (so that α is a function, and $\alpha(x) = f(x)$ for all standard x). Let β be a pre-distribution, which is standard at the standard point x_0. Let $\gamma = \alpha\beta$. Then γ is standard at x_0 and $\gamma(x_0) = \alpha(x_0)\beta(x_0)$.

PROOF. It follows from the assumptions of the theorem that $f(x)$ is S-continuous at x_0, and $f(x_0) = \alpha(x_0)$; also, that there exists a function $g(x) \in \beta$ such that $g(x)$ is S-continuous at x_0 and ${}^0(g(x_0)) = \beta(x_0)$. Consider the function $h(x) = f(x)g(x)$. This function belongs to γ. Also, for any point x_1 in the monad of x_0,

$$h(x_1) - h(x_0) = f(x_1)g(x_1) - f(x_0)g(x_0) = f(x_1)(g(x_1) - g(x_0))$$
$$+ (f(x_1) - f(x_0))g(x_0).$$

But $f(x_1)(g(x_1)-g(x_0))$ and $(f(x_1)-f(x_0))g(x_0)$ are both infinitesimal since they are products of two factors one of which is finite and the other infinitesimal. Hence, $h(x_1)-h(x_0)$ is infinitesimal, $h(x)$ is S-continuous at x_0. It follows that γ is standard at x_0 and $\gamma(x_0)={}^0(h(x_0))=f(x_0)\,{}^0(g(x_0))=\alpha(x_0)\beta(x_0)$. This proves the theorem.

5.3.17   THEOREM.   Let α be a pre-distribution and let $f(x)$ be a standard continuous function. Suppose that α is standard at all standard points x of a closed interval $a\leq x\leq b$, a and b standard, $a<b$, and that for such x, $\alpha(x)=f(x)$. Then ${}^0(h,g)=(f,g)$ for any $g(x)\in F$ whose support is included in the interval $a\leq x\leq b$ (i.e., which vanishes for $x<a$ and $x>b$) and for any $h(x)\in\alpha$.

Let β be the pre-distribution which contains $f(x)$ and let $\gamma=\alpha-\beta$. Then γ is standard at all standard points of the interval $a\leq x\leq b$, and

$$\gamma(x)=\alpha(x)-\beta(x)=f(x)-f(x)=0$$

for the standard points of that interval. If we can show that for any function $k(x)\in\gamma$ and for any $g(x)$ as described in the statement of the theorem, ${}^0(k,g)=0$ then this will entail the theorem since $h-f\in\gamma$. And, in order to show that ${}^0(k,g)=0$ in this case, i.e., that $(k,g)\simeq 0$, we only have to use the method employed in the proof of 5.3.15.

We have seen already that if the pre-distribution α is a function, i.e., if α contains a continuous standard function $f(x)$, then α is a Q-distribution. The same is true if α is the kth derivative of a function for some finite k. For let $\beta=\alpha^{(k)}$, where α is a function which contains the continuous standard function $f(x)$. Let $\{g_n(x)\}$ be a sequence of functions from F all of which vanish outside some interval $-c\leq x\leq c$, where c is a positive standard number, and suppose that $\lim_{n\to\infty}g_n^{(k)}(x)=0$ uniformly in x, for all finite k. Choose a function $h(x)\in{}^*C^\infty$ which belongs to α. We established previously that such a function exists. Then $h^{(k)}(x)$ belongs to β and we have to verify that ${}^0(h^{(k)},g_n)$ tends to zero as n tends to infinity (in R). Integration by parts shows that

$$(h^{(k)},g_n)=\int_{-\infty}^{\infty}h^{(k)}(x)g_n(x)\,dx=(-1)^k\int_{-\infty}^{\infty}h(x)g_n^{(k)}(x)\,dx=(-1)^k(h,g_n^{(k)}).$$

At the same time, f and h both belong to α, and so $(h,g_n^{(k)}) \simeq (f,g_n^{(k)})$. But $g_n^{(k)}(x) \to 0$ uniformly in x and so

$$(f,g_n^{(k)}) = \int_{-\infty}^{\infty} f(x)g_n^{(k)}(x)\,dx = \int_{-c}^{c} f(x)g_n^{(k)}(x)\,dx \to 0, \qquad \text{as } n \to \infty$$

It follows that

$$^0(h^{(k)},g_n) = (-1)^k \,{}^0(h,g_n^{(k)}) = (-1)^k (f,g_n^{(k)})$$

tends to zero as n tends to infinity, as required.

5.3.18 THEOREM. In order that a pre-distribution α be a Q-distribution, it is sufficient that for every standard positive c there exist a finite natural number k and a pre-distribution β which is a function such that the distribution $\gamma = \alpha - \beta^{(k)}$ is standard at the standard points of the interval $-c \le x \le c$ and $\gamma(x) = 0$ for such points.

PROOF. Suppose that the condition of the theorem is satisfied and let $\{g_n(x)\}$ be a sequence of functions from F such that, for some standard positive c, $g_n(x) = 0$ for $|x| \ge c$ and $\lim_{n \to \infty} g^{(k)}(x) = 0$ uniformly in x for finite k. Let $f(x) \in \alpha$ then we have to show that $^0(f,g_n)$ tends to zero as n tends to infinity (in R). By assumption, there exists a pre-distribution β which is a function such that for some finite k, $\gamma = \alpha - \beta^{(k)}$ is standard, and equals zero, at all standard points of the interval $-c \le x \le c$. Let $h(x)$ be a function of class $*C^\infty$ which is contained in β. Then $f(x) - h^{(k)}(x)$ belongs to γ. Applying 5.3.17, with $-c$ and c for the a and b in the statement of that theorem, with the zero function for the $f(x)$ mentioned there, and with $f(x) - h^{(k)}(x)$ for $h(x)$, we find that $^0(f-h^{(k)},g_n) = 0$ and hence $^0(f,g_n) = {}^0(h^{(k)},g_n)$. But, as shown in the discussion preceding the statement of 5.3.18, $^0(h^{(k)},g_n)$ tends to zero as n tends to infinity, in R, and the same is therefore true of $^0(f,g_n)$. This proves the theorem.

5.4 Remarks and references. The non-standard theory of Riemann integration is developed in ROBINSON [1963] within the framework of the Lower Predicate Calculus. For the non-standard approach to Lebesgue integration the use of a higher order language appears to be essential.

Non-standard functions which represent the Dirac Delta functions are detailed in ROBINSON [1961]. LUXEMBURG [1962] contains a non-standard theory of distributions which is based on the approach given in MIKU-ŠINSKI and SIKORSKI [1957]. The theory presented here is closer to the point of view of L. Schwartz (SCHWARTZ [1950, 1951]).

CHAPTER VI

FUNCTIONS OF A COMPLEX
VARIABLE

6.1 Analytic theory of polynomials. Let Z be the two-dimensional Euclidean space with the algebraic structure of the complex numbers. The points of Z may be regarded as pairs of real numbers in R and the formal statements that can be made about Z can all be expressed as statements about R. Let $*Z$ and $*R$ be corresponding enlargements. For any entity (set, relation, function) in Z or R there is a corresponding entity in $*Z$ or $*R$. Since $*Z$ is included in $*R$ in the sense explained we shall usually think of such a correspondence as relating to R and $*R$.

A complex-valued function of a complex variable $w=f(z)$ may be represented in the formal language in various ways, e.g., as a binary relation whose arguments are pairs of real numbers, or as a quaternary relation where $\Phi_{(0,0,0,0)}(\psi,x,y,u,v)$ stands for $u+iv=f(x+iy)$, $x+iy=z$, $u+iv=w$. Although the particular choice of the representation is irrelevant to our arguments the one just mentioned is quite suitable if we wish to express a given statement within the formal language Λ.

Let Π be the set of complex-valued polynomials of one complex variable so that the elements of Π map Z into Z. Let $*\Pi$ be the corresponding set in $*R$, so that the elements of $*\Pi$ map $*Z$ into $*Z$. Following our custom, we call the elements of $*\Pi$ also polynomials.

With every polynomial $p(z)\in\Pi$ and with every natural number $k\in N$ there is associated a complex number a_k, which is *the kth coefficient* of $p(z)$. For non-zero $p(z)$ there is a greatest k for which $a_k\neq0$, the *degree* of $p(x)$. For any $p(z)$, the number of coefficients which are different from 0 will be called the *rank r* of $p(z)$ so that $r\geq0$. All these functionals extend to $*\Pi$ automatically, although the degree and the rank of a polynomial may now be infinite natural numbers. We shall be interested chiefly in non-zero polynomials of $*\Pi$ whose rank is *finite*, $r\in N$. Such polynomials can be written in the form

6.1.1 $$p(z)=c_1z^{n_1}+c_2z^{n_2}+\cdots+c_rz^{n_r}, \qquad r\geq1,$$

where $n_1 < n_2 < \cdots < n_r$, and $c_j \neq 0$, $1 \leq j \leq r$. In this expression, neither the n_j nor the c_j need be standard although the number of terms, r, is standard, by assumption. 6.1.1 implies that for any particular complex number in $*Z$ the value of $p(z)$ can actually be obtained by the evaluation and addition of the individual terms on the right hand side.

Another functional which extends automatically to the non-zero polynomials of $*\Pi$ is *the number of zeros of a polynomial $p(z)$ (taking into account multiplicities) such that $|z| < R_0$.* We denote this functional by $\Omega = \Omega[p(z),R_0]$, where R_0 ranges over the positive real numbers and Ω ranges over the natural numbers. If $\Omega[p(z),R_0] = v$, where v is a finite natural number then this fact can be expressed equivalently by the semi-formal statement

6.1.2 $(\exists z_1)...(\exists z_v)[[|z_1| < R_0 \wedge |z_2| < R_0 \wedge \cdots \wedge |z_n| < R_0]$
$\wedge [(\exists q)[\Phi_\tau(\pi,q) \wedge p(z) = (z - z_1)...(z - z_v)q$
$\wedge [(\forall z)[q(z) = 0 \supset |z| \geq R_0]]]]]$

where π is the constant which denotes Π in R and $*\Pi$ in $*R$, and τ is the type of π, $\tau = ((0,0,0,0))$. It is not difficult to see that 6.1.2 can be translated into a sentence of K. If $\Omega[p(z),R_0]$ is infinite there is no corresponding single sentence quantifying the individual roots as in 6.1.2. However, in that case the statements

6.1.3 $(\exists z_1)...(\exists z_v)[[|z_1| < R_0 \wedge |z_2| < R_0 \wedge \cdots \wedge |z_v| < R_0]$
$\wedge [(\exists q)[\Phi_\tau(\pi,q) \wedge p(z) = (z - z_1)...(z - z_v)q]]]$

hold for all finite natural numbers v.

A monomial $p(z) = az^m$ is a polynomial of rank 1. Thus, it is implied in the definition of a monomial that $a \neq 0$. Let az^m and bz^n be two monomials in $*\Pi$ such that $m \neq n$. We shall say that bz^n *dominates* az^m, in symbols $bz^n \gg az^m$, if there exists a finite positive number ϱ such that $|bz^n| > |az^m|$ for all finite z such that $|z| > \varrho$. Clearly, this is equivalent to the condition $|b/a|\sigma^{n-m} > 1$ for all $\sigma > \varrho$, σ finite.

Let az^m, bz^n, cz^k be three monomials such that $bz^n \gg az^m$, $cz^k \gg bz^n$. (This implies that $n \neq m$ and $k \neq n$). Suppose that $k \neq m$. Then we claim that $cz^k \gg az^m$.

By assumption, there exist finite positive numbers ϱ, ϱ' such that for all finite σ, $\sigma > \varrho$ entails $|b/a|\sigma^{n-m} > 1$ and $\sigma > \varrho'$ entails $|c/b|\sigma^{k-n} > 1$.

Hence, for all finite $\sigma > \max(\varrho, \varrho')$

$$\left|\frac{c}{a}\right|\sigma^{k-m} = \left|\frac{c}{b}\right|\sigma^{k-n}\left|\frac{b}{a}\right|\sigma^{n-m} > 1 \cdot 1 = 1 ,$$

showing that cz^k dominates az^m.

For any two monomials az^m, bz^n, $n \neq m$, the function $f(\sigma) = |b/a|\sigma^{n-m}$, $\sigma > 0$, is strictly increasing for $n > m$ and strictly decreasing for $n < m$. In the former case, it is smaller than 1 for sufficiently small σ and greater than 1 for sufficiently large σ, in the latter case, greater than 1 for sufficiently small σ, smaller than 1 for sufficiently large σ. In any case, there is a single $\sigma = \sigma_0$ such that $|b/a|\sigma^{n-m} = 1$. If $n > m$ and σ_0 is finite then $bz^n \gg az^m$. If $n > m$ and σ_0 is infinite then $|a/b|\sigma^{m-n} > 1$ for all finite $\sigma > 0$ and so $az^m \gg bz^n$. On the other hand, if $m > n$ and σ_0 is finite then $az^m \gg bz^n$; while if $m > n$ and σ_0 is infinite then $bz^n \gg az^m$. This exhausts all possible relations between az^m and bz^n and shows that (for $n \neq m$) either $bz^n \gg az^m$ or $az^m \gg bz^n$. It will be seen that two monomials cannot dominate each other.

Now let $p(z)$ be a non-zero polynomial of finite rank as given by 8.1.1. The terms of $p(z)$ as they appear on the right hand side of 8.1.1 constitute a set of r monomials of different degrees, which are totally ordered by the relation $\gg$. Accordingly, one of the terms of $p(z)$, $c_i z^{n_i}$, dominates all others. Its degree, n_i, will be called the *order* of $p(z)$, in symbols $n_i = \text{ord } p(z)$.

We now come to the principal result of this section.

6.1.4 THEOREM. Let $p(z)$ be a non-zero polynomial of finite rank, r. If $\text{ord } p(z) = \mu$ is finite (i.e., if $\mu \in N$) then the number of finite zeros of $p(z)$ is μ. If $\text{ord } p(z) = \mu$ is infinite (i.e., $\mu \in {}^*N - N$) then the number of finite zeros of $p(z)$ is *infinite*.

The multiplicities of the zeros have to be taken into account in the interpretation of this theorem. For finite μ, this does not present any problem. For infinite μ, we shall take the italicized word *infinite* in the conclusion to mean that for every finite natural number ν we can find (equal or different) finite complex numbers $z_1, \ldots, z_\nu$, and a polynomial $q(z) \in {}^*\Pi$ such that $p(z) = (z - z_1) \cdots (z - z_\nu) q(z)$.

In the proof of 6.1.4, we shall make use of Rouché's theorem –

6.1.5 THEOREM (Standard. Rouché). If the functions $f_1(z)$ and $f_2(z)$ are analytic on and inside a Jordan curve Γ and $|f_2(z)| < |f_1(z)|$ on Γ then $f_1(z)$ and $f_1(z) + f_2(z)$ have the same number of zeros inside Γ.

We shall require this theorem for the particular case that $f_1(z)$ and $f_2(z)$ are polynomials and the curve Γ is a circle. For this case the transfer of the theorem from R to $*R$ presents no problem.

PROOF of 6.1.4. Let $c_i z^{n_i}$ be the term of $p(z)$ which dominates all other terms, so that $n_i = \mu = \operatorname{ord} p(z)$. Supposing that $p(z)$ is given by 6.1.1., consider any term $c_j z^{n_j}$ of $p(z)$ such that $j < i$. By assumption, there exists a finite positive number ϱ_j such that

$$\left|\frac{c_i}{c_j}\right| |z|^{n_i - n_j} > 1 \qquad \text{for } |z| > \varrho_j.$$

Let $\varrho_0 = r \max_{1 \le j < i} \varrho_j$, where r is the rank of the polynomial then for all finite z such that $|z| > \varrho_0$,

6.1.6 $$\left|\frac{c_j z^{n_j}}{c_i z^{n_i}}\right| = \left|\frac{c_j (z/r)^{n_j}}{c_i (z/r)^{n_i}}\right| r^{n_j - n_i} < \frac{1}{r}.$$

On the other hand, if $j > i$ then we claim that the ratio $|c_j z^{n_j}/c_i z^{n_i}|$ is infinitesimal for all finite z. Suppose, on the contrary, that $|c_j z_0^{n_j}/c_i z_0^{n_i}| \ge \varepsilon$ for some finite z_0 where ε is a standard positive number. Put $\varrho' = |z_0| \varepsilon^{-1/(n_j - n_i)}$ then for $|z| > \varrho'$,

$$\left|\frac{c_j z^{n_j}}{c_i z^{n_i}}\right| = \left|\frac{c_j}{c_i}\right| |z|^{n_j - n_i} > \left|\frac{c_j}{c_i}\right| |z_0|^{n_j - n_i} \varepsilon^{-1} \ge \varepsilon \varepsilon^{-1} = 1,$$

which contradicts the assumption that $c_i z^{n_i}$ dominates $c_j z^{n_j}$. Hence, $|c_j z^{n_j}/c_i z^{n_i}|$ is infinitesimal for all finite z and hence, a fortiori,

$$|c_j z^{n_j}/c_i z^{n_i}| < 1/r \quad \text{for all finite } z.$$

Put

$$p(z) = c_i z^{n_i} + q(z) \qquad \text{where } q(z) = \sum_{j \ne i}^{r} c_j z^{n_j}.$$

Then, for $|z| > \varrho_0$,

$$|q(z)| \le \sum_{j \ne i}^{r} |c_j z^{n_j}| \le (r-1) \cdot \frac{1}{r} |c_i z^{n_i}|,$$

and so

6.1.7 $$|q(z)| < |c_i z^{n_i}| \qquad \text{for } |z| > \varrho_0.$$

Suppose first that $n_i = \mu = \operatorname{ord} p(z)$ is a finite natural number. Then $c_i z^{n_i}$ has just n_i zeros, all at the origin. Hence, by the special case of Rouché's theorem mentioned above, and taking into account 6.1.7, $p(z)$ has just n_i zeros in any circle of finite radius $\varrho' > \varrho_0$. This shows that $p(z)$ possesses just n_i finite zeros, taking into account multiplicities.

Suppose next that $n_i = \mu$ is an infinite natural number. It is then a consequence of Rouché's theorem that for every $\varrho' > \varrho_0$ and for every finite natural number v, there exist z_k, $k = 1,\dots,v$, $|z_k| < \varrho'$ such that $p(z)$ is divisible by $(z - z_1)\dots(z - z_v)$. This completes the proof of 6.1.4.

6.1.4 enables us to establish the following theorem on lacunary polynomials.

6.1.8 THEOREM (Standard. Montel). Let

6.1.9
$$p(z) = 1 + a_1 z + a_2 z^2 + \dots + a_k z^k + a_{k+1} z^{n_{k+1}} + \dots + a_{k+l} z^{n_{k+l}}$$

where $0 < k < n_{k+1} < \dots < n_{k+l}$, $l \geq 1$, $a_k \neq 0$. Then there exists a positive ϱ which depends only on $a_1, a_2, \dots, a_k$, and l (and not on the particular $n_{k+j}, a_{k+j}, j = 1, \dots, l$) such that $p(z)$ possesses at least k roots in the circle $|z| \leq \varrho$.

PROOF. Suppose, contrary to the assertion of the theorem, that no such ϱ exists for a given set $a_1, a_2, \dots, a_k$, and l. Then the following statement can be formulated as a sentence of K, which holds in R and hence in $*R$.

6.1.10 'For every $\varrho > 0$ there exists a polynomial $p(z)$ of the form 6.1.9 such that $p(z)$ possesses not more than $k - 1$ zeros in the circle $|z| \leq \varrho$.'

Working in $*R$ and choosing ϱ as an infinite positive number we deduce from 6.1.10 that there exists a polynomial $p(z) \in *\Pi$ of the form 6.1.9 which possesses at most $k - 1$ finite roots. Let us consider the order of this particular polynomial. Since the numbers $1, a_1, a_2, \dots, a_k$ are standard, and since $a_k \neq 0$, the term $a_k z^k$ in $p(z)$ dominates 1 and all $a_j z^j$ for $j < k$ for which $a_j \neq 0$. All the remaining terms of $p(z), a_{k+j} z^{n_{k+j}}, j = 1, \dots, l$ are of degree greater than k. Accordingly, $\operatorname{ord} p(z) \geq k$, where $\operatorname{ord} p(z)$ may be finite or infinite. This shows that $p(z)$ has at least k finite roots, by 6.1.4. We have obtained a contradiction and the theorem is proved.

We shall now give a generalization of Montel's theorem in a new direction. Two polynomials will be called *disjoint* if they do not contain any terms (monomials) of the same degree. For example, if one polynomial is odd and the other one even then the two polynomials are disjoint.

6.1.11   THEOREM (Standard).   There exists a function $\psi(r,s,k)$ whose arguments range over the positive integers and whose functional values are real positive such that the following condition is satisfied. If $p(z)$ and $q(z)$ are two disjoint polynomials of ranks not exceeding r and s respectively, and if both $p(z)$ and $q(z)$ have at least k roots whose moduli do not exceed any given $\varrho > 0$ then the sum $p(z)+q(z)$ has at least k roots whose moduli do not exceed $\varrho\psi(r,s,k)$.

PROOF.   It is sufficient to prove the existence of $\psi(r,s,k)$ for the particular case $\varrho=1$ since the general case can be reduced to this by putting $z=\varrho z'$.

Suppose that, for a particular set of finite natural numbers r, s, and k, there does not exist a $\psi=\psi(r,s,k)$ as described in the theorem. Then the following statement holds in R and can be expressed within K and hence, holds also in *R.

6.1.12   'For every $\sigma>0$, there exists two disjoint polynomials $p(z)$ and $q(z)$ of ranks not exceeding r and s respectively, each possessing at least k roots in the unit circle, $|z|\leq 1$, and such that the sum $p(z)+q(z)$ possesses at most $k-1$ roots in the circle $|z|\leq\sigma$.'

Working in *R, and choosing for σ any infinite positive number, we find that there exist disjoint polynomials $p(z)$ and $q(z)$ in *$\varPi$, of ranks not exceeding r and s respectively, each with at least k roots in the unit circle such that $p(z)+q(z)$ possesses at most $k-1$ finite roots. Clearly, in this case, $p(z)+q(z)$ cannot reduce to the zero polynomial, and neither $p(z)$ nor $q(z)$ can reduce to that polynomial. Since $p(z)$ and $q(z)$ are disjoint, the set of all terms of $p(z)+q(z)$ is the union of the set of terms of $p(z)$ and the set of terms of $q(z)$. Suppose that the term of $p(z)$ which dominates all others is $a_i z^{n_i}$ while the corresponding term of $q(z)$ is $b_j z^{m_j}$. Thus ord $p(z)=n_i\geq k$, ord $q(z)=m_j\geq k$. Evidently, the dominant term of $p(z)+q(z)$ is then either $a_i z^{n_i}$ or $b_j z^{m_j}$. This shows that ord $(p(z)+q(z))\geq k$, implying that $p(z)+q(z)$ has at least k finite roots. We have obtained a contradiction, which proves the theorem.

Montel's theorem, 6.1.8, follows from 6.1.11, by taking $r=k+1$, $s=l$ and by choosing ϱ in 6.1.11 so that all roots of the polynomial $1+a_1z+a_2z^2+\cdots+a_kz^k$ have moduli $|z|\leq\varrho$.

We now turn to the proof of some results which are related to Kakeya's theorem on the zeros of the derivative of a polynomial.

6.1.13 THEOREM (Standard. Kakeya). Let $p(z)$ be a polynomial of degree not greater than n such that at least $k\geq 1$ roots of $p(z)$ are situated in the circle $|z|\leq\varrho$. Then there exists a number θ which depends only on n and k, $\theta=\theta(n,k)$ such that $p'(z)=dp/dz$ has at least $k-1$ roots in the circle $|z|\leq\varrho\theta$.

Clearly, one obtains an equivalent result by stating in the conclusion that $zp'(z)$ has at least k roots in the circle $|z|\leq\varrho\theta$. In this form Kakeya's theorem is a special case of the following result which is considerably more general.

6.1.14 THEOREM (Standard). There exists a function $\phi(r,k)$ whose arguments range over the positive integers and whose functional values are real positive such that the following condition is satisfied. For any polynomial of rank not exceeding r,

$$p(z)=c_1z^{n_1}+c_2z^{n_2}+\cdots+c_rz^{n_r},\quad 0\leq n_1<n_2<\cdots<n_r,$$

such that $p(z)$ possesses at least k roots in the circle $|z|\leq\varrho$, and for any set of complex numbers λ_i such that

$$0\leq|\lambda_1|\leq|\lambda_2|\leq\cdots\leq|\lambda_r|,$$

the polynomial

$$q(z)=\lambda_1c_1z^{n_1}+\lambda_2c_2z^{n_2}+\cdots+\lambda_rc_rz^{n_r}$$

possesses at least k roots in the circle $|z|\leq\varrho\phi(r,k)$.

PROOF. Again, the substitution $z=\varrho z'$ shows that it is sufficient to consider only the case $\varrho=1$. If for particular r and k, no $\phi(r,k)$ as described in the theorem existed then the following statement would hold in R.

6.1.15 'For every $\sigma > 0$, there exists a polynomial $p(z)$ and constants λ which satisfy the conditions of 6.1.14 with $\varrho = 1$, such that the corresponding $q(z)$ possesses not more than $k-1$ roots in the circle $|z| \leq \sigma$.'

Applying this statement to $*R$ and choosing for σ an infinite positive number, we should then obtain $p(z)$ and λ_i which satisfy the conditions of 6.1.14 while $q(z)$ would have less than k finite roots. We are going to show that this is impossible. At any rate it cannot happen if $p(z)$ vanishes identically and we may rule out this case.

Suppose that

$$p(z) = c_1 z^{n_1} + c_2 z^{n_2} + \cdots + c_r z^{n_r},$$

where $c_j \neq 0$ for at least one j, has at least k roots in the unit circle. Then ord $p(z) \geq k$. Thus, if $c_i z^{n_i}$ is the dominant term in $p(z)$, then $n_i \geq k$. The terms of $q(z)$ are those $\lambda_j c_j z^{n_j}$ which do not vanish identically. Since $q(z)$ has less than k finite roots it cannot vanish identically and has a dominant term, $\lambda_m c_m z^{n_m}$, $n_m \leq k-1$, and so $n_m < n_i$. Then $|\lambda_i| \geq |\lambda_m|$ and so

6.1.16
$$\left| \frac{\lambda_i c_i z^{n_i}}{\lambda_m c_m z^{n_m}} \right| = \left| \frac{\lambda_i}{\lambda_m} \right| \left| \frac{c_i z^{n_i}}{c_m z^{n_m}} \right| \geq \left| \frac{c_i z^{n_i}}{c_m z^{n_m}} \right|.$$

But $c_i z^{n_i} \gg c_m z^{n_m}$ and so the right hand side of 6.1.16 is greater than 1 for sufficiently large finite $|z|$. This shows that $\lambda_i c_i z^{n_i} \gg \lambda_m c_m z^{n_m}$, a contradiction which proves 6.1.14.

6.1.13 is obtained from 6.1.14 for $\lambda_j = n_j = j-1$, $j = 1, 2, \ldots, r$. The following theorem can be proved by the same method.

6.1.17 THEOREM (Standard).   There exists a function $\chi(r, k)$, defined for all positive integers and taking real positive values such that if

$$p(z) = c_1 z^{n_1} + c_2 z^{n_2} + \cdots + c_r z^{n_r}, \qquad 0 \leq n_1 < n_2 < \cdots < n_r$$

and

$$0 \leq |\lambda_1| \leq |\lambda_2| \leq \cdots \leq |\lambda_r|,$$

and moreover,

$$0 \leq k_1 \leq k_2 \leq \cdots \leq k_r,$$

where the k_i are natural numbers, then – if $p(z)$ possesses at least k roots in the circle $|z| \leq \varrho$, the polynomial

$$q(z) = \lambda_1 c_1 z^{n_1 + k_1} + \lambda_2 c_2 z^{n_2 + k_2} + \cdots + \lambda_r c_r z^{n_r + k_r}$$

possesses at least k roots in the circle $|z| \leq \varrho \chi(r, k)$.

6.2 Analytic functions. Let Ψ be the set of all complex valued functions
of a complex variable within R, i.e., of mappings from subsets of Z into
Z. Let Σ be the binary relation which holds between an element $f(z)$ of
Ψ and an open set of points D in Z if $f(z)$ is analytic in D, i.e., if $f(z)$ is
differentiable at all points of D. Passing to $*R$, the relation holds between
functions $f(z)$ which belong to $*\Psi$ and open internal sets D in $*Z$ if $f(z)$
is differentiable at all points z_0 of D according to the classical definition
as applied within $*R$. Thus, $f'(z_0)$ is the derivative of $f(z)$ at z_0 if for
every $\varepsilon>0$ in $*R$ there exists a $\delta>0$ in $*R$ such that

$$\left|\frac{f(z_0+h)-f(z_0)}{h}-f'(z_0)\right|<\varepsilon$$

provided $|h|<\delta$.

If B is a set of points in $*Z$, which may be internal or external, and
$f(z)\epsilon*\Psi$ then we shall say that $f(z)$ is analytic in B if there exists an open
internal set of points D in $*Z$ such that $B\subset D$ and such that $\Sigma(f(z),D)$
holds in $*R$.

6.2.1 THEOREM. Suppose that the function $f(z)$ is analytic in an S-open
set of points B in $*Z$, and that $f(z)$ is finite (i.e., $|f(z)|$ is a finite real
number) everywhere in B. Then $f(z)$ is S-continuous in B.

PROOF. Let z_0 be a point of B. Then there exists a circle with center z_0
and radius ϱ, ϱ a standard positive number, such that the set D which
consists of the points z in the interior of the circle and on its circumfer-
ence, $|z-z_0|\leq\varrho$, is included in B. Since $f(z)$ is an analytic internal
function in B, we may make use of known properties of analytic functions,
with the appropriate interpretation for $*R$. In particular, $f(z)$ attains its
maximum for the circumference of the circle at some point z_1, $|z_1-z_0|=\varrho$,
and since $f(z)$ is finite everywhere in B, we have $|f(z_1)|=m$, a finite
number. Now let z' be any point which belongs to the monad of z_0.
Then $|z'-z_0|$ is infinitesimal and

$$|f(z')-f(z_0)|\leq|z'-z_0|\frac{2m}{\varrho}$$

by Schwarz's lemma. This shows that $|f(z')-f(z_0)|$ is infinitesimal,
$f(z')\epsilon\mu(f(z_0))$. It now follows directly from 4.5.7 that $f(z)$ is S-con-
tinuous at z_0. This proves 6.2.1.

6.2.2 THEOREM. Suppose that $g(z)$ is analytic in a set of points $B \subset {}^*Z$. Suppose further that, for a point z_0 which belongs to the S-interior of B, the difference $g(z) - g(z_0)$ is finite for all points z that belong to the monad of z_0. Then $g(z)$ is S-continuous at z_0.

PROOF. Let ϱ be a standard positive number such that the set $D = \{z \mid |z - z_0| \leq \varrho\}$ is a subset of B. Then the function $f(z) = g(z) - g(z_0)$ is defined and analytic in D, and is finite in the monad $\mu(z_0)$. It then follows from 4.5.5 that $f(z)$ is finite for all z such that $|z - z_0| \leq \sigma$ where σ is some standard number $0 < \sigma \leq \varrho$. We now conclude from 6.2.1 (taking as the S-open set mentioned there the set $\{z \mid^0 |z - z_0| < \sigma\}$) that $f(z)$, and hence $g(z)$, is S-continuous at z_0. This proves the theorem.

Let B be an S-open set which contains only finite points. In agreement with a notation introduced previously, we write ${}^0B$ for the set of points z_0 of Z such that $\mu(z_0)$ has points in common with B. If $z' \in \mu(z_0) \cap B$ then $|z' - z_0| \simeq 0$. At the same time, since B is S-open, $\mu(z') \subset B$. Hence, $z_0 \in B$, and hence, ${}^0B \subset B$. It is not difficult to see that ${}^0B$ is open. Also, if ${}^0B$ is extended to ${}^{*0}B$ on passing from R to ${}^*R$ then $B \subset {}^{*0}B$ since the monad of every point of ${}^0B$ belongs to ${}^{*0}B$ and since every point of B belongs to the monad of some point of ${}^0B$.

Now let $f(z)$ be an internal function which is analytic and finite in B. As we have shown (see 6.2.1) it follows that $f(z)$ is S-continuous at all points of B. Accordingly, the function ${}^0f(z)$ can be defined for all points $z_0 \in {}^0B$ as the joint value of the standard parts of the values of $f(z)$ for $z \in \mu(z_0) \cap B = \mu(z_0)$. Theorem 4.5.10 now shows that ${}^0f(z)$ is continuous at all points of ${}^0B$ (where we may take for the set D in 4.5.10 any circle which, together with its interior, belongs entirely to B). Moreover, with the stated assumptions, we are going to prove

6.2.3 THEOREM. ${}^0f(z)$ is analytic in ${}^0B$.

PROOF. Put $g(z) = {}^0f(z)$ so that $g(z)$ is defined in ${}^0B$ and, passing to ${}^*R$, is defined in ${}^{*0}B$, and hence in B. Let $z_0 \in {}^0B$. Then there exists a positive standard ϱ such that the set $D = \{z \mid |z - z_0| \leq \varrho\}$ in ${}^*R$ belongs to B. Then a point z belongs to ${}^0D$ if and only if $|z - z_0| \leq \varrho$, for z in R. Hence D and ${}^0D$ have the same definition in ${}^*R$ and R respectively and hence, $D = {}^{*0}D$. Let D_1 be the S-interior of the circle D and D_2 its circumference. Thus, $D_2 = \{z \mid |z - z_0| = \varrho\}$. It is not difficult to see that $D_2 = {}^{*0}D_2$.

Since $g(z)$ is continuous in ${}^0D$ it is continuous also in $D = {}^{*0}D$. In particular, $g(z)$ is continuous, hence S-continuous, on D_2. At the same time, $f(z)$ also is S-continuous on D_2 and, at all standard points of D_2, $g(z) = {}^0f(z)$ and hence $g(z) \simeq f(z)$. It follows that $|f(z) - g(z)|$ is infinitesimal everywhere on D_2 and hence, that $|f(z) - g(z)| \leq \eta$ for all $z \in D_2$, for some infinitesimal η.

Now in order to prove that $g(z)$ is an analytic function in ${}^0D_1$ it is sufficient to show that for all $z_1 \in {}^0D_1$, i.e., for any standard z_1 such that $|z_1 - z_0| = \varrho_1 < \varrho_0$,

6.2.4
$$g(z_1) = \frac{1}{2\pi i} \int_{{}^0D_2} \frac{g(z)}{z - z_1} \, dz .$$

But both sides of the equation 6.2.4 are standard numbers. Hence, in order to prove that they are equal it is sufficient to show that they differ only by infinitesimal amounts. As for the left hand side, we have

6.2.5
$$g(z_1) \simeq f(z_1) = \frac{1}{2\pi i} \int_{D_2} \frac{f(z)}{z - z_1} \, dz .$$

On the right hand side of 6.2.4, we may pass from R to $*R$, without changing the value of the integral. Thus, it remains to be shown that

6.2.6
$$\frac{1}{2\pi i} \int_{D_2} \frac{g(z)}{z - z_1} \, dz \simeq \frac{1}{2\pi i} \int_{D_2} \frac{f(z)}{z - z_1} \, dz .$$

Now the difference between the left and right hand sides of 6.2.6 is equal to
$$\frac{1}{2\pi i} \int_{D_2} \frac{g(z) - f(z)}{z - z_1} \, dz$$

and
$$\left| \frac{1}{2\pi i} \int_{D_2} \frac{g(z) - f(z)}{z - z_1} \, dz \right| \leq \frac{1}{2\pi} \int_{D_2} \frac{|g(z) - f(z)|}{|z| - |z_1|} |dz| \leq$$

$$\frac{1}{2\pi} \int_{D_2} \frac{\eta}{\varrho - \varrho_1} |dz| = \eta \frac{\varrho}{\varrho - \varrho_1} ,$$

which is infinitesimal.

This proves that $g(z)$ is analytic in $^0D_1$ (within R), and, in particular, that it is differentiable at z_0. Since z_0 was chosen as an arbitrary point of ^0B the theorem is proved.

The precise rendering of an integral within the formal language Λ is to some extent a matter of choice. For example, we may think of the integral as a functional of three variables one of which is the integrand, while the other two are functions which describe the contour of integration. Whatever the formal representation, an integral in $*Z$ has the same properties as an integral in Z to the extent to which these properties can be expressed within Λ.

6.2.7 THEOREM. With the assumptions of 6.2.3, let z be any point in ^0B. Then
$$(^0f(z))^{(n)} = {}^0(f^{(n)}(z)) \qquad \text{for } n=1,2,\dots,\ n\in N.$$

In order to prove 6.2.7, we have to replace Cauchy's formulae in 6.2.4 and 6.2.5 by Cauchy's formulae for the first and higher derivatives.

6.2.8 THEOREM. With the assumptions of 6.2.3, suppose that $^0f(z)$ is not a constant. If $^0f(z_0)=b$ for some point $z_0\in{}^0B$, then $f(z)$ takes the value b somewhere in the monad of z_0. Conversely, if $f(z)=b$ for $z\in\mu(z_0)$, where b and z_0 are standard, and $z_0\in{}^0B$, then $^0f(z_0)=b$.

PROOF. To prove the first part of the theorem suppose that $^0f(z_0)=b$ where $z_0\in{}^0B$. Let C be the circle $|z-z_0|=\varrho$ in Z where the positive standard ϱ is chosen in such a way that C and its interior belong to ^0B and $g(z)={}^0f(z)-b$ does not vanish on or inside C except at z_0. Since $|g(z)|$ does not vanish on C it attains its minimum $m\neq 0$ somewhere on C. On passing from R to $*R$ we find that m is also the minimum of $g(z)$ on $*C$, which is the circle $|z-z_0|=\varrho$ in $*Z$. Moreover, $g(z)\neq 0$ for all z such that $0<|z-z_0|\leq\varrho$, for this also is a fact which is preserved on passing from R to $*R$. At the same time, $g(z_0)={}^0f(z_0)-b=0$.

Similarly as in the proof of 6.2.3, $|f(z)-b-g(z)|$ is infinitesimal on $*C$. Also, m is standard and positive and so
$$|f(z)-b-g(z)|<m\leq g(z) \qquad \text{on } *C.$$

Hence, by Rouché's theorem as applied to $*R$ there exists a point z_1, $|z_1-z_0|<\varrho$, such that

$$(f(z)-b-g(z))+g(z)=f(z)-b$$

vanishes for $z=z_1$, i.e., such that $f(z_1)=b$.

In order to show that z_1 may be supposed to belong to the monad of z_0, we use a slightly more precise argument. Suppose that $^0f(z)-b$ vanishes at $z=z_0$ with multiplicity n, where n is by necessity finite. The above argument then shows that $f(z)-b$ has just n zeros z_j, $j=1,\dots,n$, counting multiplicities, such that $|z_j-z_0|<\varrho$ where ϱ is a sufficiently small positive number, otherwise arbitrary. This shows that the z_j, $j=1,\dots,n$, all belong to the monad of z_0. For otherwise there would exist a standard positive ϱ such that for some z_j, $0<\varrho<|z_j|$ and the number of zeros of $f(z)-b$, $|z|<\varrho$, would then have to be less than n. This proves the first part of the theorem. In order to verify the second part, we observe that $^0f(z_0)$ is, by definition the standard part of $f(z)$ for an arbitrary z in the monad of z_0.

We remark that a function may be analytic and finite on an S-open set B and yet possess an infinite number of zeros in a monad which belongs to B. For example, the function $\exp(-\omega^2)\sin\omega z$ where ω is positive infinite, has an infinite number of zeros in the monad of 0. This is possible because the set of these zeros does not possess a limit point (in the sense of $*R$) anywhere in $*Z$. However,

6.2.9 THEOREM. With the assumptions of 6.2.3, suppose that $^0f(z)$ is not constant in ^0B. Then the set of zeros of $f(z)$ cannot possess an S-limit point in B.

PROOF. Suppose on the contrary that z_1 is an S-limit point of the zeros of $f(z)$. Choose a standard positive number ϱ such that $^0f(z)$ has no zeros on the circle $|z-{}^0z_1|=\varrho$. Then the function $^0f(z)$ can have only a finite number, n, of zeros inside this circle, $n\geq0$.

The argument used in the proof of the first part of 6.2.8 shows that the number of zeros of $f(z)$ inside the circle $|z-{}^0z_1|=\varrho$ also is precisely n. This shows that z_1 cannot be an S-limit point of the set of zeros of $f(z)$. Putting the result in a different way, no monad in B can contain an infinite number of zeros of $f(z)$.

6.2.10 THEOREM. With the assumptions of 6.2.3, suppose in addition, that $^0f(z)$ is not constant. If $f(z)$ is univalent (schlicht) in B then $^0f(z)$ is univalent in ^0B.

PROOF. Suppose that $^0f(z)$ is not univalent in ^0B so that $^0f(z_1)=\,^0f(z_2)$ where z_1 and z_2 are two distinct points in ^0B. By 6.2.8, $f(z_1')=b$ for some $z_1'\in\mu(z_1)$ and $f(z_2')=b$ for some $z_2'\in\mu(z_2)$. But the monads of z_1 and z_2 are disjoint and so $z_1' \neq z_2'$. This shows that $f(z)$ is not univalent in B and proves the theorem.

We may also consider meromorphic functions on the Riemann sphere in $*R$. Identifying this sphere with the unit sphere in $*R^3$ and relating it to the complex plane by stereographic projection in the usual way, we find that the monad of $z=\infty$ on the Riemann sphere contains, in addition to $z=\infty$, precisely those points which correspond to the infinite points of $*Z$.

6.3 Picard's theorems and Julia's directions

6.3.1 THEOREM. Let $f(z)$ be an internal function which is analytic for $|z-z_0|<\varrho$ where ϱ is standard and positive. Let a, b, and c be three finite complex numbers such that a and b do not belong to the same monad. Suppose that $f(z)=c$ for some $z\in\mu(z_0)$ but that $f(z)$ takes neither the value a nor the value b in $\mu(z_0)$. Then $f(z)$ is S-continuous at z_0.

PROOF. We have to show, on the assumptions of the theorem that if z', z'', are any two points in $\mu(z_0)$ then $|f(z')-f(z'')|$ is infinitesimal. It is sufficient to prove this fact for $a=0$, $b=1$. For if a and b are more general but still subject to the restrictions of the theorem, consider $g(z)=(f(z)-a)/(b-a)$. Then $g(z)\neq0$, $g(z)\neq1$ for all z in $\mu(z_0)$. Moreover, if $f(z_1)=c$ then $g(z_1)=(c-a)/(b-a)$ which is again a finite number since $|b-a|$ is not infinitesimal. On the other hand,

$$f(z)=(b-a)g(z)+a,$$

so if z', z'' are any two points in $\mu(z_0)$ then

$$f(z')-f(z'')=(b-a)(g(z')-g(z'')).$$

But $|b-a|$ is finite and so, supposing that we have already proved that $|g(z')-g(z'')|$ is infinitesimal, we obtain the same conclusion for

$|f(z')-f(z'')|$. Accordingly, we shall suppose from now on that $a=0$, $b=1$.

The assumption that $f(z)$ does not take the values 0 or 1 in $\mu(z_0)$ entails that there even exists a circle with center z_0 and radius $r<\varrho$, r a standard positive number, such that $f(z)$ does not take the values 0 or 1 in the interior of the circle. For, at any rate, there exists a first positive integer $n=n_0$ such that $f(z)$ does not take the values 0 or 1 for $|z-z_0|<1/n$ and $1/n<\varrho$. No infinite n can be the first positive integer of this kind. Hence, n_0 is finite. We set $r=1/n_0$.

Let $c=f(z_1)$ where $z_1\in\mu(z_0)$ and c is the number specified in the statement of the theorem. We map the circle $|z-z_0|<r$ into the unit circle in the complex ζ-plane by means of the function $\zeta=F(z)=\lambda(f(z))$ where $w=v(\zeta)$ is the elliptic modular function which maps the equilateral cusped triangle with vertices $\exp\frac{1}{2}i\pi$, $\exp\frac{7}{6}i\pi$, $\exp\frac{11}{6}i\pi$ into the upper half of the w-plane, and $\zeta=\lambda(w)$ is its inverse. $v(\zeta)$ is analytic for $|\zeta|<1$ while $\lambda(w)$ is analytic but many-valued for all $w\neq0$, 1 in the w-plane and has branch points at $w=0,1$. Observe that these descriptions apply both in R and in $^*R$.

We determine $F(z)$ uniquely by requiring that $\gamma=F(z_1)=\lambda(c)$ be situated in the fundamental triangle if $\text{Im}(c)\geq0$, or else in one of the adjacent triangles. For R, the fact that this definition determines $F(z)$ as a one-valued function for $|z-z_0|<r$ is established by means of the monodromy theorem. Without going into a detailed discussion of the meaning of this theorem in $^*R$ we may infer that the conclusion applies also in $^*R$.

$|F(z)|$ is bounded by 1 for $|z-z_0|<r$. Accordingly, by 6.2.1, $F(z)$ is S-continuous in the S-interior of this circle. Hence, $F(z)\in\mu(F(z_0))$ for all $z\in\mu(z_0)$.

Suppose first that $^0c=0$ so that c belongs to the monad of 0. If $f(z)\in\mu(0)$ also for all other $z\in\mu(z_0)$ then the theorem is proved. Suppose then that for some $z_2\in\mu(z_0)$, $d_2=f(z_2)$ does not belong to the monad of 0. Suppose that d_2 is not finite so that $|d_2|=|f(z_2)|>3$. On the other hand, $|c|=|f(z_1)|<1$. Consider the variation of $f(z)$ along the straight segment from z_1 to z_2 in $^*Z$. This segment is contained entirely within $\mu(z_0)$. By the theorem of Bolzano–Weierstrass on continuous functions (3.4.6) there exists a z_3 on the segment such that if $d_3=f(z_3)$ then $|d_3|=2$. d_3 is a finite number which belongs neither to the monad of 0 nor to the monad of 1.

Suppose next that $d_2 = f(z_2)$ belongs to the monad of 1. By considering again the variation of $f(z)$ along the segment from z_1 to z_2, we find that for some z_3 on the segment, $|d_3| = \frac{1}{2}$ where $d_3 = f(z_3)$. Thus, d_3 is again a finite number which belongs neither to the monad of zero nor to the monad of 1. Finally if d_2 is finite but $^0d_2 \neq 1$, put $d_3 = d_2$.

If $c \in \mu(1)$ then we may obtain a $z_3 \in \mu(z_0)$ such that $d_3 = f(z_3)$ is finite but belongs neither to the monad of 0 nor to the monad of 1 by an exactly similar procedure. If c belongs neither to the monad of 0 nor to the monad of 1, we take $z_3 = z_1$ so that $d_3 = f(z_3) = c$ is again finite.

Now let $^0\zeta_3$ be the standard part of $\zeta_3 = F(z_3) = \lambda(d_3)$, in accordance with our usual notation. Then $^0\zeta_3 = \lambda(^0d_3)$ and so $|^0\zeta_3| < 1$. It follows that ζ_3 belongs to the S-interior of the unit circle, so that $v(\zeta)$ is S-continuous at $\zeta = \zeta_3$. This shows that $f(z) = v(F(z))$ is S-continuous at $z = z_3$. But z_3 belongs to the monad of z_0 and so $f(z)$ is S-continuous at z_0. This completes the proof.

6.3.2 Theorem. Let B be an S-open and S-connected set (*an S-domain*) in the $^*Z$-plane and let $f(z)$ be a function which is analytic in B. Let a and b be two finite points in $^*Z$ which do not belong to the same monad. Suppose that $f(z)$ takes neither the value a nor the value b in B but that it takes a finite value at some point of B. Then $f(z)$ is finite and S-continuous in B.

Proof.   We may suppose that $a = 0$, $b = 1$ as in the proof of 6.3.1. By 6.3.1, $f(z)$ is S-continuous at all points of B in whose monad $f(z)$ takes at least one finite value, and there are such points, by the assumptions of the theorem to be proved. Let $D \subset B$ be the set of points of B in whose monad $f(z)$ takes a finite value and let $D' = B - D$. For any $z_1 \in D_1$, all other points of the monad of z_1 also belong to D, $\mu(z_1) \subset D$. Similarly, $z_1 \in D'$ entails $\mu(z_1) \subset D'$. We are going to show that D and D' are S-open.

Indeed, let $z_0 \in D$ then $f(z)$ is finite in the entire monad of z_0, by 6.3.1. But $f(z)$ is defined in some S-circle with center z_0, i.e., for $|z - z_0| < \varrho$ where ϱ is positive standard. Hence, by 4.5.5, there exists a positive standard σ, $\sigma < \varrho$, such that $f(z)$ is finite for $^0|z - z_0| < \sigma$. This shows that D is S-open.

D' also is S-open. For let $z_0 \in D'$ and consider the function $g(z) = 1/f(z)$. This function is analytic in D and is infinitesimal in the monad of z_0. It follows, again by 4.5.5 that $g(z)$ is finite for $|z-z_0|<\varrho$ for some positive standard ϱ. Consider the function $h(z)=g(z_0+z)$. $h(z)$ is analytic and finite for $|z|<\varrho$ and so it is S-continuous in the S-interior of that circle, i.e., for the points of the set $E=\{z\ ||^0z|<\varrho\}$. By 6.2.3 $^0h(z)$ exists and is analytic (within R) in the set ^0E, where it is not difficult to verify that ^0E is given by $|z|<\varrho$. Also, $^0h(0)=0$ since $h(0)=g(z_0)$ is infinitesimal. But $h(z)\neq0$ for $|^0z|<\varrho$, and by 6.2.8, this is possible only if $^0h(z)$ is constant in ^0E, and hence $^0h(z)=0$ for $|z|<\varrho$ in R. Since $h(z)$ is S-continuous for $|^0z|<\varrho$ we conclude that $h(z)$ is infinitesimal in that set and, in particular, is infinitesimal for $|z|<\frac{1}{2}\varrho$. It follows that $f(z)$ is infinite for $|z-z_0|<\frac{1}{2}\varrho$ and this proves that D' is S-open.

We have now shown that both D and D' are S-open. It follows that these sets are S-open also relative to the set B of which they are subsets. But D is not empty, and B is S-connected, by assumption, so D' must be empty. This shows (see 6.2.1) that $f(z)$ is finite and S-continuous everywhere in B, and completes the proof of 6.3.2.

6.3.3 THEOREM (Standard. Picard's first theorem).   An entire analytic function which is not a constant takes all complex values except, possibly, just one.

PROOF.   Suppose, contrary to the assertion of the theorem that some non-constant entire function $f(z)$ does not take the values a and b anywhere, where a and b are two different complex numbers. It is understood that we contemplate this possibility in the first instance within R. Going on to $*R$, consider the function $F(z)=f(\omega z)$ where ω is an infinite positive number. $f(z)$ does not become equal to a or b in $*Z$ either, by transfer from R to $*R$. It follows that $F(z)\neq a$, $F(z)\neq b$ for all z such that $|z|<1$. On the other hand, $F(0)=f(0)$ is a finite number. Hence, by 6.3.1, $F(z)$ is S-continuous at the point $z=0$. But $f(z)$ is not a constant and so, for some standard z_1 such that $|z_1|<1$ (more precisely, for all such z_1 except for a finite number), $f(z_1)\neq f(0)$. Put $z_2=z_1\omega^{-1}$, so that $z_2\in\mu(0)$. Then $F(z_2)=f(\omega z_2)=f(\omega z_1\omega^{-1})=f(z_1)$ and so $|F(z_2)-F(0)|$ is not infinitesimal. This contradicts the fact that $F(z)$ is S-continuous at the origin and proves the theorem.

6.3.4 THEOREM (Standard. Landau).   For given complex a_0 and a_1,

$a_1 \neq 0$, there exists a positive number ϱ_0 such that if $f(z)$ is any function which is analytic in a circle $|z| < \varrho$ and such that $f(0) = a_0$, $f'(0) = a_1$ and, moreover, $f(z) \neq 0$, $f(z) \neq 1$ for all $|z| < \varrho$ – then $\varrho \leq \varrho_0$.

PROOF.   Suppose that no such ϱ_0 exists in R for given complex standard numbers a_0 and a_1. Then 'for every $\sigma > 0$, there exists an $f(z)$ which is analytic in a circle $|z| < \varrho$, $\varrho > \sigma$, such that $f(z) \neq 0$ and $f(z) \neq 1$ in that circle and such that $f(0) = a_0$ and $f'(0) = a_1$'. The statement in quotation marks is true equally in R and $*R$ and so, working in $*R$ and choosing σ infinite, we find that there exists a function $f(z)$ which is analytic for $|z| < \sigma$ and does not take the values 0 and 1 for such z in $*Z$ and for which $f(0) = a_0$, $f'(0) = a_1$. Consider the function $g(z) = f(\tfrac{1}{2}\sigma z)$. This function is analytic in the S-interior of the unit circle and does not take the values 0 and 1 in that set. Also, $g(0) = a_0$ and $g'(0) = \tfrac{1}{2}\sigma f'(0) = \tfrac{1}{2}\sigma a_1$ so that $g'(0)$ is infinite. $g(0)$ is finite and $|1 - 0|$ is not infinitesimal. Hence, $g(z)$ is S-continuous at the origin, by 6.3.1. It follows that there exists a positive standard r such that $|g(z)| < |a_0| + 1$ for $|z| \leq r$. Hence, $|g'(0)| \leq (|a_0| + 1)/r$ by one of Cauchy's inequalities. This contradicts the fact that $g'(0)$ is infinite and proves the theorem.

Other versions of Landau's theorem can be proved in the same way. For example, as a slight variation of a theorem of Montel, we may establish the conclusion of Landau's theorem for families of functions which satisfy, for given complex a, $b \neq 0$ and positive c, that $f(0) = a$ and that there exists a z_0, $|z_0| \leq c$ such that $f(z_0) = b$.

Next we prove,

6.3.5   THEOREM (Standard. Schottky).   Let α and ϱ be specified positive numbers and consider the family of functions $f(z)$ which are analytic for $|z| < \varrho$ such that $|f(0)| \leq \alpha$ and such that $f(z) \neq 0$, $f(z) \neq 1$ for $|z| < \varrho$. Let θ be any positive number less than 1. Then there exists a constant λ which depends only on θ and α such that for all functions of the family in question $|f(z)| \leq \lambda$ for $|z| \leq \theta \varrho$.

PROOF.   Suppose first that $\varrho = 1$ and that the theorem is untrue for some particular α and θ. Then 'for every $\lambda > 0$, there exists an $f(z)$ which satisfies the assumptions of 6.3.5 for $\varrho = 1$ and for the specified α and θ and such that $|f(z_0)| > \lambda$ for some $|z_0| \leq \theta$.' The statement in quotation marks must be true equally in R and $*R$ and so, for infinite λ in $*R$, there exists a particular $f(z)$ which satisfies the assumptions of the theorem but such that $f(z_0)$ is infinite for some z_0, $|z_0| \leq \theta$. Let B be the S-interior of the unit circle, so that B contains 0 and z_0. Since $f(0)$ is finite it follows from 6.3.2 that $f(z)$ is finite throughout B. This contradicts the fact that $f(z_0)$ is infinite and proves the existence of a $\lambda = \lambda(\alpha, \theta)$ as required, for $\varrho = 1$. If, to begin with, $\varrho \neq 1$, we only have to consider the function $g(z) = f(\varrho z)$ in place of $f(z)$.

We shall now give some new results which imply, among other things, the existence of a Julia direction for an analytic function with an isolated essential singularity at infinity and hence also Picard's second theorem. We begin with some auxiliary considerations of a geometrical nature.

As pointed out at the end of section 4.7, a straight segment $z_1 z_2$ in the $*Z$-plane is S-connected if z_1 and z_2 are both finite so that all points of the segment are finite. A related result is the following.

6.3.6 THEOREM. Let $B = z_1 z_2$ be a circular arc in the $*Z$-plane whose measure in radians is a, $0 < a \leq 2\pi$, and which consists of finite points only. Then B is S-connected.

PROOF. We may write the equation of the arc in the form

6.3.7
$$z = z_0 + \varrho e^{i(\theta_1 + a\lambda)}, \qquad 0 \leq \lambda \leq 1$$

where z_0 and ϱ may be finite or infinite. Let z_3, z_4 be any two points on the arc, corresponding to λ_3, λ_4 respectively. Then

$$z_4 - z_3 = \varrho e^{i(\theta_1 + a\lambda_4)} - \varrho e^{i(\theta_1 + a\lambda_3)}$$
$$= \varrho e^{i(\theta_1 + a(\lambda_4 - \lambda_3)/2)} \left(e^{ia(\lambda_4 - \lambda_3)/2} - e^{-ia(\lambda_4 - \lambda_3)/2} \right)$$

and so

6.3.8
$$|z_4 - z_3| = 2\varrho |\sin \tfrac{1}{2} a (\lambda_4 - \lambda_3)| \leq \varrho a |\lambda_4 - \lambda_3|.$$

Now ϱa is the length of the arc B. All points of the arc are finite, and since the points of a circular arc in R attain their maximum modulus for some point on the arc, the same applies to all circular arcs in $*R$. Accordingly, that maximum, for B, must be a finite number, ϱ_0. Then B is included in a circle B_1, with radius ϱ_1 about the origin, where ϱ_1 is a positive standard number. Elementary considerations show that the length of B cannot exceed the length of B_1, $\varrho a \leq 2\pi \varrho_1$. Hence, 6.3.8 implies

6.3.9 $|z_4 - z_3| \leq 2\pi \varrho_1 |\lambda_4 - \lambda_3|$.

This shows that the mapping ψ defined by 6.3.7 is S-continuous, for in order to make $|z_4 - z_3|$ smaller than any given positive standard δ we only have to take $|\lambda_4 - \lambda_3|$ smaller than the positive standard number $\delta/2\pi \varrho_1$. It follows that the inverse mapping, ψ^{-1}, from B onto $0 \leq \lambda \leq 1$ maps S-open sets into S-open sets. Thus, if it were possible to decompose B into two non-empty S-open sets, the same would be true of the segment $0 \leq \lambda \leq 1$. But this segment is S-connected and so our claim follows.

6.3.10   THEOREM.   Let B be an S-connected set which is contained in an S-open set, D. Let $f(z)$ be an internal function which is defined and analytic in D. Suppose that there exists points z_1, z_2 which belong to B such that $f(z_1)$ is finite and $f(z_2)$ is infinite. Then there exists a point $z_0 \in B$ such that all finite values which are not taken by $f(z)$ for some $z \in \mu(z_0)$ all belong to the same monad.

PROOF.   We show first that there exists a point $z_0 \in B$ such that $f(z)$ takes both finite and infinite values in $\mu(z_0)$. Let D' be the set of points $z \in D$ such that $f(z)$ takes only finite values in $\mu(z)$. Then D' is S-open by 4.5.5. We may suppose that D' contains the point z_1 for if $f(z)$ is not finite everywhere in the monad of z_1 then we may take $z_0 = z_1$. Let $B' = D' \cap B$ so that B' is not empty and is S-open in B. Next, let D'' be the set of points $z \in D$ such that $f(z)$ takes only infinite values in $\mu(z)$. We may suppose that D'' contains the point z_2, for if this is not the case then $f(z)$ is not infinite everywhere in $\mu(z_2)$ although $f(z_2)$ is infinite, so that we may set $z_0 = z_2$.

D'' also is S-open. For let $z_3 \in D''$ then $f(z)$ is infinite everywhere in $\mu(z_3)$, and hence, does not vanish anywhere in that set. There exists a first natural number n such that $f(z) \neq 0$ for all z for which $|z - z_3| < 1/n$. This n cannot be infinite (for in that case it would not be the first number with the property in question) so n is finite, $f(z) \neq 0$ for $|z - z_3| < \varepsilon_0$ where $\varepsilon_0 = n^{-1}$ is positive standard. The function $g(z) = 1/f(z)$ is analytic for $|z - z_3| < \varepsilon_0$ and is infinitesimal in $\mu(z_3)$. Moreover, $g(z)$ is finite for

$|z - z_3| < \varepsilon_1$ where $\varepsilon_1 \leq \varepsilon_0$ is some positive standard number (compare 4.5.4, 4.5.5). Hence, the function $h(z) = g(z_3 + z)$ is analytic and finite for $|z| < \varepsilon_1$ and so $h(z)$ is S-continuous for $|^0z| < \varepsilon_1$, by 6.2.1. At the same time, $^0h(0) = 0$ since $h(0) = g(z_0)$ is infinitesimal, and $^0h(z)$ is analytic (in R, and hence in $*R$) for $|z| < \varepsilon_1$. But $h(z) \neq 0$ for $|z| < \varepsilon_1$ and by 6.2.8 this is possible only if $^0h(z) = 0$ for all $|z| < \varepsilon_1$. This shows that $h(z)$ is infinitesimal for $|^0z| < \varepsilon_1$ and in particular for $|z| < \frac{1}{2}\varepsilon_1$. It follows that $f(z)$ is infinite for $|z - z_3| < \frac{1}{2}\varepsilon_1$ and so D'' is S-open and the set $B'' = D'' \cap B$ is not empty and is S-open in B. But B is S-connected, by assumption and D' and D'' are disjoint. This shows that $D' \cup D''$ cannot exhaust B. We conclude that $B - (D' \cup D'')$ is not empty, there exists a point $z_0 \in B$ such that $f(z)$ takes both finite and infinite values in $\mu(z_0)$.

Now suppose that there are finite values a and b which do not belong to the same monad such that $f(z) \neq a$ and $f(z) \neq b$ for all $z \in \mu(z_0)$. At the same time, $f(z)$ takes some finite value, c say, in the monad of z_0. Then the conditions of 6.3.1 are satisfied and so $f(z)$ is S-continuous at z_0. But this is impossible since $f(z)$ takes both finite and infinite values in $\mu(z_0)$. We conclude that the finite complex numbers which are not taken as values by $f(z)$ in $\mu(z_0)$ – if there are such numbers – all belong to the same monad. This proves 6.3.10.

Let z_0 be a complex number, and define B by

6.3.11

$B = \{z \mid$ there exists a positive infinitesimal δ such that $|z - z_0| < \delta|z_0|\}$.

Thus, B is non-empty only if $z_0 \neq 0$ and it then consists of all z such that $|(z/z_0) - 1|$ is infinitesimal, i.e., such that z/z_0 belongs to the monad of 1. Then $z_0 \neq 0$ will be called a *P-center for a given internal function* $f(z)$ if $f(z)$ is defined and analytic in B and if the finite points which are not taken as values by $f(z)$ in B (if any) all belong to the same monad. Equivalently, a point z_0 is a P-center for $f(z)$ if $z_0 \neq 0$ and if the function $g(z) = f(z/z_0)$ is defined for the monad of 1, $\mu(1)$, and if $g(z)$ takes in $\mu(1)$ all finite values with the possible exception of a set of values contained in a single monad. The set B introduced above will be called the *P-circle with center* z_0.

If z_0 and z_1 are two complex numbers different from 0 (in $*Z$) then z_1 belongs to the P-circle with center z_0 if and only if $(z_1/z_0) \simeq 1$. It is not difficult to see that this condition defines an equivalence relation between z_0 and z_1. The equivalence classes with respect to this relation are precisely the P-circles. Thus, if $z_1 \neq 0$ belong to a P-circle with center z_0 then it is itself a center for that circle. It follows that if z_0 is a P-center for $f(z)$ then any point z_1 such that $z_1 \neq 0$ which belongs to the P-circle with center z_0 also is a P-center for $f(z)$.

The following definition will be used in connection with some *standard* theorems, i.e., theorems within classical Function Theory. Thus, although we do not say so explicitly in the definition it is supposed to refer to numbers and functions within R. Of course it can (and will) be re-interpreted also for $*R$.

6.3.12 DEFINITION (Standard).   The set of complex numbers J will be called a *P-set for the function* $w = f(z)$ if –

'There exists a complex number w_0 such that for every positive r, δ, and ε, there exist a complex number $z_0 \in J$ and a positive number δ_0, $|z_0| > r$ and $\delta_0 < \delta$ such that the circular disk $C(z_0, \delta_0)$ which is given by $|z - z_0| < \delta_0 |z_0|$ belongs to the domain of definition of $f(z)$ and such that $f(z)$ takes in $C(z_0, \delta_0)$ all values w which satisfy the inequalities $|w| < 1/\varepsilon$ and $|w - w_0| > \varepsilon$.'

Observe that w_0 is supposed to be the same for all circles $C(z_0, \delta_0)$. The definition also includes the possibility that w_0 is actually redundant, i.e., that all values w such that $|w| < 1/\varepsilon$ are taken by $f(z)$ in the circular disks $C(z_0, \delta_0)$. In order to satisfy the definition formally in this case, we may introduce an arbitrary complex number as w_0, e.g., $w_0 = 0$. The circles $C(z_0, \delta_0)$ were introduced by Milloux and are known as 'cercles de remplissage'.

6.3.13 THEOREM.   Let $f(z)$ be a standard analytic function in the set D which is given by $|z| > \varrho$, ϱ standard, $\varrho \geq 0$, and let J be a standard set of complex numbers. Then J is a P-set for $f(z)$ if and only if there exists a point z_0 in $*J$, $|z_0|$ infinite, such that z_0 is a P-center for $f(z)$.

PROOF.   The condition is necessary. Suppose that J is a P-set for $f(z)$. That is to say, the statement given in quotation marks in 6.3.12 holds in R. We fix the standard number w_0 and we re-interpret the remainder of the statement ('for every positive r, δ, and ε, etc.') in $*R$, so that the symbol J refers to $*J$. Choose r infinite and δ and ε infinitesimal. Then the point z_0 which exists according to the assertion of the statement is infinite while δ_0 is infinitesimal. Let B be the P-circle with center z_0. Then $B \subset {}^*D$. For if z' is any point in B then $|z'| > |z_0|(1-\delta') > r(1-\delta')$ for some infinitesimal δ' and so $|z'| > \varrho$. Also, the function $f(z)$ takes any finite value in the circular disk $|z - z_0| < \delta_0 |z_0|$, with the possible exception of values which belong to the monad of w_0. This shows that z_0 is a P-center for $f(z)$.

The condition is also sufficient. Let $z_0 \in {}^*J$ be a P-center for $f(z)$ such that z_0 is infinite and let B be the P-circle with center z_0. $f(z)$ takes all finite complex numbers as values, with the possible exception of a set of numbers which all belong to the same monad. If there are such numbers, let w_0 be their joint standard part. If no such numbers exist, put $w_0 = 0$. Choose any three positive standard numbers, r, δ and ε. Choose δ_0 as the smaller one of the two numbers $\frac{1}{2}$, $\frac{1}{2}\delta$. Then the circular disk $|z - z_0| < \delta_0 |z_0|$ belongs to ${}^*D$ and includes B as a subset. Thus, the following statement holds in ${}^*R$.

6.3.14 'There exist a complex number $z_0 \in {}^*J$ and a positive number δ_0, $|z_0| > r$, $\delta_0 < \delta$, such that the circular disk $C(z_0, \delta_0)$ which is given by $|z - z_0| < \delta_0 |z_0|$ belongs to ${}^*D$ and $f(z)$ takes in $C(z_0, \delta_0)$ all values w which satisfy the inequalities $|w| < 1/\varepsilon$, $|w - w_0| > \varepsilon$.'

Expressing 6.3.14 within the formal language Λ and re-interpreting in R, we see that 6.3.14 holds in R provided we replace ${}^*D$ by D and ${}^*J$ by J. This shows that J is a P-set for $f(z)$ and completes the proof of 6.3.13.

The argument which proves the sufficiency of the condition in 6.3.13 leads also to the more general

6.3.15 THEOREM. Let $f(z)$ and J be a standard function and a standard set of complex numbers, respectively. If the infinite point $z_0 \in {}^*J$ is a P-center for $f(z)$ then J is a P-set for $f(z)$.

We are now in a position to prove the principal new result of this section.

6.3.16 THEOREM (Standard). Let $f(z)$ be a function which is defined and analytic for $|z| > \varrho$, ϱ a specified positive constant, and let $\{B_j\}$, $j = 1,2,3,\ldots$ be a sequence of circular arcs or finite straight segments. Suppose that the angular measure of the circular arcs is less than 2π and that the end points of the arcs or segments are included in all cases. (Both arcs and segments may occur in a given sequence.)

Suppose further that the following conditions are satisfied:

(i) If $z_1^{(j)}$ and $z_2^{(j)}$ are the end points of B_j then $\lim_{j\to\infty}|z_1^{(j)}|=\infty$;

(ii) There exists a positive constant η such that for all $z\in B_j$,

$$\eta^{-1}|z_1^{(j)}|\le|z|\le\eta\,|z_1^{(j)}|,\qquad j=1,2,\dots;$$

(iii) $\lim_{j\to\infty}|f\,(z_1^{(j)})|=\infty$;

(iv) There exists a positive constant m such that

$$|f(z_2^{(j)})|\le m,\qquad j=1,2,3,\dots.$$

Then there exists a sequence of points $b_j\in B_j$, $j=1,2,3,\dots$ such that $J=\{b_j\}$ constitutes a P-set for $f(z)$.

PROOF. Let ω be an infinite natural number and consider the set B_ω, with end points $z_1^{(\omega)},z_2^{(\omega)}$. B_ω is either a circular arc or a straight segment. $z_1^{(\omega)}$ is infinite, by condition (i) while $f(z_1^{(\omega)})$ is infinite, and $f(z_2^{(\omega)})$ is finite, by conditions (iii) and (iv). Let $r=|z_1^{(\omega)}|$ and introduce a new variable z' by $z'=z/r$. Consider the function $g(z')=f(z)=f(rz')$, then $g(z')$ is defined and analytic for $|z'|>\varrho/r$. Let B_ω' be the arc or segment which is the image of B_ω in the z'-plane. Then the points z' of B_ω' satisfy the inequality $1/\eta\le|z'|\le\eta$. Since $1/\eta$ is positive standard and ϱ/r is infinitesimal the points of B_ω' belong to the S-interior of the set $\{z'\,|\,z'>\varrho/r\}$. And since the points of B_ω' are all finite, B_ω' is S-connected (see 6.3.6). It now follows from 6.3.10 that there exists a point $z_0'\in B_\omega'$ such that all finite values which are omitted (not taken) by $g(z')$ in the monad $\mu(z_0')$ all belong to a single monad, $\mu(w_0)$ say, where w_0 is standard. (If there are no such points put $w_0=0$). Let $z_0=rz_0'$ then the transformation $z=rz'$ maps $\mu(z_0')$ on the P-circle with center z_0, to be denoted by $\pi(z_0)$. Indeed, let $z\in\pi(z_0)$, then $|z-z_0|\le\delta|z_0|$ for some infinitesimal δ. But $|z_0|\le\eta r$ and so $|rz'-rz_0'|\le\delta\eta r$, and hence $|z'-z_0'|\le\delta\eta$, $|z'-z_0'|\simeq0$, $z'\in\mu(z_0')$. Conversely, if $z'\in\mu(z_0')$, then $|z'-z_0'|\le\delta$, for some infinitesimal δ, and so $|rz'-rz_0'|\le\delta r$, $|z-z_0|\le\delta\eta|z_0|$ since $r\le\eta|z_0|$ and hence, $z\in\pi(z_0)$. Since $f(z)$ takes the same values in $\pi(z_0)$ as $g(z)$ in $\mu(z_0')$ we conclude that z_0 is a P-center for $f(z)$.

For any natural number j, we may describe B_j by the parametric equation

$$z = (1 - \lambda) z_1^{(j)} + \lambda z_2^{(j)}, \qquad 0 \leq \lambda \leq 1,$$

if B_j is a straight segment, and by

$$z = z_0^{(j)} + \varrho^{(j)} e^{i(\theta_1^{(j)} + a^{(j)}\lambda)}, \qquad 0 \leq \lambda \leq 1,$$

if B_j is a circular arc (see 6.3.7). Let λ_0 be the value of the parameter λ which corresponds to z_0 and to z_0'. We may suppose that λ_0 is a standard number. Indeed, if this is not the case from the outset, then the standard part of $\lambda_0, {}^0\lambda_0$, yields a point in the z'-plane which belongs to the monad of z_0' and hence, may be taken in place of z_0'. Thus, we may suppose that z_0 and z_0' are given by a standard λ_0. We now define $J = \{b_j\}$ as the standard sequence which is obtained by choosing b_j as the point of B_j which corresponds to the parametric value $\lambda = \lambda_0$, for all finite j. Then b_j is given by λ_0 also for all infinite j and, in particular, $b_\omega = z_0$. This shows that ${}^*J$ contains a P-center for $f(z)$, i.e., z_0, and proves the theorem by 6.3.13.

If the standard function $f(z)$ possesses an isolated essential singularity at infinity then we can always find a sequence of circular arcs $\{B_j\}$ which satisfies the conditions of 6.3.16. For this purpose, define a complex valued function of the real variable $r, g(r)$, for sufficiently large $r, r > \varrho$, in such a way that $|g(r)| = r$ and $|f(g(r))| = \max_{|z|=r} |f(z)|$. Then $\lim_{r \to \infty} |g(r)| = \infty$. Since $f(x)$ has an essential singularity at infinity, there exists a standard sequence $\{z_2^{(j)}\}$, $\varrho < |z_2^{(1)}| < |z_2^{(2)}| < |z_2^{(3)}| < \cdots$, $\lim_{j \to \infty} |z_2^{(j)}| = \infty$, such that $|f(z_2^{(j)})|$ is bounded. Put $r_j = |z_2^{(j)}|$ and $z_1^{(j)} = g(r_j)$. By considering only sufficiently large r, we may suppose that $z_1^{(j)} \neq z_2^{(j)}$, $j = 1, 2, \ldots$. We now choose B_j as one of the two arcs with endpoints $z_1^{(j)}, z_2^{(j)}$ on the circle $|z| = r_j$. Then the conditions of 6.3.16 are satisfied for any $\eta > 1$.

A more arbitrary choice of circular areas is possible if there exists a $\varrho > 0$ such that $f(z)$ is analytic and omits a value for $|z| > \varrho$. We may suppose without loss of generality that this value is 0, and we may then define a complex valued function of a real variable, $h(r)$, for $r > \varrho$ in such a way that $|h(r)| = r$ and $|f(h(r))| = \min_{|z|=r} |f(z)|$. For any standard sequence $\{r_j\}$ such that $\varrho < r_1 < r_2 < \cdots$ and $\lim_{j \to \infty} r_j = \infty$, put $z_1^{(j)} = g(r_j)$, $z_2^{(j)} = h(r_j)$, $j = 1, 2, 3, \ldots$ and choose B_j as one of the two arcs on $|z| = r_j$

which connects $z_1^{(j)}$ with $z_2^{(j)}$. Then the conditions of 6.3.16 are again satisfied for any $\eta > 1$.

In any case, the condition of 6.3.16 has now established the following result.

6.3.17 THEOREM. Let $f(z)$ be a standard function which is analytic for $|z| > \varrho$, ϱ standard, and which possesses an isolated essential singularity of infinity. Then there exists an infinite complex number z_0 which is a P-center for $f(z)$.

Under the conditions of 6.3.17, let $\theta_0 = \arg z_0$, $0 \le \theta_0 < 2\pi$, and let θ_1 be the standard part of θ_0, $\theta_1 = {}^0\theta_0$. Then $z_1 = |z_0| e^{i\theta_1}$ belongs to the P-circle with center z_0 since

$$\left| \frac{z_1}{z_0} - 1 \right| = |e^{i(\theta_1 - \theta_0)} - 1| = 2 \left| \sin \frac{\theta_1 - \theta_0}{2} \right|$$

is infinitesimal. It follows that z_1 also is a P-center for $f(z)$. Since the ray given by $\arg z = \theta_1$ includes z_1 we have, taking into account 6.3.13,

6.3.18 THEOREM (Standard. Julia, strengthened by Milloux). Let $f(z)$ be analytic for $|z| > \varrho$, with an essential singularity at infinity. Then the line $\arg z = \theta_1$ is a P-set for at least one value of θ_1.

Let $\theta = \gamma(r)$ be a standard real function which is defined for $r > \varrho$, ϱ positive standard, and which may or may not be continuous. $\gamma(r)$ determines a set of points Γ in the complex plane, which is defined by $\Gamma = \{z | z = r\, e^{i\gamma(r)}\}$. Consider the family of sets $\{\Gamma_t\}$, $0 \le t < 2\pi$, which is given by $\Gamma_t = \{z | z = r\, e^{i(t + \gamma(r))}\}$. Then $\Gamma_0 = \Gamma$, and the sets Γ_t are disjoint for different t, $0 \le t < 2\pi$, and together they exhaust the set $\{z | |z| > \varrho\}$ in the Z-plane if t and r take standard values, and in the *Z-plane if t and r are supposed to take also non-standard values. Strengthening results by Julia and Valiron, we prove as our last theorem in this section:

6.3.19 THEOREM (Standard). Let $f(z)$ be analytic for $|z| > \varrho$ with an essential singularity at infinity and, for given $\gamma(r)$, let the sets Γ_t be obtained as detailed above. Then there exists a t_1, $0 \le t_1 < 2\pi$, such that Γ_{t_1} is a P-set for $f(z)$.

PROOF. As an exception to our general practice, we shall find it convenient to denote corresponding sets of the family $*\{\Gamma_t\}$ in the Z and $*Z$ planes (for standard t) by the same symbol Γ_t.

$f(z)$ possesses a P-center z_0, by 6.3.17. Since the sets Γ_t, as interpreted in the $*Z$-plane, cover the set $\{z | |z| > \varrho\}$, there exists a t_0, standard or non-standard, $0 \le t_0 < 2\pi$ such that Γ_{t_0} contains z_0. Still working in $*R$, consider the set Γ_{t_1} where $t_1 = {}^0t_0$, unless ${}^0t_0 = 2\pi$, in which case we put $t_1 = 0$. Let z_1 be the point of Γ_{t_1} for which $|z_1| = |z_0|$. Then $\arg z_1 - \arg z_0$ is infinitesimal (provided the arguments are taken with the appropriate determination) and so z_1 belongs to the P-circle with center z_0 and is itself a P-center for $f(z)$. It now follows from 6.3.13 that I_{t_1}, taken within R, is a P-set for $f(z)$. This proves the theorem.

6.4 Compactness arguments in classical Function Theory. When comparing the theory of the preceding section with the method of normal families one observes, informally speaking, that the notions of a normal family, of equicontinuity, and of the oscillation of a function at a point in the latter order of ideas are related to our notions of an internal analytic function which is S-continuous where it is finite, of S-continuity, and of the behavior of an internal analytic function in a monad. A precise correlation of this kind is not possible. Thus, a given standard normal family of functions does not give rise to a single internal function. It may, however, give rise to a set of internal analytic functions in a way which is illustrated by the following theorem.

6.4.1 THEOREM. Let F be a family of meromorphic functions which are defined in a domain (open connected set) D on the Riemann sphere, all notions referring to R. Then F is normal in D if and only if all the elements of $*F$ are S-continuous in the S-interior of $*D$ on the Riemann sphere in $*R$.

In this connection, $*F$ is the extension of the family F from R to $*R$, as usual. The functions of this family are defined in the set $*D$ which is the extension of D from R to $*R$. The notions of S-continuity and S-interior are to be interpreted in terms of the chordal metric on the Riemann sphere.

PROOF OF 6.4.1. It is known that a necessary and sufficient condition for a family of functions to be normal in a domain D on the Riemann sphere is that it be equicontinuous at all points of D. Now let z_0 be any standard point which belongs to $*D$ and hence, to D. If F is equicontinuous at z_0 then for any positive standard δ there exists a positive standard ε such that $\chi(z,z_0)<\varepsilon$ implies $\chi(f(z),f(z_0))<\delta$ for all $f(z)\epsilon F$ and for all $z\epsilon D$, where χ stands for chordal distance. But this must be true also when re-interpreted in $*R$ and so we conclude that *any* function which belongs to the family $*F$ is S-continuous at z_0. Next, let z_1 be any point of $*D$ which belongs to the monad of z_0. Then for any $z\epsilon*D$ such that $\chi(z,z_1)<\frac{1}{2}\varepsilon$, we have at the same time, $\chi(z,z_0)<\varepsilon$ – since $\chi(z_0,z_1)$ is infinitesimal – and so

$$\chi(f(z),f(z_1))\leq\chi(f(z),f(z_0))+\chi(f(z_0),f(z_1))<2\delta .$$

Thus, $f(z)$ is S-continuous also at z_1. But the Riemann sphere is compact and so every point which belongs to the S-interior of $*D$ is in the monad of some standard point of $*D$. This shows that $f(z)$ is S-continuous at all points of the S-interior of $*D$, the condition of the theorem is necessary.

The condition is also sufficient. Suppose that it is satisfied but that at some point $z_0\epsilon D$, F is not equicontinuous. Thus, there exists a positive standard δ such that

'for every positive ε there exist a $z\epsilon D$ and an $f(z)\epsilon F$ such that $\chi(f(z), f(z_0))\geq\delta$ while $\chi(z, z_0)<\varepsilon$.'

Formulating the statement in quotation marks as a sentence of K, re-interpreting it in $*R$, and taking for ε a positive infinitesimal number, we find that there exist a point $z\epsilon*D$ and a function $f(z)\epsilon*F$ such that $\chi(f(z),f(z_0))\geq\delta$ although $\chi(z, z_0)<\varepsilon$, and hence, although z belongs to the monad of z_0. This shows that $f(z)$ is not S-continuous at z_0, although z_0 belongs to the S-interior of $*D$. We conclude that the condition of the theorem is sufficient.

However, as we have seen, the theory of generalized complex functions provided by Non-standard Analysis actually replaces the theory of normal families in some cases and is more effective in others. At the same time, we may use our methods also in order to define compactness arguments regarding families of functions and related results. We prove as an example

6.4.2 THEOREM (Standard. Vitali). Let $\{f_n(z)\}$ be a sequence of functions which are analytic and uniformly bounded in the interior of (i.e., on every compact subset of) a domain D. Suppose that $\{f_n(z)\}$ converges for an infinite set of points z_k in D, $k = 1, 2, \ldots$ with a limit point z_0 in D. Then $\{f_n(z)\}$ converges everywhere in D and the convergence is uniform in the interior of D.

PROOF. Let ω be an infinite natural number and consider the function $f_\omega(z)$, which belongs to $*\{f_n(z)\}$. Let z' be any standard point which belongs to $*D$ and hence to D, and let $\delta > 0$ be a standard number such that the points for which $|z - z'| \le \delta$ all belong to $*D$. Such a number δ exists since D is open. By assumption, there exists a positive standard m such that it is true in R that '$|f_n(z)| \le m$ for $|z - z'| \le \delta$, $m = 0, 1, 2, 3, \ldots$'. Accordingly, the statement in quotation marks is true also in $*R$ and, in particular, $|f_\omega(z)| \le m$ for $|z - z'| \le \delta$. This shows that $f_\omega(z)$ is finite in the monads of the standard points of $*D$, i.e., in the monads of the points of D. Denoting the union of these monads by B, we have $D = {}^0B$. B is S-open and so $f_\omega(z)$ is S-continuous in B, by 6.2.1. Moreover, the function $F(z) = {}^0f_\omega(z)$ is analytic in $D = {}^0B$, by 6.2.3.

The function $F(z)$ is independent of the particular choice of the infinite natural number ω. For since $\lim_{n \to \infty} f_n(z_k)$ exists for $k = 1, 2, \ldots$, we have, in R, that $|f_n(z_k) - f_m(z_k)| < \delta$, for arbitrary positive δ and $n > \nu(\delta)$, $m > \nu(\delta)$. It follows that, in $*R$, $|f_\omega(z_k) - f_\eta(z_k)|$ is infinitesimal for all finite natural numbers k and for all infinite natural ω and η. Hence ${}^0f_\omega(z_k) = {}^0f_\eta(z_k)$ for all finite natural k, and hence ${}^0f_\omega(z) = {}^0f_\eta(z)$ by the identity theorem for standard analytic functions.

For any standard $z \in D$ and for any positive standard δ it is true in $*R$ that 'there exists a natural number ν such that $|f_n(z) - F(z)| < \delta$ for all $n > \nu$'. Indeed, this is true in $*R$ for any infinite natural ν since $|f_n(z) - F(z)|$ is then infinitesimal. But the statement in quotation marks can be formalized within the vocabulary of K and so it is true also in R. This proves that $\lim_{n \to \infty} f_n(z) = F(z)$ for all $z \in D$, so that the sequence $\{f_n(z)\}$ converges in D.

We still have to show that the convergence is uniform on any compact subset of D. Let $E \subset D$, E compact, i.e., closed and bounded, and suppose that the convergence of $\{f_n(z)\}$ in E is not uniform. Then there exists a positive standard δ such that it is true in R that for any natural number m there exist a natural $l > m$ and a $z' \in E$ such that $|f_l(z') - F(z')| \ge \delta$. Passing to $*R$ and choosing m infinite we then find that there exists an infinite natural number l and a point $z'\in{}^*E$ such that $|f_l(z')-F(z')|\geq\delta$. Now z' is finite since E is bounded and so it possesses a standard part $z''={}^0z'$. z'' belongs to $^*E$ since E is closed and so $z''\in E$ and $z''\in D$. It follows that $F(z'')$ is defined and that $f_l(z'')-F(z'')$ is infinitesimal. At the same time, $F(z'')-F(z')$ is infinitesimal since $F(z)$ is analytic, hence continuous on D; and $f_l(z')-f(z'')$ is infinitesimal, by 6.2.1. Accordingly,

$$f_l(z')-F(z')=(f_l(z')-f_l(z''))+(f_l(z'')-F(z''))+(F(z'')-F(z'))$$

is infinitesimal. This contradicts the relation $|f_l(z')-F(z')|\geq\delta$ and completes the proof of Vitali's theorem.

6.5   Remarks and references.   The theory and results described in this chapter were reported previously in ROBINSON [1962], in slightly different form. For the classical background consult CARTWRIGHT [1957], JULIA [1923], KAKEYA [1917], MARDEN [1949], MILLOUX [1924, 1928], MONTEL [1923, 1927, 1933], OSTROWSKI [1926], VALIRON [1923]. Related results will be found in ROBINSON [1965a].

CHAPTER VII

LINEAR SPACES

7.1 Normed spaces. Let B be a finite or infinite dimensional normed linear space whose field of scalars are either the real or the complex numbers. In either case, the definition and discussion of B involve B together with the real numbers, R, so we assume as in previous cases that B and R are embedded, simultaneously, in some full structure M and we develop the non-standard theory of B in an enlargement $*M$ of M.

Let $\|a\|$ be the norm of any point a of B. Since B is a metric space under the definition $\varrho(a,b) = \|a-b\|$ the theory of sections 4.3 through 4.6 is applicable. The finite points of B in the sense of Chapter IV are precisely the points of B whose norm is finite.

Let T be an operator on B which is linear in the algebraic sense, i.e., $T(a+b) = Ta + Tb$ and $T\lambda a = \lambda Ta$ for all a and b in B and for all numbers λ in the appropriate field of scalars. Then the extension of T to $*B$ (which will still be denoted by T, as usual) is linear in the algebraic sense. We consider operators with domain and range in the same space, B.

7.1.1 THEOREM. The standard operator T is bounded if and only if T transforms every finite point a in $*B$ into a finite point.

PROOF. The condition of the theorem is necessary. For if T is bounded, and $\|T\|$ is the norm of T, so that $\|T\|$ is a standard real number, then for any point a in B or in $*B$,

$$\|Ta\| \le \|T\| \, \|a\| \, .$$

This shows that if a is finite then Ta also is finite.

The condition is also sufficient. Suppose T is not bounded, then there exists a standard sequence $\{a_n\}$ of elements of B such that $\|a_n\| = 1$, $n = 0,1,2,\ldots$, but $\|Ta_n\|$ is unbounded in M. Passing to $*M$ we see that this implies that $\|Ta_\omega\|$ is infinite, for some infinite ω, although $\|a_\omega\| = 1$ so that a_ω is finite.

7.1.2 THEOREM. The standard operator T is bounded if and only if T transforms every point in the monad of the origin, 0, into a point in the monad of 0.

PROOF. A point is in the monad of 0 if and only if its norm is infinitesimal. The necessity of the condition is thus obvious. To prove its sufficiency, choose $\{a_n\}$ as in the proof of 7.1.1, so that $\|a_n\| = 1$ while $\{Ta_n\}$ is unbounded and $\|Ta_\omega\|$ is infinite for a particular infinite ω. Consider the point $b = a_\omega/\|Ta_\omega\|$. Then $\|b\| = 1/\|Ta_\omega\|$ so that $\|b\|$ is infinitesimal, b is in the monad of 0. At the same time $\|Tb\| = \|Ta_\omega\|/\|Ta_\omega\| = 1$ so that Tb is not in the monad of 0. This completes the proof of 7.1.2.

7.1.3 THEOREM. The standard operator T is bounded if and only if T transforms every near-standard point into a near-standard point.

PROOF. The condition of the theorem is necessary. Suppose a is near-standard, then $a = {}^0a + b$ where $\|b\|$ is infinitesimal. Then $Ta = T({}^0a + b) = T({}^0a) + Tb$. Suppose T is bounded, then $T({}^0a)$ is a standard point while $\|Tb\|$ is infinitesimal, by 7.1.2. This shows that Ta is near-standard, as required.

The condition is also sufficient. Suppose that T is not bounded and choose $\{a_n\}$ and ω as in the proofs of 7.1.1 and 7.1.2 so that $\|Ta_\omega\|$ is infinite. Put $b = a_\omega/\sqrt{\|Ta_\omega\|}$, where the positive square root is to be taken in the denominator. Then b is near-standard since it belongs to the monad of 0, $\|b\| = 1/\sqrt{\|Ta_\omega\|}$ is infinitesimal. On the other hand, $\|Tb\| = \sqrt{\|Ta_\omega\|}$ is infinite and so $\|Tb\|$ cannot be near-standard. This completes the proof of the theorem.

The notion of a compact mapping from a metric space into a metric space has been dealt with in section 4.6. Specializing 4.6.4 to the case of an algebraically linear operator from B into B, we obtain

7.1.4 THEOREM. The standard operator T is compact if and only if it maps every finite point into a near-standard point.

In the theory of normed linear spaces a compact operator is also said to be *completely continuous*. Comparing 7.1.4 with 7.1.3 we see that a compact operator is by necessity bounded (and hence, continuous).

Now suppose that B is complete, i.e., a Banach space. Making use of 4.6.5, we obtain the well known

7.1.5 THEOREM (Standard). Let $\{T_n\}$ be a sequence of compact linear operators from a Banach space B into B. Suppose that the sequence converges in norm to an operator T. Then T is compact.

In order to reduce this theorem to 4.6.5, we take as the space T of that theorem the unit ball U in B, which is given by $\|a\| \leq 1$, while S is identified with B. Then it is not difficult to verify that the convergence in norm of $\{T_n\}$ to T, $\lim_{n\to\infty} \|T_n - T\| = 0$ implies that $\lim_{n\to\infty} T_n a = Ta$ uniformly on U. We conclude that every point of U is mapped by T on a near-standard point, and hence, that *every* finite point of B is mapped by T on a near-standard point.

As an example of the application of our order of ideas to a concrete function space, we shall establish the compactness of certain integral operators.

Let B be the space of real continuous functions of one variable defined in the closed interval $0 \leq x \leq 1$, with the norm $\|\phi(x)\| = \max_{0 \leq x \leq 1} |\phi(x)|$. It is not difficult to verify that B is a normed linear space over the field of real numbers. (B is actually a Banach space, but we shall not require the completeness of B.)

Define the operator T on B by

7.1.6
$$\psi(x) = T\phi = \int_0^1 K(x,t)\,\phi(t)\,dt,$$

where $K(x,t)$ is continuous in the closed square $0 \leq x \leq 1$, $0 \leq t \leq 1$. It is not difficult to see that $\psi(x)$ is continuous, i.e., T maps B into B. Moreover, T is clearly linear in the algebraic sense.

7.1.7 THEOREM (Standard). The operator T is compact.

PROOF. In the present case, we may identify M with R, i.e., with the full structure based on the real numbers as set of individuals. Then the points of B are given by certain functions (i.e., formally, by certain relations) within M. On passing from $M=R$ to an enlargement $*M=*R$, B is carried into an enlargement $*B$, and the preceding theory is applicable. Thus, in order to prove that T is compact, we only have to show that every finite ϕ in $*B$ is mapped by T into a near-standard ψ. Now if ϕ is finite there exists a standard positive number m such that $|\phi(x)|\leq m$ for $0\leq x\leq 1$. At the same time, $K(x,t)$ is bounded by a standard positive k, and this bound applies both in R and in $*R$. Hence,

$$|\psi(x)|=\left|\int_0^1 K(x,t)\phi(t)\,dt\right|\leq km$$

so that $\psi(x)$ is finite for $0\leq x\leq 1$ in $*R$. We propose to show that $\psi(x)$ is S-continuous at all points of this interval.

Indeed, for any x_1,x_2 in the interval of definition of $\psi(x)$,

$$\psi(x_2)-\psi(x_1)=\int_0^1 (K(x_2,t)-K(x_1,t))\phi(t)\,dt.$$

Now suppose that $x_1\simeq x_2$ so that x_1 and x_2 belong to the same monad. Then $K(x_2,t)-K(x_1,t)\simeq 0$ since K is continuous. Putting

$$|K(x_1,t)-K(x_2,t)|=\eta(t),$$

we then have

$$|\psi(x_2)-\psi(x_1)|\leq m\int_0^1 \eta(t)\,dt$$

where $\eta(t)\simeq 0$ for $0\leq t\leq 1$. Clearly $\int_0^1\eta(t)\,dt\geq 0$. Moreover, this integral must be infinitesimal for if ε is any positive standard number then

$$0\leq\int_0^1\eta(t)\,dt\leq\int_0^1\varepsilon\,dt=\varepsilon.$$

We conclude that $\psi(x_2)-\psi(x_1)$ is infinitesimal for infinitesimal x_2-x_1 and hence, that $\psi(x)$ is S-continuous in the closed interval $0\leq x\leq 1$. It follows that the function $^0\psi(x)$ is defined and continuous and that $\psi(x)\simeq {}^0\psi(x)$, for all x in this interval. Thus, $^0\psi(x)$ is a point in B and $\psi(x)$ belongs to the monad of $^0\psi(x)$ since $\|\psi(x)-{}^0\psi(x)\|=\max_{0\leq x\leq 1}$

$|\psi(x) - {}^0\psi(x)|$ is infinitesimal. We conclude that $\psi(x)$ is near-standard, T is compact.

In section 4.4, we showed how to define a space T_μ whose points are the monads of an enlargement $*T$ of a given metric space T. We may adapt this construction to the case when $T = B$ is a normed linear space. The norm of a point $\xi = \mu(p)$ in $B_\mu = T_\mu$ will now be defined provided one, and so, all of the elements of ξ (which are points of B) are finite, and we then put $\|\xi\| = {}^0\|p\|$. Defining addition and multiplication by a scalar in B_μ in a similar way, we find that the points of B_μ which, as monads, belong to the principal galaxy of B constitute a normed linear space B_G. If we denote by B_S the subspace of B_G whose points, taken as monads, contain points of B, then B_S is isometric and algebraically isomorphic to B and (compare 4.4.6) the closure of B_S in B_G is isometric and algebraically isomorphic to the completion of B.

7.2 Hilbert space. Let H be an abstract Hilbert space with the complex numbers as scalar field and suppose, specifically, that H is infinite-dimensional, separable, and complete. While we shall thus consider the case of complex Hilbert space, our analysis is applicable also to the real case, with minor verbal or formal alterations.

We take for granted some of the more elementary facts about Hilbert space. Denoting the inner product in H by (x, y), the definition $\|x\| = \sqrt{(x, x)}$ turns H into a Banach space, so that the theory of section 7.1 becomes applicable. Let $\{e_n\}$, $n = 1, 2, 3, \ldots$ be an arbitrary but definite orthonormal basis of H. Then any point a in H has a unique representation $a = \sum_{n=1}^{\infty} a_n e_n$ where the coordinates a_n are complex numbers. We shall indicate this relation also by $a = (a_1, a_2, a_3, \ldots)$. In terms of its coordinates, the norm of a is given by $\|a\| = \sqrt{(a_1 \bar{a}_1 + a_2 \bar{a}_2 + a_3 \bar{a}_3 + \cdots)}$, and the inner product of any two points $a = (a_1, a_2, a_3, \ldots)$ and $b = (b_1, b_2, b_3, \ldots)$ equals $(a, b) = a_1 \bar{b}_1 + a_2 \bar{b}_2 + a_3 \bar{b}_3 + \cdots$.

Following our usual practice we suppose that both H and the real numbers R are contained in some full structure M, and we develop the Non-standard Analysis of the subject in an enlargement $*M$ of M. On passing to $*M$, H and R are carried into $*H$ and $*R$, respectively. At the same time, the basis $\{a_n\}$ of H is extended to an internal sequence of points in $*H$, which is defined for all positive integers in $*N$, finite or infinite. As usual, we retain the notation $\{e_n\}$ also for the enlarged sequence. The scalar products (e_n, e_m) are equal to 1 or 0 according as $n = m$ or $n \neq m$, also within $*M$, as we see immediately by transfer from M. Every point a in $*H$ has a unique representation,

7.2.1
$$a = \sum_{n=1}^{\infty} a_n e_n,$$

where the a_n are complex numbers within $*M$ and n varies over all positive integers, finite or infinite. The equation 7.2.1 is to be interpreted in terms of the classical definition as applied to $*M$. That is to say, for every positive ε in $*R$ there exists a natural number $m = m(\varepsilon)$ in $*N$ such that

$$\left\| a - \sum_{n=1}^{k} a_n e_n \right\| < \varepsilon \qquad \text{for all } k > m.$$

In this connection, an expression $\sum_{n=1}^{k} a_n e_n$ though not necessarily a finite sum in the ordinary sense is nevertheless defined in $*M$ by transfer from M.

Let $a = (a_1, a_2, a_3, \ldots)$ be any point in $*H$. Then $\|a\|^2 = \sum_{n=1}^{\infty} a_n \bar{a}_n$, so that a is finite in the sense defined previously for metric spaces and more particularly for normed linear spaces if and only if $\sum_{n=1}^{\infty} a_n \bar{a}_n$ is finite.

7.2.2 THEOREM.   A point $a = (a_1, a_2, a_3, \ldots)$ in $*H$ is near-standard if and only if a is finite and the sum $\sum_{n=k+1}^{\infty} a_n \bar{a}_n$ is infinitesimal for all infinite k.

PROOF.   The condition is necessary. Suppose first that a is standard, a is a point of H. Then a is finite. Also, the sequence $\{s_k\}$ which is defined by $s_k = \sum_{n=1}^{k} a_n \bar{a}_n$ is a standard sequence which converges to $s = \sum_{n=1}^{\infty} a_n \bar{a}_n$ and so $s - s_k = \sum_{n=k+1}^{\infty} a_n \bar{a}_n$ is infinitesimal for all infinite k.

Suppose next that a is any near-standard point in $*H$ and let $b = {}^0a = (b_1, b_2, b_3, \ldots)$ be the standard part of a. Then $\|a - b\|$ is infinitesimal. But $\|a - b\|^2 = \sum_{n=1}^{\infty} (a_n - b_n)(\bar{a}_n - \bar{b}_n)$ where the terms of the infinite series on the right hand side are non-negative. Hence, $\sum_{n=k+1}^{\infty} (a_n - b_n)(\bar{a}_n - \bar{b}_n)$ is infinitesimal for all natural numbers k, finite or infinite. At the same time, as shown already, $\sum_{n=k+1}^{\infty} b_n \bar{b}_n$ is infinitesimal for all infinite k. Now, by Minkowski's inequality, which is valid by transfer from M,

7.2.3
$$\left(\sum_{n=k+1}^{\infty} a_n \bar{a}_n \right)^{\frac{1}{2}} \leq \left(\sum_{n=k+1}^{\infty} (a_n - b_n)(\bar{a}_n - \bar{b}_n) \right)^{\frac{1}{2}} + \left(\sum_{n=k+1}^{\infty} b_n \bar{b}_n \right)^{\frac{1}{2}}.$$

(This is simply a special case of the triangle inequality.) For infinite k, both sums on the right hand side of 7.2.3 are infinitesimal. Hence $\sum_{n=k+1}^{\infty} a_n \bar{a}_n$ is infinitesimal, the condition is necessary.

The condition is also sufficient. Suppose that it is satisfied, so that $\sum_{n=1}^{\infty} a_n \bar{a}_n$ is finite, while $\sum_{n=k+1}^{\infty} a_n \bar{a}_n$ is infinitesimal for all infinite k. Then $a_n \bar{a}_n = |a_n|^2$ is finite for all n, and a_n possesses a standard part, ${}^0 a_n$. We define a standard point $b = (b_1, b_2, ...)$ by setting $b_n = {}^0 a_n$ for all finite positive integers n. In order to prove that b is indeed a point in H, we have to show that the sums $\sum_{n=1}^{k} b_n \bar{b}_n$ are uniformly bounded by a standard positive number m as k varies over the finite positive integers. Now for all such k the difference $\sum_{n=1}^{k} b_n \bar{b}_n - \sum_{n=1}^{k} a_n \bar{a}_n = \sum_{n=1}^{k} {}^0 a_n {}^0 \bar{a}_n - \sum_{n=1}^{k} a_n \bar{a}_n$ is infinitesimal, while $\sum_{n=1}^{\infty} a_n \bar{a}_n$ is known to be finite. It follows that if $m-1$ is any standard positive number greater than $\sum_{n=1}^{\infty} a_n \bar{a}_n$ then $\sum_{n=1}^{k} a_n \bar{a}_n < m-1$ and hence $\sum_{n=1}^{k} b_n \bar{b}_n < m$. This shows that $b = (b_1, b_2, b_3, ...)$ is a point in H, i.e., equivalently, that b is a standard point in $*H$. We claim that a belongs to the monad of b or, which is the same, that $\|a - b\|$ is infinitesimal. Now

7.2.4
$$\|a - b\|^2 = \sum_{n=1}^{\infty} (a_n - b_n)(\bar{a}_n - \bar{b}_n) = \sum_{n=1}^{k} (a_n - b_n)(\bar{a}_n - \bar{b}_n)$$
$$+ \sum_{n=k+1}^{\infty} (a_n - b_n)(\bar{a}_n - \bar{b}_n).$$

The first sum on the right hand side of 7.2.4 is infinitesimal for all finite k since, for finite n, $(a_n - b_n)(\bar{a}_n - \bar{b}_n) = (a_n - {}^0 a_n)(\bar{a}_n - {}^0 \bar{a}_n)$ is infinitesimal. In view of 3.3.20, we may therefore conclude that $\sum_{n=1}^{k} (a_n - b_n)(\bar{a}_n - \bar{b}_n)$ is infinitesimal also for some *infinite* k. Choosing such a k, we obtain for the second sum on the right hand side of 7.2.4,

$$\left(\sum_{n=k+1}^{\infty} (a_n - b_n)(\bar{a}_n - \bar{b}_n) \right)^{\frac{1}{2}} \leq \left(\sum_{n=k+1}^{\infty} a_n \bar{a}_n \right)^{\frac{1}{2}} + \left(\sum_{n=k+1}^{\infty} b_n \bar{b}_n \right)^{\frac{1}{2}}$$

by Minkowski's inequality, and so the second sum on the right hand side of 7.2.4 also is infinitesimal. This shows that a is near-standard and completes the proof of the theorem.

Now let $\{a_n\}$ be an internal sequence of points of $*H$ which is defined for $n = 1, 2, ..., \omega$, where ω is an infinite natural number. According to the terminology of section 5.1, $\{a_n\}$ is thus a Q-finite sequence.

7.2.5 THEOREM. Suppose that there exists a positive standard number ε such that $\|a_i - a_j\| \geq \varepsilon$ for all finite i and j, $i \neq j$. Then there exists an $n \leq \omega$ such that a_n is not near-standard.

PROOF. Suppose a_n is near-standard for all n for which it is defined, and consider the standard infinite sequence $\{b_n\}$ which is determined by $b_n = {}^0a_n$ for all finite n. Then $\|a_n - b_n\|$ is infinitesimal for all finite n. It now follows from 3.3.20 that there exists an infinite natural number $m \leq \omega$ such that $\|a_n - b_n\|$ is infinitesimal for all $n \leq m$. Put $b = {}^0b_k = {}^0a_k$ for some infinite k which is not greater than m, otherwise arbitrary. Then b is a limit point of the sequence $\{b_n\}$ in M. This implies that there exist distinct finite natural numbers i and j such that $\|b - b_i\| < \frac{1}{4}\varepsilon$ and $\|b - b_j\| < \frac{1}{4}\varepsilon$. We conclude that $\|b_i - b_j\| < \frac{1}{2}\varepsilon$. But this contradicts the assumption $\|a_i - a_j\| \geq \varepsilon$ since the difference between $\|a_i - a_j\|$ and $\|b_i - b_j\|$ is infinitesimal, and completes the proof of 7.2.5.

Now let $a = \sum_{k=1}^{\infty} a_n e_n$ be an arbitrary point in $*H$. For any positive integer k, finite of infinite, we consider the projection operator P_k which is given by

$$P_k : a \rightarrow b = \sum_{n=1}^{k} a_n e_n .$$

P_k is an internal entity, and if k is finite then P_k is even standard. P_k is linear in the algebraic sense. Moreover, P_k is bounded, with norm $\|P_k\| = 1$, since for any a in $*H$,

$$\|P_k a\| = \left\| \sum_{n=1}^{k} a_n e_n \right\| = \left(\sum_{n=1}^{k} a_n a_n \right)^{\frac{1}{2}} \leq \left(\sum_{n=1}^{\infty} a_n a_n \right)^{\frac{1}{2}} = \|a\|$$

while $P_k e_1 = e_1$.

Also, if a is near-standard (in particular, if a is standard) and k is infinite then it follows from

$$\|(I - P_k) a\| = \|a - P_k a\| \leq \left(\sum_{n=k+1}^{\infty} a_n \bar{a}_n \right)^{\frac{1}{2}}$$

and from 7.2.2 that $\|(I - P_k)a\|$ is infinitesimal, where I is the identity operator. Suppose now that T is a compact standard operator, so that Ta is near-standard for any finite a, by 7.1.4. Then $\|(T - P_k T)a\| = \|(I - P_k)Ta\|$ is infinitesimal for all infinite k and all finite a. In particular, this is the case for $\|a\| = 1$, from which we may conclude that $\|T - P_k T\|$ is infinitesimal for all infinite k.

Now let ε be any standard positive number. Then there exists a first natural number m such that $\|T - P_m T\| < \varepsilon$, and m must be finite since $\|T - P_k T\| < \varepsilon$ for all infinite k. Transferring from *M to M, we conclude that $\|T - P_m T\| < \varepsilon$ for some (finite) natural number m. But the operator $T' = P_m T$ is of finite rank, that is to say, it maps H into a finite dimensional space (i.e., the space spanned by $e_1, e_2, e_3, ..., e_m$). Accordingly, we have proved the following theorem.

7.2.6 THEOREM (Standard).   Let T be a compact operator on H. For every $\varepsilon > 0$ there exists a bounded operator of finite rank T', such that $\|T - T'\| < \varepsilon$.

It is well known that this result can be used as a starting point for the development of the Fredholm theory for compact operators.

7.3 Spectral theory for compact operators.   We turn to the study of a self-adjoint compact operator T in H. Thus, we suppose that T is defined and compact on H and that

$$(Ta, b) = (a, Tb)$$

for all a and b in H and hence, for all a and b in *H.

Let ω be an arbitrary but fixed infinite natural number. We denote by H_ω the subspace of *H which is spanned by $e_1, e_2, ..., e_\omega$. Thus H_ω consists of all points which are linear combinations of $e_1, e_2, ..., e_\omega$ with complex coefficients in *M. Since H_ω is *finite-dimensional* in the sense of *M, all the standard results of the theory of linear transformations in finite-dimensional spaces apply to H_ω, by transfer from M.

Let P_ω be the projection operator from *H onto H_ω,

$$P_\omega(a_1 e_1 + a_2 e_2 + \cdots) = a_1 e_1 + a_2 e_2 + \cdots + a_\omega e_\omega.$$

Then $\|P_\omega\| = 1$, and $\|(I - P_\omega)a\|$ is infinitesimal for all near-standard a, as shown above. For any compact self-adjoint operator T on H, we define an internal operator, T_ω, in *H by

$$T_\omega = P_\omega T P_\omega.$$

Clearly, T_ω is linear and bounded, and $\|T_\omega\| \le \|P_\omega\| \|T\| \|P_\omega\| = \|T\|$.

Moreover, if a is any finite point in *H, $T_\omega a$ is near-standard. For $T_\omega a = P_\omega T(P_\omega a)$ and $\|P_\omega a\| \leq \|a\|$ so that $P_\omega a$ is also finite. It follows that $T(P_\omega a)$ is near-standard, $T(P_\omega a) = b + \eta$ where b is standard and $\|\eta\|$ is infinitesimal. Now

$$T_\omega a = P_\omega(b + \eta) = b - (I - P_\omega)b + P_\omega \eta$$

and so

$$\|T_\omega a - b\| \leq \|(I - P_\omega)b\| + \|P_\omega \eta\|.$$

$\|P_\omega \eta\| \leq \|\eta\|$ is infinitesimal, and $\|(I - P_\omega)b\|$ is infinitesimal, as shown above. Hence, $\|T_\omega a - b\|$ is infinitesimal, $T_\omega a$ is near-standard, as asserted.

T_ω maps *H into H_ω and so, in particular, maps H_ω into itself. Finally, T_ω is self-adjoint, by transfer from M (for all projection operators in H are self-adjoint) and

$$(P_\omega T P_\omega a, b) = (T P_\omega a, P_\omega b) = (a, P_\omega T P_\omega b).$$

Let T' be the restriction of T_ω to H_ω. Then it follows from the above that T' is a self-adjoint linear operator on H_ω and $\|T'\| \leq \|T_\omega\| \leq \|T\|$. T' can be represented by a Hermitian matrix with respect to the basis $\{e_1, e_2, \ldots, e_\omega\}$ of H_ω.

Consider the characteristic equation

7.3.1 $(\lambda I' - T')x = 0,$

where I' is the identity operator in H_ω. By standard theory, transferred to *M, there exist real numbers $\lambda_1, \lambda_2, \ldots, \lambda_\omega$ (the *eigenvalues* of T'), equal or different, and arranged in non-increasing order of their absolute values, and corresponding non-zero points of H_ω (*eigenvectors* of T') $r_1, r_2, \ldots, r_\omega$, which may be taken to have norm $\|r_j\| = 1$, such that $(r_j, r_k) = 0$ for $j \neq k$ and .

7.3.2 $(\lambda_j I' - T')r_j = 0, \qquad j = 1, 2, \ldots, \omega.$

Equation 7.3.2 shows that λ_j is finite for $j = 1, 2, \ldots, \omega$ since it implies

$$|\lambda_j| \, \|r_j\| = \|T' r_j\| \leq \|T'\| \, \|r_j\| \leq \|T\|$$

and so $|\lambda_j| \leq \|T\|$. We are going to prove

7.3.3 THEOREM. The eigenvector r_j is near-standard for any eigenvalue λ_j that is not infinitesimal.

Indeed, we may rewrite 7.3.2 as

7.3.4
$$\lambda_j r_j = T_\omega r_j,$$

since $T_\omega a = T'a$ for all a in H_ω. But r_j is finite and so $T_\omega r_j$ is near-standard. Thus, $\lambda_j r_j$ is near-standard, and so long as λ_j is not infinitesimal, this implies that $r_j = \lambda_j^{-1}(\lambda_j r_j)$ is near-standard, the theorem is proved.

7.3.5 THEOREM. Let ε be positive standard and suppose that $|\lambda_j| \geq \varepsilon$ for $j=1,2,\ldots,k$. Then k is finite.

PROOF. It follows from 7.3.3 that, for ε and k which satisfy the assumptions of the theorem under consideration, r_n is near-standard for $n=1,2,\ldots,k$. On the other hand, $\|r_j - r_l\| = (r_j - r_l, r_j - r_l)^{\frac{1}{2}} = \sqrt{2}$. Hence, by 7.2.5, r_j cannot be near-standard for all $n=1,2,\ldots,k$ if k is infinite. This proves the theorem.

7.3.6 COROLLARY. For any λ_j which is not infinitesimal, the number of eigenvectors r_k for which $\lambda_k = \lambda_j$ is finite (i.e., is a standard natural number). In other words, the dimension of the eigenspace which corresponds to λ_j is finite (a standard natural number).

Now renumber the eigenvalues so as to omit repetition, still in non-increasing order of absolute values. Call the result $\{v_1, v_2, \ldots, v_\mu\}$ so that $\mu \leq \omega$ and for each v_i there exists a $j \geq i$ such that $v_i = \lambda_j$. Let $H_1', H_2', \ldots, H_\mu'$ be the subspaces spanned by the corresponding eigenvectors and let $P_1', P_2', \ldots, P_\mu'$ be the projection operators from H_ω to $H_1, H_2, \ldots, H_\mu$, respectively. Then by standard theory (spectral theorem for finite-dimensional spaces)

7.3.7
$$H_1' \oplus H_2' \oplus \cdots \oplus H_\mu' = H_\omega,$$

7.3.8
$$P_1' + P_2' + \cdots + P_\mu' = I',$$

7.3.9
$$v_1 P_1' + v_2 P_2' + \cdots + v_\mu P_\mu' = T',$$

where $\oplus$ indicates the direct sum.

We are going to show how to carry over the results of spectral theory from H_ω to the original Hilbert space H.

Let V be any internal operator on H_ω which is linear in the algebraic sense. We have, by transfer from M, that V possesses a norm which is defined by

$$\|V\| = \max \|Va\|, \qquad a \in H_\omega, \qquad \|a\| = 1.$$

If V is self-adjoint, then 7.3.7 – 7.3.9 apply, with V for T'.

7.3.10 THEOREM. If the eigenvalues, $v_1,\ldots,v_\mu$ of V are all infinitesimal then $\|V\|$ is infinitesimal.

PROOF. Let $v = |v_1| = \max_{1\leq i \leq \mu}|v_i|$, then v is infinitesimal. Let a be any point in H_ω such that $\|a\| = 1$. Then

$$\|a\| = (\|a_1'\|^2 + \cdots + \|a_\mu'\|^2)^{\frac{1}{2}} = 1,$$

where $a_1',\ldots,a_\mu'$ are the projections of a onto $H_1',\ldots,H_\mu'$ respectively, and

$$\|Va\| = (\|v_1 a_1'\|^2 + \cdots + \|v_\mu a_\mu'\|^2)^{\frac{1}{2}} \leq (v^2(\|a_1'\|^2 + \cdots + \|a_\mu'\|^2)^{\frac{1}{2}} \leq v$$

and so $\|Va\| \leq v$. This proves 7.3.10.

If the number $\|V\|$ is finite then we call the operator V *finite*. If Va is near-standard for every near-standard point a in H_ω then we call V *near-standard*. It is not difficult to see that in that case V is finite (compare 7.1.3). For every near-standard linear operator V on H_ω, we define a bounded linear operator ^0V on H as follows.

Let a be a point in H, $a = (a_1, a_2, \ldots)$. Then there exist points in H_ω which belong to the monad of a, e.g., the point $P_\omega a = (a_1, a_2, \ldots, a_\omega, 0, 0, \ldots)$. Thus the set $\mu(a) \cap H_\omega$ is non-empty, where $\mu(a)$ is the monad of a, as usual. Moreover, the points $b = Va'$ for $a' \in \mu(a) \cap H_\omega$ are infinitely close to each other since V is finite, and they are near-standard since V is near-standard. We define $(^0V)a$ as the joint standard part of these points b. In particular, therefore, $(^0V)a = {}^0(VP_\omega a)$. From this fact it is easy to deduce that ^0V is linear in the algebraic sense. Also, if $\|a\| = 1$, a in H, then $\|P_\omega a\| \leq 1$. Hence $\|^0Va\| = \|^0(VP_\omega a)\| = {}^0\|VP_\omega a\| \leq {}^0\|V\|$. This shows that ^0V is bounded and $\|^0V\| \leq {}^0\|V\|$.

If a is any near-standard point in H_ω, with standard part ^0a then $^0V{}^0a = {}^0(Va)$, since $a \in \mu(^0a) \cap H_\omega$.

We return to a self-adjoint compact operator T on H and consider the related operators $T_\omega = P_\omega T P_\omega$ and T', where T' is the restriction of T_ω to H_ω. Since the norm of T' cannot exceed $\|T_\omega\| \leq \|P_\omega\| \|T\| \|P_\omega\| \leq \|T\|$, the operator T' is finite. T' is even near-standard for if a is a near-standard point in H_ω then $T'a = T_\omega a = P_\omega TP_\omega a = P_\omega Ta$. Now Ta is near-standard since a is at any rate finite and so $P_\omega(Ta)$ also is near-standard, as asserted.

7.3.11 THEOREM. The operator $^0(T')$ exists and is equal to T.

PROOF. $^0(T')$ exists since T' is near-standard. Now let a be any point in H then we have to show that $^0(T')a = Ta$ or, in other words that $^0(T'P_\omega a) = Ta$. But $T'P_\omega a = P_\omega TP_\omega a$ and $P_\omega a \simeq a$, $TP_\omega a \simeq Ta$, and so $P_\omega TP_\omega a \simeq P_\omega Ta \simeq Ta$, $^0(T'P_\omega a) = Ta$, as asserted.

Now let λ_j be an eigenvalue of T', and r_j a corresponding eigenvector as in 7.3.2, such that λ_j is not infinitesimal and $\|r_j\| = 1$. As we have seen, λ_j is then finite and r_j is near-standard.

7.3.12 THEOREM. $^0\lambda_j$, the standard part of λ_j, is an eigenvalue of T (in H) and ^0r_j, the standard part of r_j is a corresponding eigenvector.

PROOF. 7.3.2 can be written in the form

$$\lambda_j r_j - T'r_j = 0.$$

This entails

7.3.13 $^0(\lambda_j r_j) - {}^0(T'r_j) = 0.$

But $^0(T'r_j) = (^0T')(^0r_j)$, as shown above, and so $^0(T'r_j) = T(^0r_j)$, while $^0(\lambda_j r_j) = {}^0\lambda_j{}^0r_j$. Hence, from 7.3.13,

7.3.14 $^0\lambda_j{}^0r_j - T(^0r_j) = 0,$

which proves the theorem.

Let $\{v_1, v_2, \ldots, v_\mu\}$ be the eigenvalues of T' ranged as before in nonincreasing order of absolute values, without repetitions, and let W be the set of natural numbers i such that v_i is not infinitesimal. W is not necessarily an internal set and may be finite or infinite in the ordinary sense. For any v_i such that $i \in W$, ^0v_i is an eigenvalue of T, by 7.3.12. Let m be the number of λ_j which coincide with v_i; m is finite, by 7.3.5. Also, m is the dimension of the subspace H_i' of H_ω (see 7.3.7), and H_i' is spanned by m of the vectors r_j. These will be denoted for convenience also by $r_{i1}, \ldots, r_{im}$.

For any λ_j which is not infinitesimal, ^0r_j exists and $\|^0r_j\| = \|r_j\| = 1$. And, for $j \neq l$,

$$|(^0r_j, {}^0r_l)| = |(^0r_j, {}^0r_l) - (r_j, r_l)| =$$
$$|(^0r_j - r_j, {}^0r_l) + (r_j, {}^0r_l - r_l)| \le |(^0r_j - r_j, {}^0r_l)| + |(r_j, {}^0r_l - r_l)| \le$$
$$\|^0r_j - r_j\| \, \|^0r_l\| + \|r_j\| \, \|^0r_l - r_l\| = \|^0r_j - r_j\| + \|^0r_l - r_l\| \simeq 0,$$

so that $(^0r_j, {}^0r_l) \simeq 0$. But $(^0r_j, {}^0r_l)$ is standard and so $(^0r_j, {}^0r_l) = 0$.

For any given $i \in W$, let $^0H_i'$ be the subspace of H which is spanned by $^0r_{i1}, \ldots, {}^0r_{im}$. We have just shown that the vectors are orthogonal and so $^0H_i'$ is m-dimensional, like H_i. The same argument shows that for distinct $i \in W$, the corresponding $^0H_i'$ are orthogonal to each other.

Let P_i be the projection operator from H to $^0H_i'$. We are going to show that $P_i = {}^0P_i'$ where P_i' is the projection operator from H_ω to H_i' as introduced previously (see 7.3.8). To see this, let a be any point in H. Then $^0(P_\omega a) = a$. Moreover there exist finite complex numbers μ_i such that

7.3.15 $$P_\omega a = \mu_1 r_{j1} + \cdots + \mu_m r_{im} + a', \qquad \text{where}$$

7.3.16 $$P_i'(P_\omega a) = \mu_1 r_{i1} + \cdots + \mu_m r_{im},$$

and a' is a point in H_ω which is orthogonal to H_i'. Taking standard parts in 7.3.16 we obtain

$$^0(P_i'(P_\omega a)) = {}^0\mu_1 {}^0r_{i1} + \cdots + {}^0\mu_m {}^0r_{im}$$

This shows that $^0P_i'$ exists and

$$^0P_i' a = {}^0\mu_1 {}^0r_{i1} + \cdots + {}^0\mu_m {}^0r_{im}.$$

Thus, $^0P_i'$ maps a into $^0H_i'$. On the other hand, by 7.3.15,

$$a = {}^0(P_\omega a) = {}^0\mu_1 {}^0r_{i1} + \cdots + {}^0\mu_m {}^0r_{im} + {}^0a',$$

where $^0a'$ exists since $P_\omega a$ and $\mu_1 r_{i1} + \cdots + \mu_m r_{im}$ are near-standard. As before, $^0a'$ is orthogonal to $^0r_{i1}, \ldots, {}^0r_{im}$ and hence, to $^0H_i'$. This shows that

$$P_i a = {}^0\mu_1 {}^0r_{i1} + \cdots + {}^0\mu_m {}^0r_{im},$$

i.e., $P_i = {}^0P_i'$, as asserted.

Now suppose that $T \neq 0$. Then there exists a point a in H such that $\|a\| = 1$ and $\|Ta\| = \alpha > 0$. Consider the point $b = P_\omega a$, which belongs to H_ω. We have $P_\omega b = b$ and $\|TP_\omega b\| = \|Tb\| \ge \frac{1}{2}\alpha$ since $\|b - a\| \simeq 0$. Moreover, $c = Tb$ is near-standard since T is compact and so $\|c - P_\omega c\| = \|(I - P_\omega)c\| \simeq 0$

and $\|T'b\| = \|P_\omega TP_\omega b\| = \|P_\omega c\| > \frac{1}{4}\alpha$ while at the same time $\|b\| \le \|a\| = 1$. This shows that $\|T'\|$ is not infinitesimal. We may therefore conclude from 7.3.10 that the eigenvalues of T' are not all infinitesimal and so the corresponding set of positive integers W is not empty.

W contains, together with any positive integer i also all positive integers which are smaller than i. Now if $m \in W$ for some infinite m, then $|v_m|$ is not infinitesimal and so $|v_m| \ge \varepsilon$ for some positive standard ε. But $|v_i| \ge |v_m|$ for all finite i and so $|v_i| \ge \varepsilon$ for all finite i. This contradicts 7.3.5 and shows that W cannot contain any infinite positive integers. Thus W is either finite, or it coincides with the set of all finite positive integers $N - \{0\}$.

Suppose first that W is finite, $W = \{1,...,k\}$ where k is a finite natural number. Then the space $^0H_1' + {}^0H_2' + \cdots + {}^0H_k'$ is finite-dimensional. Let H_0 be the orthogonal complement of this space in H, and let P_0 be the projection operator from H onto H_0.

7.3.17 THEOREM. On the assumptions just stated,

7.3.18 $$^0H_1' \oplus {}^0H_2' \oplus \cdots \oplus {}^0H_k' \oplus H_0 = H,$$

7.3.19 $$P_1 + P_2 + \cdots + P_k + P_0 = I,$$

and

7.3.20 $$^0v_1 P_1 + {}^0v_2 P_2 + \cdots + {}^0v_k P_k = T.$$

PROOF. 7.3.18 follows from the definition of H_0, where we recall that the several $^0H_i'$ are orthogonal. 7.3.19 is an immediate consequence of 7.3.18. In order to prove 7.3.20, we make use of 7.3.9 and 7.3.11, so

7.3.21 $$\begin{aligned} T = {}^0(T') &= {}^0(v_1 P_1' + v_2 P_2' + \cdots + v_\mu P_\mu') \\ &= {}^0(v_1 P_1' + \cdots + v_k P_k') + {}^0(v_{k+1} P_{k+1}' + \cdots + v_\mu P_\mu') \\ &= {}^0v_1 P_1 + \cdots + {}^0v_k P_k + {}^0(v_{k+1} P_{k+1}' + \cdots + v_\mu P_\mu') \end{aligned}$$

since $^0P_i' = P_i$, $i = 1,...,k$. These equations are meaningful since T' and $v_1 P_1' + \cdots + v_k P_k'$ are near-standard operators in H_ω. As far as T' is concerned, this fact was established earlier and is implicit in 7.3.11. Again, if a is any finite point in H_ω, with projections $a_1,...,a_k$ into $H_1',...,H_k'$ respectively, then

$$a = a_1 + \cdots + a_k + a'$$

for some a' in H_ω that is orthogonal to $a_1,\dots,a_k$. For any i, $1 \le i \le k$, a_i may be written as

$$a_i = \mu_{i1} r_{i1} + \cdots + \mu_{im(i)} r_{im(i)},$$

where $r_{i1},\dots,r_{im(i)}$ are the eigenvectors for eigenvalue v_i, as introduced previously, and the μ_{ij} are finite since a and the points a_i are finite. But the r_{ij} are near-standard, by 7.3.3, and so the points $\mu_{ij} r_{ij}$ are near-standard for these r_{ij} and finally, the a_i are near-standard, being finite sums of near-standard points. Now

$$(v_1 P_1' + \cdots + v_k P_k') a = v_1 P_1' a + \cdots + v_k P_k' a = v_1 a_1 + \cdots + v_k a_k,$$

and this sum also is near-standard since $v_1,\dots,v_k$ are finite. It follows that $v_1 P_1' + \cdots + v_k P_k'$ is near-standard, as asserted.

Since $v_{k+1} P_{k+1}' + \cdots + v_\mu P_\mu'$ is the difference between T' and

$$v_1 P_1' + \cdots + v_k P_k',$$

it is also near-standard, and the term $^0(v_{k+1} P_{k+1}' + \cdots + v_\mu P_\mu')$ which was introduced in 7.3.21 is indeed meaningful. We shall show presently that

7.3.22 $^0(v_{k+1} P_{k+1}' + \cdots + v_\mu P_\mu') = 0.$

Combining 7.3.22 with 7.3.21, last line, we obtain

$$^0 v_1 P_1 + \cdots + {}^0 v_k P_k = T,$$

and this is 7.3.20.

It remains to prove 7.3.22. For any finite a in H_ω,

$$\begin{aligned}
\|(v_{k+1} P_{k+1}' + \cdots + v_\mu P_\mu') a\| &= \|v_{k+1} P_{k+1}' a + \cdots + v_\mu P_\mu' a\| \\
&\le (\|v_{k+1} P_{k+1}' a\|^2 + \cdots + \|v_\mu P_\mu' a\|^2)^{\frac{1}{2}} \\
&\le |v_{k+1}| (\|P_{k+1}' a\|^2 + \cdots + \|P_\mu' a\|^2)^{\frac{1}{2}} \le |v_{k+1}| \, \|a\|
\end{aligned}$$

where $|v_{k+1}| \, \|a\|$ is infinitesimal. This shows that $\|v_{k+1} P_{k+1}' + \cdots + v_\mu P_\mu'\|$ is infinitesimal, entailing 7.3.22.

Suppose next that W is infinite so that W coincides with the set of finite positive integers. Then the v_i are finite but not infinitesimal for all finite i and the corresponding subspaces $^0 H_i'$ of H are finite-dimensional and mutually orthogonal. Thus, $H' = \sum_{i=1}^{\infty} {}^0H_i'$ exists as a direct sum. Let H_0 be the orthogonal complement of H' in H where we put $H_0 = \{0\}$ if $H' = H$. Then

7.3.23
$$H_0 \oplus \sum_{i=1}^{\infty} {}^0H_i' = H .$$

If P_i is the projection operator from H to ${}^0H_i'$ as before, $i = 1,2,\ldots, i$ finite, and P_0 is the projection operator from H to H_0 then it is an immediate consequence of 7.3.23 that

7.3.24
$$\sum_{i=0}^{\infty} P_i = I$$

provided we understand this equation in the sense that

7.3.25
$$\sum_{i=0}^{\infty} P_i a = a \qquad \text{for all } a \in H$$

Equations 7.3.23 and 7.3.24 or 7.3.25, correspond to 7.3.18 and 7.3.19 for the case of infinite W. The counterpart of 7.3.20 for this case is

7.3.26
$$\sum_{i=1}^{\infty} {}^0 v_i P_i a = T a \qquad \text{for all points } a \text{ in } H.$$

In order to prove 7.3.26, we have to show that $s_n = \|T a - \sum_{i=1}^{n} {}^0 v_i P_i a\|$ tends to zero as n tends to infinity in M, for any point a in H. Now

$$Ta = ({}^0T')a = {}^0(T'P_\omega a)$$

by 7.3.11, and

$$\sum_{i=1}^{n} {}^0 v_i P_i a = \sum_{i=1}^{n} {}^0 v_i {}^0 P_i' a = \sum_{i=1}^{n} {}^0 v_i {}^0(P_i' P_\omega a) = {}^0\left(\left(\sum_{i=1}^{n} v_i P_i'\right) P_\omega a\right).$$

Hence,

7.3.27 $\qquad s_n = \left\| {}^0(T' P_\omega a) - {}^0\left(\left(\sum_{i=1}^{n} v_i P_i'\right) P_\omega a\right) \right\| = \left\| {}^0\left(\sum_{i=n+1}^{\mu} v_i P_i'\right) P_\omega a \right\|,$

where we have taken into account 7.3.9. Let α be a standard real number which is greater than $\|a\|$. Then $\|P_\omega a\| < \alpha$ and (compare the proof of 7.3.22),

$$\left\|\left(\sum_{i=n+1}^{\mu} v_i P_i'\right) P_\omega a\right\| \le |v_{n+1}|\, \|P_\omega a\| < |v_{n+1}|\,\alpha.$$

Hence, by 7.3.27, $s_n < |^0 v_{n+1}|\alpha + \eta$ for any positive standard η, and so $s_n \le |^0 v_{n+1}|\alpha$. But the sequence of $|v_i|$ is non-increasing, and for every positive standard r there exist only a finite number of i for which $|v_i| \ge r$, by 7.3.5. It follows that the same applies to $|^0 v_i|$ and so $\lim_{i \to \infty} |^0 v_i| = 0$ in M. Hence, $\lim_{n \to \infty} s_n = 0$, as required in order to establish the truth of 7.3.26.

Summing up, we have proved

7.3.28   THEOREM.   If W is infinite, then

7.3.29 $$H_0 \oplus {}^0 H_1' \oplus {}^0 H_2' \oplus \cdots = H$$

7.3.30 $$P_0 a + P_1 a + P_2 a + \cdots = a$$

and

7.3.31 $${}^0 v_1 P_1 a + {}^0 v_2 P_2 a + \cdots = Ta$$

for any a in H.

The possibility that ${}^0 v_i = {}^0 v_j$ for $i \neq j$ where both i and j belong to W, has not been excluded. However, in this case, the corresponding spaces ${}^0 H_i'$ and ${}^0 H_j'$ are still orthogonal and the total dimension of the space spanned by all the eigenvectors ${}^0 r_{ik}$ whose eigenvalues are equal to ${}^0 v_i$ is finite, again by 7.3.5. On the other hand, all non-zero eigenvalues of T appear among the ${}^0 v_i$ for $i \in W$. For (using a familiar argument), if $Ta = \lambda a$ for some a in H with $\|a\| = 1$ and $\lambda \neq {}^0 v_i$ for all $i \in W$, and $\lambda \neq 0$ then, multiplying 7.3.20 or 7.3.31 (for finite or infinite W, respectively) by P_i or P_0, we obtain

$${}^0 v_i P_i a = \lambda P_i a$$

and

$$0 = \lambda P_0 a.$$

This implies $P_0 a = P_1 a = P_2 a = \cdots = 0$ and hence $a = 0$, contrary to assumption. As for $\lambda = 0$, this is an eigenvalue if W is finite, or if W is infinite and $H_0 \neq \{0\}$. However, on this point no further details are provided by our analysis.

7.4   An invariant subspace problem.   The main purpose of this section is the proof of the following theorem.

7.4.1   THEOREM (Standard).   Let T be a bounded linear operator on H and let $p(z) \neq 0$ be a polynomial with complex coefficients such that $p(T)$ is compact. Then T leaves invariant at least one closed linear subspace of H other than H or $\{0\}$.

   T is said to *leave invariant* a subspace E of H if $TE \subset E$.

   Some auxiliary considerations will be prefaced to the proof of 7.4.1.

   Let $T = (a_{jk})$, $j, k = 1, 2, 3, \ldots$ be the matrix representation of the bounded linear operator T with respect to a given basis $\{e_1, e_2, e_3, \ldots\}$. Then it is well known, and not difficult to verify, that

7.4.2
$$\sum_{k=1}^{\infty} |a_{jk}|^2 < \infty, \qquad j = 1, 2, 3, \ldots$$

and

7.4.3
$$\sum_{j=1}^{\infty} |a_{jk}|^2 < \infty, \qquad k = 1, 2, 3, \ldots .$$

   Within $^*M$, the subscripts of a_{jk} vary also over the infinite natural numbers. 7.4.2 implies that $\sum_{k=\omega}^{\infty} |a_{jk}|^2$ is infinitesimal for all infinite ω, provided j is finite. This is not necessarily true for infinite j. In fact, $a_{j\omega}$ need not be infinitesimal for infinite j. For example, if $T = I$, the identity matrix, then $a_{j\omega} = 1$ for $j = \omega$. However,

7.4.4   THEOREM.   If $T = (a_{jk})$ is a compact linear operator on H then a_{jk} is infinitesimal for all infinite k (j finite or infinite).

PROOF.   Suppose first that j is finite, while k is infinite. Then

$$|a_{jk}|^2 < \sum_{n=k}^{\infty} |a_{jn}|^2 .$$

The right hand side of this inequality is infinitesimal, since k is infinite, so a_{jk} is infinitesimal.

   Suppose next that both j and k are infinite and define a point $a = (a_1, a_2, a_3, \ldots)$ in $^*H$ by $a_n = 0$ for $n \neq k$ and by $a_k = 1$. Then $\|a\| = 1$. Since a is finite and T is compact, $b = Ta$ must be near-standard. But $b =$

$(b_1, b_2, b_3, \ldots)$ where $b_j = \sum_{n=1}^{\infty} a_{jn} a_n = a_{jk}$. Hence, taking into account 7.2.2, $a_{jk} \simeq 0$ for infinite j. This completes the proof of 7.4.4.

The operator $T = (a_{jk})$ will be called *almost superdiagonal* with respect to a given basis if $a_{jk} = 0$ for $j > k + 1$, $k = 1, 2, 3, \ldots$.

7.4.5 THEOREM. Let $T = (a_{jk})$ be an almost superdiagonal standard bounded linear operator on H such that $Q = p(T) = (b_{jk})$ is compact for a standard polynomial with complex coefficients,

$$p(z) = c_0 + c_1 z + \cdots + c_m z^m, \qquad c_m \neq 0, \qquad m \geq 1.$$

Then there exists an infinite natural number ω such that $a_{\omega+1, \omega}$ is infinitesimal.

PROOF. We show by direct computation that for any positive integer h,

7.4.6 $b_{h+m,h} = c_m a_{h+1,h} a_{h+2,h+1} \cdots a_{h+m, h+m-1}$.

Now by 7.4.4, $b_{h+m,h}$ is infinitesimal for all infinite h. Fixing h as an infinite natural number, we conclude that one of the factors on the right hand side of 7.4.6 must be infinitesimal. But c_m is standard and different from zero, hence, is not infinitesimal. Thus, for some j, $0 \leq j < m$, $a_{h+j+1, h+j}$ is infinitesimal, and the conclusion of the theorem is satisfied by $\omega = h + j$.

Let E be any internal linear subspace of $*H$. In accordance with a previous definition, the set of points ${}^0E$ in H consists of all standard points a whose monad has a non-empty intersection with E, $\mu(a) \cap E \neq \emptyset$. Theorem 4.3.3 shows that ${}^0E$ is closed irrespective of whether or not E is closed. Moreover, if a and b are points of ${}^0E$ and λ is an arbitrary standard complex number, then there exist points a' and b' in E such that $a' \in \mu(a) \cap E$, $b' \in \mu(b) \cap E$. It follows that $a' + b' \in E$ and $\lambda a' \in E$; and that $a' + b' \in \mu(a+b)$ and $\lambda a' \in \mu(\lambda a)$. We conclude that $a+b$ and λa belong to ${}^0E$, so that ${}^0E$ is a closed linear subspace of H.

An internal linear subspace E of $*H$ may be finite-dimensional in the sense of $*M$ (i.e., Q-finite-dimensional), though not in the standard sense. The dimension of E will be denoted by $\dim(E)$. For example, in the case of the space H_ω considered in the preceding section, $\dim(H_\omega) = \omega$.

7.4.7 THEOREM. Let E and E_1 be two internal linear subspaces of $*H$

such that $E \subset E_1$ and $\dim(E_1) = \dim(E) + 1$. Then $^0E \subset {}^0E_1$, and any two points of $^0E_1$ are linearly dependent modulo ^0E.

PROOF. $^0E \subset {}^0E_1$ is trivially true since $E \subset E_1$. Now let σ_1 and σ_2 be any two points in $^0E_1$. Then there exist points τ_1, τ_2 in E_1 such that $\sigma_1 \simeq \tau_1$, $\sigma_2 \simeq \tau_2$. Since the dimension of E_1 exceeds that of E by one, there exists a representation

7.4.8 $\tau_2 = \lambda \tau_1 + \tau,$

or vice versa, where τ belongs to E and λ is a complex number. Suppose now that σ_1 and σ_2 are linearly independent modulo ^0E. If $\lambda \simeq 0$, then $\lambda \tau_1 \simeq 0$ since τ_1 is finite, and so $\tau_2 \simeq \tau$ and $\sigma_2 \simeq \tau$, which implies that σ_2 belongs to ^0E. This contradicts the assumption that σ_1 and σ_2 are linearly independent modulo ^0E. Again, if λ were infinite, then the equation

7.4.9 $\tau_1 = \lambda^{-1}\tau_2 - \lambda^{-1}\tau,$

which is a consequence of 7.4.8, would show that σ_1 belongs to E (since $\lambda^{-1} \simeq 0$ and $\lambda^{-1}\tau \in E$). We conclude that λ is finite and possesses a standard part $^0\lambda \neq 0$. Also $\tau = \tau_2 - \lambda\tau_1 \simeq \sigma$, where $\sigma = \sigma_2 - {}^0\lambda\sigma_1$ since

$$\tau - \sigma = \tau_2 - \lambda\tau_1 - (\sigma_2 - {}^0\lambda\sigma_1) = \tau_2 - \sigma_2 + \lambda(\tau_1 - \sigma_1) + (\lambda - {}^0\lambda)\sigma_1.$$

On the right hand side of this equation $\tau_2 - \sigma_2 \simeq 0$, $\lambda(\tau_1 - \sigma_1) \simeq 0$, and $(\lambda - {}^0\lambda)\sigma_1 \simeq 0$. Hence, $\tau - \sigma \simeq 0$, as asserted. It follows that $\sigma \in {}^0E$, so that σ_1 and σ_2 are linearly dependent modulo ^0E. This contradiction proves 7.4.7.

Let T be a bounded linear operator on H and let H_ω and T_ω be defined as in section 7.3. Let E be an internal linear subspace of H_ω. E is Q-finite-dimensional since H_ω is Q-finite-dimensional.

7.4.10   THEOREM.   If E is invariant for T_ω, i.e., $T_\omega E \subset E$, then ^0E is invariant for T, $T^0E \subset {}^0E$.

PROOF.   Let $\sigma \in {}^0E$, then we have to show that $T\sigma \in {}^0E$. By assumption, there exists a $\tau \in E$ such that $\sigma \simeq \tau$. Then $T_\omega \tau \in E$. Now

$$T\sigma - T_\omega\tau = T\sigma - P_\omega T P_\omega \tau = T\sigma - P_\omega T\tau = (T - P_\omega T)\sigma + P_\omega T(\sigma - \tau).$$

On the right hand side of this equation $T\sigma$ is standard and so

$$\|(T - P_\omega T)\sigma\| = \|(I - P_\omega)T\sigma\|$$

is infinitesimal. And $\|\sigma - \tau\|$ is infinitesimal, so $\|P_\omega T(\sigma - \tau)\|$ also is infinitesimal. This shows that $\|T\sigma - T_\omega \tau\|$ is infinitesimal, and so $T\sigma \simeq T_\omega \tau$, $T\sigma \in {}^0E$, as asserted.

PROOF of 7.4.1.   For a given T, which satisfies the assumptions of the theorem, consider the set of points $A = \{\sigma, T\sigma, T^2\sigma, ...\}$ in H, where σ is any point of H such that $\|\sigma\| = 1$. If A is linearly dependent algebraically, then $T^{n+1}\sigma = \sum_{k=0}^{n} \lambda_k T^k \sigma$ for some (finite) natural number n. In that case, the finite-dimensional subspace of H which is generated by $\sigma, T\sigma, ..., T^n\sigma$ is invariant for T, and we have finished. Similarly, if the closed linear subspace E generated by A is a proper subspace of H, then E may serve as the required invariant subspace for T. Accordingly, we shall suppose from now on that A is linearly independent algebraically and generates the entire space H. We now choose a special basis $\{e_1, e_2, e_3, ...\}$ for H in the following way.

Put $e_1 = \sigma$. By one of the assumptions just made, the space H_2 generated by e_1 and $Te_1 = T\sigma$ must be two-dimensional. Choose e_2 in H_2 such that $\|e_2\| = 1$ and $(e_1, e_2) = 0$. Then e_1 and e_2 form an orthonormal basis for H_2. Next consider the space H_3, generated by e_1, e_2, and Te_2. H_3 is also the space generated by σ, $T\sigma$, and $T^2\sigma$, and is therefore strictly three-dimensional. Choose e_3 in H_3 such that $\|e_3\| = 1$, $(e_1, e_3) = (e_2, e_3) = 0$. Then e_1, e_2, e_3 form an orthonormal basis for H_3. Continuing in this way, we obtain an orthonormal sequence of points $\{e_n\}$ in H, $n = 1, 2, 3, ...$ such that for every $n \geq 1$, the space generated by $e_1, e_2, e_3, ..., e_n$ coincides with the space generated by $\sigma, T\sigma, ..., T^{n-1}\sigma$. In more detail $\{e_n\}$ may be obtained from $\{T^n\sigma\}$, $n = 0, 1, 2, ...$ by the Gram–Schmidt procedure.

Te_n is linearly dependent on $e_1, e_2, ..., e_{n+1}$ and may be written in the form

7.4.11 $Te_n = a_{1n}e_1 + a_{2n}e_2 + \cdots + a_{n+1,n}e_{n+1}, \quad n = 1, 2, ...$

Since the closed linear subspace generated by A coincides with H, $\{e_n\}$ constitutes a basis for H. 7.4.11 now shows that the matrix (a_{jk}) which represents T with respect to this basis is almost superdiagonal.

Passing to $^*H$, we see from 7.4.5 that there exists an infinite natural number ω such that $a_{\omega+1,\omega}$ is infinitesimal. For such an ω, we consider the space H_ω and the operator $T_\omega = P_\omega T P_\omega$. Let $\xi = (x_1, x_2, ...)$ be any finite point in $*H$ and let $\zeta=(z_1,z_2,\ldots)=(TP_\omega-T_\omega)\xi=(I-P_\omega)TP_\omega\xi$. We find by direct computation that $z_{\omega+1}=a_{\omega+1,\omega}x_\omega$ and $z_n=0$ for $n\neq\omega+1$, Hence, $\|\zeta\|\leq|a_{\omega+1,\omega}|\,\|\xi\|$ so that ζ is infinitesimal. Thus, $(TP_\omega-T_\omega)\xi\simeq0$, and $TP_\omega\xi\simeq T_\omega\xi$ for all finite points ξ in $*H$. More generally, we prove by induction that for all finite ξ in $*H$ and for all finite positive integers r,

7.4.12
$$T^rP_\omega\xi\simeq T_\omega^r\xi.$$

We have just seen that 7.4.12 is true for $r=1$. Suppose 7.4.12 has been proved for $r-1$, $r\geq2$. Then

$$T^rP_\omega\xi\simeq TT_\omega^{r-1}\xi=TP_\omega T_\omega^{r-1}\xi\simeq T_\omega T_\omega^{r-1}\xi=T_\omega^r\xi.$$

This proves the assertion for r and hence, establishes the truth of 7.4.12 for all finite positive integers r.

We apply 7.4.12 to the monomials of $p(T)$ and take into account that $P_\omega\xi=\xi$ for $\xi\in H_\omega$. This yields

7.4.13
$$p(T)\xi\simeq p(T_\omega)\xi \qquad \text{for finite } \xi\in H_\omega.$$

Let T' be the restriction of T_ω to H_ω, as before. Since H_ω is Q-finite-dimensional, i.e., finite-dimensional in the sense of $*M$, the standard theory of linear operators (or matrices) for finite-dimensional spaces, transferred from M to $*M$, shows that T' possesses a chain of invariant subspaces

7.4.14
$$E_0\subset E_1\subset E_2\subset\cdots\subset E_\omega,$$

where $\dim(E_j)=j$, $0\leq j\leq\omega$, so that $E_0=\{0\}$, $E_\omega=H_\omega$. Thus, $T'E_j\subset E_j$ $j=0,1,2,\ldots,\omega$. Let P_j' be the projection operator from $*H$ onto E_j, $j=0,1,2,\ldots,\omega$, so that $P_\omega'=P_\omega$.

Choose ξ in H such that $\|\xi\|=1$. Then we may suppose that $p(T)\xi\neq0$, for otherwise $\xi,T\xi,\ldots,T^m\xi$ would be linearly dependent, proving the theorem, as pointed out previously. Since $\xi\simeq P_\omega\xi$, we have $p(T)\xi\simeq p(T)P_\omega\xi$, so $p(T)P_\omega\xi$ is not infinitesimal. It now follows from 7.4.13 that $p(T_\omega)P_\omega\xi$ is not infinitesimal. But $p(T_\omega)P_\omega\xi\simeq p(T_\omega)\xi$, and so $\|p(T_\omega)\xi\|>\varrho$ for some standard positive ϱ. Define ϱ_j for $j=0,1,2,\ldots,\omega$ by

7.4.15
$$\varrho_j=\|p(T_\omega)\xi-p(T_\omega)P_j'\xi\|.$$

Then $\varrho_j\leq\|p(T_\omega)\|\,\|\xi-P_j'\xi\|$, and $\varrho_0=\|p(T_\omega)\xi\|$. Hence, $\varrho_0>\varrho$. Also, $\|\xi-P_\omega'\xi\|=\|\xi-P_\omega\xi\|$ is infinitesimal, hence $\varrho_\omega<\frac12\varrho$. It follows that there exists a smallest positive integer λ, which may be finite or infinite, such that $\varrho_\lambda < \frac{1}{2}\varrho$ but $\varrho_{\lambda-1} \geq \frac{1}{2}\varrho$.

With every E_j there is associated a closed linear subspace ${}^0E_j$ of H, as introduced earlier in this section. We propose to show that

7.4.16 ${}^0E_{\lambda-1} \neq H$ and ${}^0E_\lambda \neq \{0\}$.

For if ${}^0E_{\lambda-1} = H$ then $\xi \in {}^0E_{\lambda-1}$, and so $\|\xi - P'_{\lambda-1}\xi\| \simeq 0$. But

$$\varrho_{\lambda-1} \leq \|p(T_\omega)\| \, \|\xi - P'_{\lambda-1}\xi\|,$$

so $\varrho_{\lambda-1} \simeq 0$, which is contrary to the choice of λ. Also, ${}^0E_\lambda$ cannot reduce to $\{0\}$. For consider the point $\eta = p(T_\omega)P'_\lambda\xi$. η belongs to E_λ since $P'_\lambda\xi \in E_\lambda$ and E_λ is invariant under $p(T')$ and, equivalently, under $p(T_\omega)$. We claim that η is near-standard.

Indeed, since $P'_\lambda\xi \in H_\omega$, $\eta = p(T_\omega)P'_\lambda\xi \simeq p(T)P'_\lambda\xi$, by 7.4.13, and $p(T)P'_\lambda\xi$ is near-standard since $P'_\lambda\xi$ is finite and $p(T)$ is compact. Hence, η is near-standard, and ${}^0\eta$ exists and belongs to ${}^0E_\lambda$. But η cannot be infinitesimal. For if $\eta \simeq 0$ then, by 7.4.15

$$\varrho_\lambda \geq \|p(T_\omega)\xi\| - \|p(T_\omega)P'_\lambda\xi\| > \varrho - \zeta,$$

where $\zeta = \|\eta\| \simeq 0$. Hence $\varrho_\lambda > \frac{1}{2}\varrho$, which is again contrary to the choice of λ. We conclude that η is not infinitesimal and hence that ${}^0\eta \neq 0$. Thus, ${}^0E_\lambda \neq \{0\}$ since it contains ${}^0\eta$.

Now by 7.4.10, both ${}^0E_{\lambda-1}$ and ${}^0E_\lambda$ are invariant for T. If neither of these spaces were a non-trivial closed invariant subspace of H for T we should have ${}^0E_{\lambda-1} = \{0\}$ and ${}^0E_\lambda = H$. This contradicts 7.4.7 and completes the proof of 7.4.1.

Throughout this chapter, we have considered only separable H. But 7.4.1 is actually trivially true for non-separable H for in that case the set $A = \{\sigma, T\sigma, T^2\sigma, \ldots\}$ generates a non-trivial closed invariant subspace of H for T, for any $\sigma \neq 0$.

7.5   Remarks and references.   Going beyond 7.3, A. R. BERNSTEIN has, in his dissertation (1965, unpublished), given the spectral theory of self-adjoint bounded linear operators within the framework of Non-standard Analysis. For the background to 7.4, consult ARONSZAJN and SMITH [1954] and HALMOS [1963]. The fact that any compact linear operator in Hilbert space possesses a non-trivial invariant closed linear subspace was first proved by J. von Neumann and N. Aronszajn, and the corresponding result for Banach spaces is due to N. Aronszajn and K. T. Smith. For the result given in 7.4, compare BERNSTEIN and ROBINSON [1966]. Bernstein has shown that this result, also, can be generalized to arbitrary Banach spaces.

Further methods and results in Non-standard Functional Analysis will be found in LUXEMBURG [1962], TAKEUTI [1962] and ROBINSON [1964].

CHAPTER VIII

TOPOLOGICAL GROUPS AND
LIE GROUPS

8.1 Topological groups. A topological group G is (i) a topological space and (ii) a group, such that the group multiplication $ab = c$ is a continuous function from $G \times G$ into G and the operation of inversion, $a^{-1} = b$ is a continuous function from G into G. That is to say, by one of the characteristic properties of a continuous function in a topological space, if $ab = c$ and W is an open neighborhood of c then there exist open neighborhoods U and V of a and b respectively, such that $UV \subset W$. In this connection, UV is defined, as usual, as the set of all products $a'b'$ where $a' \in U$, $b' \in V$. Similarly, if $a^{-1} = b$ then for any open neighborhood V of b there exists an open neighborhood U of a such that $U^{-1} \subset V$ where U^{-1} consists of all elements c^{-1} of the group such that $c \in U$. For the time being, we do not stipulate that G is a Hausdorff space.

In order to carry out the Non-standard Analysis of G, we may consider in the first instance a full structure M whose individuals are the group elements of G. However, we shall presently consider situations which involve at the same time the natural numbers, or the real numbers, or even a general metric or normed space. In all these cases we shall suppose tacitly that these structures also are contained in M. Passing to an enlargement $*M$ of M, we then find that $*M$ contains in the first instance an enlargement $*G$ of G and, where required, also enlargements of the natural numbers, of the real numbers, and of the metric or normed space under consideration.

The monad $\mu(a)$ of a standard point a in $*G$ is defined, as usual, as the intersection of all standard open neighborhoods of a. An element $a \in \mu(e)$ is called *infinitesimal*, where e is the identity (neutral element).

8.1.1 THEOREM. Let a and b be any two standard points in $*G$ (i.e., a and b belong to G). Then $\mu(a)\mu(b) \subset \mu(ab)$.

PROOF. Let $a' \in \mu(a)$ and $b' \in \mu(b)$. We have to show that for any open neighborhood W of ab in G, $a'b' \in {}^*W$. By assumption, there exist open sets U and V in G such that $a \in U$, $b \in V$ and $UV \subset W$. Then $a' \in {}^*U$, $b' \in {}^*V$ and ${}^*U {}^*V \subset {}^*W$. Hence, $a'b' \in {}^*W$, proving the theorem.

8.1.2 THEOREM. Let a be a standard point in ${}^*G$. Then $\mu(a^{-1}) = (\mu(a))^{-1}$.

PROOF. Given any open neighborhood V of a^{-1} in G, we show similarly as in the proof of 8.1.1, that $c^{-1} \in {}^*V$ for all $c \in \mu(a)$. We conclude that $(\mu(a))^{-1} \subset \mu(a^{-1})$. Applying the same argument to a^{-1} in place of a we find that $(\mu(a^{-1}))^{-1} \subset \mu((a^{-1})^{-1}) = \mu(a)$. Hence, $\mu(a^{-1}) \subset (\mu(a))^{-1}$ proving $\mu(a^{-1}) = (\mu(a))^{-1}$, as asserted.

Reinforcing 8.1.1, we now prove

8.1.3 THEOREM. Let a and b be any two standard points in G. Then $\mu(a)\mu(b) = \mu(ab)$.

PROOF. In view of 8.1.1, we only have to show that $\mu(a)\mu(b) \supset \mu(ab)$. Let $c' \in \mu(ab)$. Since $a^{-1} \in \mu(a^{-1})$, we then have $a^{-1}c' \in \mu(a^{-1})\mu(ab)$ and so $a^{-1}c' \in \mu(a^{-1}(ab)) = \mu(b)$, by 8.1.1. Thus, $c' = a(a^{-1}c')$ where $a \in \mu(a)$, $a^{-1}c' \in \mu(b)$. This shows that $c' \in \mu(a)\mu(b)$ and proves the theorem.

8.1.4 THEOREM (Standard). Let a and W be a point, and a set of points, respectively, in G. If W is open, then Wa is open.

PROOF. Since W is open, $\mu(b) \subset {}^*W$ for any $b \in W$. Now let c be a standard point such that $c \in Wa$. Then $c = ba$ for some $b \in W$. Now $\mu(c) = \mu(b)a$. For on one hand, $\mu(b)a \subset \mu(c)$, by 8.1.1 and on the other hand, if $c' \in \mu(c)$ then $b' = c'a^{-1} \in \mu(c)\mu(a^{-1}) = \mu(b)$, so $c' = b'a \in \mu(b)a$, $\mu(c) = \mu(b)a$. But $\mu(b)a \subset {}^*Wa$ and so $\mu(c) \subset {}^*Wa = {}^*(Wa)$, Wa is open.

8.1.5 THEOREM. Suppose that the points a and b in ${}^*G$ both belong to the same monad, $a \in \mu(c)$, $b \in \mu(c)$ for some standard point c. then $ab^{-1} \in \mu(e)$ where e is the identity in G and ${}^*G$.

PROOF. $b^{-1} \in \mu(c^{-1})$, by 8.1.2. Hence $ab^{-1} \in \mu(c)\mu(c^{-1}) = \mu(cc^{-1}) = \mu(e)$, where we have made use of 8.1.3.

We now introduce a concept which has no precise counterpart in the standard theory of topological groups.

8.1.6 THEOREM. The elements of $\mu(e)$ constitute a subgroup J_1 of $*G$. J_1 will be called the *infinitesimal group* of $*G$ (or of G).

PROOF. If $a\in\mu(e)$ and $b\in\mu(e)$ then $ab\in\mu(e)\mu(e)=\mu(e)$. If $a\in\mu(e)$ then $a^{-1}\in\mu(e^{-1})=\mu(e)$. This proves 8.1.6.

8.1.7 THEOREM. The near-standard points of $*G$ (i.e., the points of $*G$ which belong to the monads of standard points) constitute a subgroup J_0 of $*G$.

PROOF. If $a'\in\mu(a)$, $b'\in\mu(b)$ then $a'b'\in\mu(a)\mu(b)=\mu(ab)$. If $a'\in\mu(a)$ then $(a')^{-1}\in(\mu(a))^{-1}=\mu(a^{-1})$. This proves 8.1.7.

8.1.8 THEOREM. J_1 is a normal subgroup of J_0.

PROOF. Evidently $J_1\subset J_0$. Also if $a\in J_1$ and $g\in J_0$, then $a\in\mu(e)$ and $g\in\mu(h)$ for some standard point h. Hence $gag^{-1}\in\mu(h)\mu(e)(\mu(h))^{-1}=\mu(heh^{-1})=\mu(e)$, $gag^{-1}\in J_1$. This proves the theorem.

8.1.9 THEOREM. Suppose the group G is a Hausdorff space. Then the quotient group J_0/J_1 is isomorphic to G.

PROOF. The cosets of J_1 within J_0 are precisely the monads of $*G$, i.e., the monads of points of G. But these cosets are also the elements of J_0/J_1 Thus, we may define a mapping ψ from J_0/J_1 onto G by putting $\psi:b\to a$ for $b\in J_0/J_1$, $a\in G$ if $b=\mu(a)$. This mapping is one-to-one since, for a Hausdorff space, the monads of distinct points of G are disjoint. It is not difficult to verify that ψ provides an isomorphism. This proves 8.1.9.

If G is compact then all points of $*G$ are near-standard, hence $J_0=*G$ and, if G is at the same time a Hausdorff space, $*G/J_1$ is isomorphic to G.

So far we have considered only the monads of standard points as provided by the topology of G. We now define the left and right monads of an arbitrary point a in $*G$ as the left and right cosets of $J_1=\mu(e)$ which contain a, so $\mu_l(a) = a\mu(e)$, $\mu_r(a) = \mu(e)a$. Thus a point b belongs to $\mu_l(a)$ if $a^{-1}b \in \mu(e)$ and b belongs to $\mu_r(a)$ if $ba^{-1} \in \mu(e)$.

If a is a standard point then $\mu(a) = \mu_l(a) = \mu_r(a)$. For in that case $\mu_l(a) = a\mu(e) \subset \mu(a)\mu(e) = \mu(a)$, and if $b \in \mu(a)$, then $a^{-1}b \in \mu(a^{-1})\mu(a) = \mu(e)$, so that $\mu_l(a) \supset \mu(a)$. Hence $\mu_l(a) = \mu(a)$ and similarly, $\mu_r(a) = \mu(a)$.

Consider any subgroup H of $*G$. H may or may not be internal. For example, if G is the additive group of real numbers, so that $*G$ is the additive group of $*R$, then $H = M_0$ and $H = M_1$ are external. In any case we then have $H^n = H$ for any finite natural number n. However, it does not follow that the same is true also for infinite n.

Let ω be an infinite natural number, and let $N(\omega)$ be the set of natural numbers n such that there exists a finite natural m for which $n \le m\omega$. Then $N(\omega)$ constitutes an initial segment of $*N$, i.e., $n \in N(\omega)$ and $k < n$ for natural k implies $k \in N(\omega)$. Evidently, $N(\omega)$ includes ω and all finite natural numbers.

For any given subgroup H of G, we now define the subgroup $H^{(\omega)}$ of $*G$ as the set of all products $a_0 a_1 \ldots a_n$ for any *internal* sequence $\{a_0, a_1, \ldots, a_n\}$ of $n+1$ elements of H, $n \in N(\omega)$ (so that the sequence is Q-finite, and the product of its elements exists, by transfer from M). It is not difficult to verify that $H^{(\omega)}$ is indeed a group. In particular, if H is the cyclic group generated by an element a of $*G$ (so that H consists of all finite integer powers of a) then $H^{(\omega)}$ consists of all powers $a^{\pm n}$ where $n \in N(\omega)$.

We define $H^{(\infty)}$ as the union of all $H^{(\omega)}$ (for infinite ω). Thus, $H^{(\infty)}$ may be defined directly as the set of all products $a_0 a_1 \ldots a_n$ for any internal sequence $\{a_0, a_1, \ldots, a_n\}$ of $n+1$ elements of H, n an arbitrary finite or infinite natural number.

8.1.10 THEOREM. Suppose H is a normal subgroup of $*G$. Then the groups $H^{(\omega)}$ and $H^{(\infty)}$ are normal subgroups of $*G$.

PROOF. Let g be any point of $*G$, and let $h \in H^{(\omega)}$ for given infinite ω. Then we have to show that $ghg^{-1} \in H^{(\omega)}$. By assumption, there exists a Q-finite internal sequence $\{a_0, a_1, \ldots, a_n\}$, $n \in N(\omega)$ such that $a_i \in H$, $i = 0, 1, \ldots, n$ and $h = a_0 a_1 \ldots a_n$. But then $\{ga_0 g^{-1}, ga_1 g^{-1}, \ldots, ga_n g^{-1}\}$ also is an internal sequence and so $ghg^{-1} = ga_0 g^{-1} ga_1 g^{-1} \ldots ga_n g^{-1} \in H^{(\omega)}$, as required. The proof for $H^{(\infty)}$ is quite similar, and for that case the conclusion follows also from the fact that $H^{(\infty)}$ is the union of the $H^{(\omega)}$.

The *component of the identity*, G_e, of a topological group G is defined as the component of G which contains e, in the ordinary topological sense. It is known that G_e is a normal subgroup of G. For let a be an arbitrary point of G_e and consider the mapping $\psi: b \rightarrow ab^{-1}$ which maps G_e on the set aG_e^{-1}. This mapping is one-to-one and is topological (e.g., since it maps monads in G_e on monads in aG_e^{-1}, and vice versa). Thus, aG_e^{-1} is a connected set, being the image of a connected set, and $e \in aG_e^{-1}$ (for $b = a^{-1}$). Hence, $aG_e^{-1} \subset G_e$, $ab^{-1} \in G_e$ for all $b \in G_e$. For $a = e$, this shows that for any $b \in G_e, b^{-1}$ also is in G_e, and further, for arbitrary $a \in G_e$, that $a(b^{-1})^{-1} = ab \in G_e$. Moreover, if a is any point in G, then $aG_e a^{-1}$ is a connected subset of G which contains e, hence $aG_e a^{-1} \subset G_e$, G_e is normal.

Another well known result, which will be required presently is

8.1.11 THEOREM (Standard). If U is an open neighborhood of e and a subset of G_e then $\bigcup_{n=1}^{\infty} U^n = G_e$.

PROOF. Let $G' = \bigcup_{n=1}^{\infty} U^n$, then G' is an open set since it is the union of open sets. We propose to show that G' is also closed. For let a be a point of G which belongs to the closure of G'. Then the monad of a, $\mu(a)$ has a point b in common with $*G'$, and hence with some $(*U)^n$ where n may be finite or infinite. But $b \in \mu(a)$ implies $ba^{-1} \in \mu(a)\mu(a^{-1}) = \mu(e)$ and so $(ba^{-1})^{-1} = ab^{-1} \in \mu(e)$, $ab^{-1} \in *U$. At the same time $b \in (*U)^n$ and so $a = ab^{-1} b \in *U(*U)^n = (*U)^{n+1}$ and so $a \in *G'$, $a \in G'$, G' is closed. But G' is a subset of G_e. Since we have shown that it is both open and closed we conclude that $G' = G_e$, proving 8.1.11.

We are now in a position to prove

8.1.12 THEOREM. Suppose G_e contains a neighborhood of e. Then $J_1^{(\infty)} = *G_e$.

PROOF. By 4.1.2 there exists an internal open set U which is contained in $\mu(e)$. Then $\bigcup_{n=1}^{\infty} U^n = *G_e$, by transferring 8.1.11 to $*M$ (where the left hand side is now taken in the sense of $*M$). But $U \subset \mu(e) = J_1$ and so we conclude that $J_1^{(\infty)} \supset *G_e$. On the other hand, the elements of $J_1^{(\infty)}$ are all products of elements of $*G_e$ and so $J_1^{(\infty)} \subset *G_e$. The theorem follows.

Suppose next that G is a *local topological group*. That is to say, G is a topological space, and a partial operation (partial function of two variables) is defined from $G \times G$ into G such that the following conditions are satisfied. (i) For any elements a,b, and c of G such that ab, $(ab)c$, bc, and $a(bc)$ are defined, the associative law holds, that is to say $(ab)c = a(bc)$. (ii) For a specific element e of G called the *identity*, the products ae and ea exist for all a in G and $ae = ea = a$. (iii) Suppose ab is defined. Then there exist open neighborhoods U and V of a and b respectively, such that $a'b'$ is defined for all $a' \in U$, $b' \in V$. Moreover, the operation $a'b'$ is continuous on $U \times V$. (iv), Suppose the element a of G possesses a right inverse, $b = a^{-1}$, $ab = e$. Then there exists an open neighborhood U of a such that every $a' \in U$ possesses a right inverse $b' = (a')^{-1}$, $a'b' = e$. Moreover, the operation $(a')^{-1}$ is continuous on U.

Passing to G^*, we consider the monad of e, $\mu(e)$. Since $ee = e$, there exists a standard open neighborhood U of e such that ab exists and is a continuous function of a and b in U. It follows that the product ab exists for all a and b in $\mu(e)$ and, similarly as in 8.1.1 that for such a and b, $ab \in \mu(e)$. And since e possesses a right inverse, i.e., itself, there exists an open neighborhood W of e such that a^{-1} exists and is a continuous function for all $a \in W$. We conclude that a^{-1} exists for all $a \in \mu(e)$ and, moreover, that $a^{-1} \in \mu(e)$ for such a. Since the associative law applies to the multiplication of elements of $\mu(e)$, we conclude that $\mu(e)$ still constitutes a group, J_1. Also, since for any $a \in J_1$, a^n exists for all finite integers n we conclude that a^n exists in $^*G$ also for sufficiently small infinite (positive or negative) integers n. The elements of J_1 are called *infinitesimal*.

8.2 Metric groups. We shall say that G is a metric group if e is a topological group such that some open neighborhood U of the identity is a metric space whose metric is compatible with the specified topology of U. Thus, we may talk of the distance $\varrho(a,b)$ between any two points a,b in U. In terms of this distance, $\mu(e) = J_1$ consists of the points a of U such that $\varrho(a,e)$ is infinitesimal.

Let η be any positive infinitesimal number. Then we define $J_0(\eta)$ as the set of elements $a \in J_1$ such that $\varrho(a,e)/\eta$ is finite; and we define $J_1(\eta)$ as the set of elements $a \in J_1$ such that $\varrho(a,e)/\eta$ is infinitesimal.

For any positive infinitesimal η, we shall consider the following conditions:

8.2.1 $J_0(\eta)$ is a group. Thus, if $a \in J_0(\eta)$ and $b \in J_0(\eta)$ then $ab \in J_0(\eta)$ and $a^{-1} \in J_0(\eta)$.

8.2.2 $J_1(\eta)$ is a group.

8.2.3   If $a \in J_0(\eta)$, $b \in J_0(\eta)$ then $aba^{-1}b^{-1} \in J_1(\eta)$.

8.2.4 $J_1(\eta)$ is a normal subgroup of $J_0(\eta)$.
It follows from 8.2.1 – 8.2.3 that

8.2.5   The commutator subgroup of $J_0(\eta)$ is contained in $J_1(\eta)$.

Let G be a topological group, B a normed linear space over the real or complex numbers and ϕ a topological mapping from an open neighborhood of the identity e in G into an open neighborhood V of the origin 0 in B, such that $\phi(e) = 0$. Then the group operation in G defines a partial function from $V \times V$ into V by the following definition

8.2.6 $$f(x,y) = \phi(\phi^{-1}(x)\,\phi^{-1}(y)).$$

This definition implies $f(0,0) = 0$. Moreover, if $f(a,b)$ is defined for points a and b in B then there exist open neighborhoods V_a and V_b of a and b respectively such that $f(a',b')$ is defined for all $a' \in V_a$, $b' \in V_b$. In particular, there exists an open neighborhood W of 0 in B such that $f(a,b)$ is defined for all a and b in W. For such a set W, we now introduce the following condition.

8.2.7   For any two points a and b in W, there exist open neighborhoods $W_a \subset W$, $W_b \subset W$ of a and b respectively, such that for all $x \in W_a$ and $y \in W_b$,

8.2.8 $$f(x,y) = f(a,b) + \kappa(x-a) + \lambda(y-b) + \psi(x,y)$$

where κ and λ are scalars that depend only on a and b and

8.2.9 $$\lim_{\substack{x \to a \\ y \to b}} \psi(x,y)/(\|x-a\| + \|y-b\|) = 0.$$

8.2.10   THEOREM.   In order that condition 8.2.7 be satisfied it is necessary and sufficient that for all standard points a and b in $^*W$ there exist standard complex numbers κ and λ such that for all $x \in \mu(a)$, $y \in \mu(b)$,

8.2.11 $$f(x,y) = f(a,b) + \kappa(x-a) + \lambda(y-b) + z,$$

where $|z| \leq r\|x-a\| + s\|y-b\|$ for infinitesimal r and s which may depend also on x and y.

PROOF. Suppose that the condition 8.2.11 is satisfied, and let ε be any positive standard number. Then

$$\|f(x,y) - f(a,b) - \kappa(x-a) - \lambda(y-b)\| = \|z\| \leq r\|x-a\| + s\|y-b\|$$

and hence

8.2.12 $$\|f(x,y) - f(a,b) - \kappa(x-a) - \lambda(y-b)\| < \varepsilon(\|x-a\| + \|y-b\|)$$

for all $x \in \mu(a)$, $y \in \mu(b)$ and hence, in particular, for all x,y such that $\|x-a\| < \delta$, $\|y-b\| < \delta$ where δ is any positive infinitesimal number. Since such a δ exists in $*M$, we conclude by transfer to M that there exists a *standard* positive δ such that 8.2.12 is satisfied for all $\|x-a\| < \delta$, $\|y-b\| < \delta$. This shows that the condition of the theorem is sufficient.

The condition is also necessary. For suppose 8.2.7 is satisfied and let $x \in \mu(a)$, $y \in \mu(b)$. Then 8.2.9 shows that $\|\psi(x,y)\|/(\|x-a\| + \|y-b\|) = r$ is infinitesimal. Hence,

$$\|f(x,y) - f(a,b) - \kappa(x-a) - \lambda(y-b)\| = \|\psi(x,y)\| = r\|x-a\| + r\|y-b\|.$$

This proves our assertion.

Supposing from now on that 8.2.7 is satisfied, put $a = x = 0$ in 8.2.11. This yields

$$f(0,y) = f(0,b) + \lambda(y-b) + z$$

for all $y \in \mu(b)$ where $\|z\| \leq s\|y-b\|$ for some infinitesimal s. But $f(0,y) = \phi(e\phi^{-1}(y)) = \phi(\phi^{-1}(y)) = y$ and in particular, $f(0,b) = b$. Hence

$$y = b + \lambda(y-b) + z,$$

where λ is a standard number which is independent of y. Then

$$(1-\lambda)(y-b) = z$$

for any $y \in \mu(b)$. Taking an arbitrary $y \in \mu(b)$ other than b and bearing in mind that $\|z\| \leq s\|y-b\|$ for some infinitesimal s (which may depend on y), we obtain

$$\|(1-\lambda)(y-b)\| = |1-\lambda|\,\|y-b\| \leq s\,\|y-b\|$$

and so $|1-\lambda| \leq s$. But s is infinitesimal and $1-\lambda$ is standard, hence $1-\lambda = 0$, $\lambda = 1$. The same method shows that $\kappa = 1$ and so 8.2.8 yields, for $a = b = 0$.

8.2.13 $$f(x,y) = x + y + \psi(x,y)$$

where $\psi(x,y)$ satisfies 8.2.9 (for $a = b = 0$), while 8.2.11 yields

8.2.14 $$f(x,y) = x + y + z$$

for $x \simeq 0$, $y \simeq 0$, where $\|z\| \leq r\|x\| + s\|y\|$ for some infinitesimal r and s. Evidently, this condition is satisfied if and only if there exists an infinitesimal r such that $\|z\| = r(\|x\| + \|y\|)$.

The metric of the linear space B induces a metric in the open subset U of G by means of the definition $\varrho(a,b) = \|\phi(a) - \phi(b)\|$. Since the mapping ϕ is topological, the given topology of U coincides with the topology induced in U by the metric ϱ. Thus, G is a metric group in the sense defined previously.

We are now going to show that 8.2.14 implies the conditions 8.2.1 – 8.2.4 and hence also 8.2.5, for arbitrary positive infinitesimal η.

The condition states that for all $a \in \mu(e)$, $b \in \mu(e)$, and $a' = \phi(a)$, $b' = \phi(b)$,

8.2.15 $$\varrho(ab,e) = \|\phi(ab)\| = \|f(a',b')\| = \|a' + b' + \psi(a',b')\|$$
$$= \|\phi(a) + \phi(b) + \psi(\phi(a), \phi(b))\|,$$

where $\|\psi(\phi(a), \phi(b))\| = r(\|\phi(a)\| + \|\phi(b)\|)$ for some positive infinitesimal r. Hence

$$\varrho(ab,e) \leq \|\phi(a)\| + \|\phi(b)\| + r(\|\phi(a)\| + \|\phi(b)\|)$$
$$= \varrho(a,e) + \varrho(b,e) + r(\varrho(a,e) + \varrho(b,e)).$$

Hence, if

$$\varrho(a,e) \leq m\eta, \qquad \varrho(b,e) \leq m\eta$$

then

$$\varrho(ab,e) \leq 2m(1+r)\eta.$$

We conclude that both $J_0(\eta)$ and $J_1(\eta)$ are closed under multiplication.

Now take $b = a^{-1}$ in 8.2.15 so that b, like a, belongs to the monad of e. This yields

$$0 = \|\phi(a) + \phi(a^{-1}) + \psi(\phi(a), \phi(a^{-1}))\|$$

and hence

8.2.16 $$\phi(a) + \phi(a^{-1}) = -\psi(\phi(a), \phi(a^{-1}))$$

and

8.2.17 $$\|\phi(a) + \phi(a^{-1})\| = \|\psi(\phi(a), \phi(a^{-1}))\| = r(\|\phi(a)\| + \|\phi(a^{-1})\|)$$

where r is infinitesimal.

Dividing 8.2.17 by η we obtain

8.2.18 $$\left\|\frac{1}{\eta}\phi(a) + \frac{1}{\eta}\phi(a^{-1})\right\| = r\left(\left\|\frac{1}{\eta}\phi(a)\right\| + \left\|\frac{1}{\eta}\phi(a^{-1})\right\|\right).$$

8.2.18 in turn implies

$$\left\|\frac{1}{\eta}\phi(a^{-1})\right\| - \left\|\frac{1}{\eta}\phi(a)\right\| \le r\left(\left\|\frac{1}{\eta}\phi(a)\right\| + \left\|\frac{1}{\eta}\phi(a^{-1})\right\|\right)$$

and hence,

$$(1-r)\left\|\frac{1}{\eta}\phi(a^{-1})\right\| \le (1+r)\left\|\frac{1}{\eta}\phi(a)\right\|$$

and so

8.2.19 $$\left\|\frac{1}{\eta}\phi(a^{-1})\right\| \le \frac{1+r}{1-r}\left\|\frac{1}{\eta}\phi(a)\right\| < 2\left\|\frac{1}{\eta}\phi(a)\right\|.$$

8.2.19 shows that if $\|(1/\eta)\phi(a)\|$ is finite, then $\|(1/\eta)\phi(a^{-1})\|$ is finite, and if $\|(1/\eta)\phi(a)\|$ is infinitesimal, then $\|(1/\eta)\phi(a^{-1})\|$ is infinitesimal. We conclude that if $a \in J_0(\eta)$ then $a^{-1} \in J_0(\eta)$; and if $a \in J_1(\eta)$ then $a^{-1} \in J_1(\eta)$. Accordingly, we have shown that both $J_0(\eta)$ and $J_1(\eta)$ are groups, as required by conditions 8.2.1 and 8.2.2.

Next we prove 8.2.3. We have, for $a \in \mu(e)$, $b \in \mu(e)$,

$$\phi(ab) = \phi(a) + \phi(b) + \psi(\phi(a), \phi(b)),$$
$$\phi(a^{-1}b^{-1}) = \phi(a^{-1}) + \phi(b^{-1}) + \psi(\phi(a^{-1}), \phi(b^{-1})),$$

where
$$\|\psi(\phi(a), \phi(b))\| = r(\|\phi(a)\| + \|\phi(b)\|)$$
and
$$\|\psi(\phi(a^{-1}), \phi(b^{-1}))\| = s(\|\phi(a^{-1})\| + \|\phi(b^{-1})\|)$$

for certain non-negative infinitesimal r and s. Furthermore,

$$\phi(aba^{-1}b^{-1}) = \phi(ab) + \phi(a^{-1}b^{-1}) + \psi(\phi(ab), \phi(a^{-1}b^{-1}))$$

where
$$\|\psi(\phi(ab), \phi(a^{-1}b^{-1}))\| = t(\|\phi(ab)\| + \|\phi(a^{-1}b^{-1})\|)$$

for some non-negative infinitesimal t. Now,

$$\phi(aba^{-1}b^{-1}) = \phi(a) + \phi(a^{-1}) + \phi(b) + \phi(b^{-1}) + \psi(\phi(a), \phi(b))$$
$$+ \psi(\phi(a^{-1}), \phi(b^{-1})) + \psi(\phi(ab), \phi(a^{-1}b^{-1}))$$

or, making use of 8.2.16,

$$\phi(aba^{-1}b^{-1}) = -\psi(\phi(a), \phi(a^{-1})) - \psi(\phi(b), \phi(b^{-1}))$$
$$+ \psi(\phi(a), \phi(b)) + \psi(\phi(a^{-1}), \phi(b^{-1})) + \psi(\phi(ab), \phi(a^{-1}b^{-1})).$$

Accordingly,

8.2.20
$$\frac{1}{\eta}\varrho(aba^{-1}b^{-1}, e) = \left\|\frac{1}{\eta}\phi(aba^{-1}b^{-1})\right\| \leq \left\|\frac{1}{\eta}\psi(\phi(a), \phi(a^{-1}))\right\|$$
$$+ \left\|\frac{1}{\eta}\psi(\phi(b), \phi(b^{-1}))\right\| + \left\|\frac{1}{\eta}\psi(\phi(a), \phi(b))\right\|$$
$$+ \left\|\frac{1}{\eta}\psi(\phi(a^{-1}), \phi(b^{-1}))\right\| + \left\|\frac{1}{\eta}\psi(\phi(ab), \phi(a^{-1}b^{-1}))\right\|.$$

Now suppose that a and b belong to $J_0(\eta)$. Then

$$\|(1/\eta)\phi(a)\|, \ \|(1/\eta)\phi(b)\|, \ \|(1/\eta)\phi(a^{-1})\|,$$
$$\|(1/\eta)\phi(b^{-1})\|, \ \|(1/\eta)\phi(ab)\|, \ \|(1/\eta)\phi(a^{-1}b^{-1})\|$$

are all finite. We conclude that the moduli on the right hand side of 8.2.20,

$\|(1/\eta)\psi(\phi(a),\phi(a^{-1}))\|$, $\|(1/\eta)\psi(\phi(b),\phi(b^{-1}))\|$, and so on, are all infinitesimal. This shows that $(1/\eta)\phi(aba^{-1}b^{-1})$ is infinitesimal and proves 8.2.3.

Finally, we prove 8.2.4. For $a\in\mu(e)$, $b\in\mu(e)$,

$$\phi(aba^{-1})=\phi(ab)+\phi(a^{-1})+\psi(\phi(ab),\phi(a^{-1}))$$
$$=\phi(a)+\phi(b)+\phi(a^{-1})+\psi(\phi(a),\phi(b))+\psi(\phi(ab),\phi(a^{-1})).$$

Making use of 8.2.16, we now obtain

$$\phi(aba^{-1})=\phi(b)+\psi(\phi(a),\phi(b))+\psi(\phi(ab),\phi(a^{-1}))$$
$$-\psi(\phi(a),\phi(a^{-1})).$$

Hence

8.2.21 $\varrho(aba^{-1},e)=\|\phi(aba^{-1})\|\leq\|\phi(b)\|+\|\psi(\phi(a),\phi(b))\|$
$$+\|\psi(\phi(ab),\phi(a^{-1}))\|+\|\psi(\phi(a),\phi(a^{-1}))\|.$$

Dividing 8.2.21 by η, we find that the terms on the right hand side of the resulting inequality, i.e.,

$$(1/\eta)\|\phi(b)\|,(1/\eta)\|\psi(\phi(a),\phi(b))\|,(1/\eta)\|\psi(\phi(ab),\phi(a^{-1}))\|,$$
$$(1/\eta)\|\psi(\phi(a),\phi(a^{-1}))\|$$

are infinitesimal for $a\in J_0(\eta)$, $b\in J_1(\eta)$. This shows that $\varrho(aba^{-1},e)/\eta$ is infinitesimal and proves 8.2.4.

Since $J_1(\eta)$ is a normal subgroup of $J_0(\eta)$ the quotient group $J_0(\eta)/J_1(\eta)$ exists, and since $J_1(\eta)$ contains the commutator subgroup of $J_0(\eta)$, $J_0(\eta)/J_1(\eta)$ must be Abelian. A more precise description of $J_0(\eta)/J_1(\eta)$ is obtained in the following way.

Let B_0 be the additive group of finite points of $*B$ and let $B_1=\mu(0)$, regarded as an additive group. (B_0 has been called previously the principal galaxy of $*B$).

8.2.22   THEOREM. $J_0(\eta)/J_1(\eta)$ is isomorphic to the additive group B_0/B_1.

PROOF.   Consider the following mapping, $x=\chi(a)$, from $J_0(\eta)$ into B.

$$\chi:\qquad a\rightarrow(1/\eta)\phi(a)=x.$$

This mapping is one-to-one and onto, for if x is any point in B_0 then the corresponding point $a = \chi^{-1}(x)$ in $J_0(\eta)$ is given by $a = \phi^{-1}(\eta x)$.

We have $\|\chi(a)\| = (1/\eta)\|\phi(a)\| = (1/\eta)\varrho(a, e)$. It follows that $\chi(a)$ belongs to B_1 if and only if a belongs to $J_1(\eta)$, i.e., χ maps $J_1(\eta)$ on B_1.

Let $a \in J_0(\eta)$, $b \in J_0(\eta)$ then

$$\phi(ab) = \phi(a) + \phi(b) + \psi(\phi(a), \phi(b)).$$

Dividing by η and taking into account that

$$\|\psi(\phi(a), \phi(b))\| = r(\|\phi(a)\| + \|\phi(b)\|),$$

where r is infinitesimal, we obtain

8.2.23 $$\chi(ab) \simeq \chi(a) + \chi(b).$$

Similarly,

$$\begin{aligned}\phi(ab^{-1}) &= \phi(a) + \phi(b^{-1}) + \psi(\phi(a), \phi(b^{-1})) \\ &= \phi(a) - \phi(b) + \psi(\phi(a), \phi(b^{-1})) - \psi(\phi(b), \phi(b^{-1})),\end{aligned}$$

where we have taken into account 8.2.16. Dividing again by η and bearing in mind that

$$\frac{1}{\eta}\psi(\phi(a), \phi(b^{-1})) \simeq 0,$$

and

$$\frac{1}{\eta}\psi(\phi(b), \phi(b^{-1})) \simeq 0,$$

we obtain

8.2.24 $$\chi(ab^{-1}) \simeq \chi(a) - \chi(b).$$

This shows that $\chi(a) - \chi(b)$ belongs to B_1 if and only if $\chi(ab^{-1})$ is infinitesimal, i.e., if and only if ab^{-1} belongs to $J_1(\eta)$. We conclude that the cosets in $J_0(\eta)$ with respect to $J_1(\eta)$ are mapped on the cosets in B_0 with respect to B_1, and hence that χ induces a mapping X from $J_0(\eta)/J_1(\eta)$ onto B_0/B_1. More particularly, $\xi = X(\alpha)$ for $\alpha \in J_0(\eta)/J_1(\eta)$ if $\chi(a) \in \xi$ for all $a \in \alpha$, ξ being an element of B_1/B_0. 8.2.23 now shows that $X(\alpha\beta) = X(\alpha) + X(\beta)$, while 8.2.24 implies that $X(\beta^{-1}) = -X(\beta)$ for any α and β in $J_0(\eta)/J_1(\eta)$. This proves 8.2.22.

If B is finite-dimensional then the points of B_0 are near-standard and so

B_0/B_1 is isomorphic to B. Thus, we have

8.2.25 COROLLARY. If B is finite-dimensional then $J_0(\eta)/J_1(\eta)$ is isomorphic to B.

Since G is locally connected it follows from 8.1.12 that $J_1^{(\infty)} = {}^*G_e$, where G_e is the component of the identity in G. For the case now under consideration, we may supplement this result as follows.

8.2.26 THEOREM. For any positive infinitesimal η,

$$(J_1(\eta))^{(\infty)} = (J_0(\eta))^{(\infty)} = {}^*G_e.$$

PROOF. Evidently, $(J_1(\eta))^{(\infty)} \subset (J_0(\eta))^{(\infty)} \subset J_1^{(\infty)}$. Thus, it only remains for us to prove that $(J_1\eta))^{(\infty)} \supset {}^*G_e$. Consider the subset U of ${}^*G$ which is defined by

$$U = \{a \,|\, \varrho(a,e) < \eta^2\}.$$

U is an open internal set. Also, $U \subset J_1(\eta)$ since $(1/\eta)\varrho(a,e)$ is infinitesimal for all $a \in U$. Transferring 8.1.11 to ${}^*M$, we see that $\bigcup_{n=1}^{\infty} U^n = {}^*G_e$. But $\bigcup_{n=1}^{\infty} U^n \subset (J_1(\eta))^{(\infty)}$. Hence $(J_1(\eta))^{(\infty)} \supset {}^*G_e$, proving the theorem.

8.3 One-parametric subgroups. Let G be a topological group as considered in section 8.1 above and let H be any (internal or external) subgroup of ${}^*G$. Let H' be the set of all near-standard elements of H.

8.3.1 THEOREM. H' is a group.

PROOF. H' is not empty since $e \in H'$. Let $a \in H'$, $b \in H'$. Then there exist standard points c and d in ${}^*G$ (i.e., $c \in G$, $d \in G$) such that $a \in \mu(c)$, $b \in \mu(d)$. But then $ab \in \mu(cd)$, $a^{-1} \in \mu(c^{-1})$, so that ab and a^{-1} and are near-standard. This proves our assertion.

In agreement with our earlier notation, let ${}^0H$ be the set of standard points whose monads have a non-empty intersection with H (or, which is the same, with H').

8.3.2 THEOREM. ${}^0H$ is a subgroup of G.

PROOF.   Evidently, ^0H is a subset of G. Let $a \in {}^0H$, $b \in {}^0H$, then $c \in \mu(a)$, $d \in \mu(b)$ for certain elements c and d of H. Hence $cd \in \mu(ab)$, $c^{-1} \in \mu(a^{-1})$. This shows that the monads of ab and a^{-1} have non-empty intersections with H, and so $ab \in {}^0H$, $a^{-1} \in {}^0H$, ^0H is a group.

If H is internal and G satisfies the first axiom of countability then ^0H is closed, by 4.3.12. This will be the case, in particular, if G is a metric group, for in this case, the sequence $\{U_n\}$ provides a base for the system of neighborhoods of the identity, where $U_n = \{b \mid \varrho(b,e) < (1/n)\}$, and the sequence $\{U_n a\}$ provides a base for the system of neighborhoods of any other $a \in G$.

Now let a be an arbitrary element of $*G$ and let $H = \{a^n\}$, where n ranges over all integers, finite or infinite. Then H is an internal set and a subgroup of $*G$ and H' is a subgroup of $*G$ while ^0H is a subgroup of G. Moreover, if G satisfies the first axiom of countability then ^0H is closed.

We shall suppose from now on that G is a Hausdorff space. This will be the case, in particular, if G is metric.

8.3.3   THEOREM. ^0H is Abelian.

PROOF.   Let $b \in {}^0H$, $c \in {}^0H$, then there exists integers m and n, finite or infinite, such that $a^m \in \mu(b)$, $a^n \in \mu(c)$. Hence $a^{m+n} = a^m a^n \in \mu(bc)$ and $a^{n+m} = a^n a^m \in \mu(cb)$. But $a^{m+n} = a^{n+m}$ and so $\mu(bc) \cap \mu(cb) \neq \emptyset$. This proves that $bc = cb$, since G is supposed to be a Hausdorff space.

Suppose now that a is infinitesimal. Then a^n is infinitesimal for all finite n. Let $H = \{a^n\}$ as before. We define a mapping from H' onto ^0H, $\Phi : b \to c$ by the condition $b \in \mu(c)$. It is easy to verify that Φ is a homomorphism. Let K' be the kernel of this homomorphism. Then $K' = H' \cap \mu(e)$. Also, $a^n \in K'$ if and only if $a^{-n} \in K'$. Let P be the set of natural numbers n such that $a^{\pm n} \in K'$. Then $P \supset N$. If $P \supset *N$, ^0H reduces to the identity.

8.3.4   THEOREM.   Suppose P is a segment (i.e., $n \in P$ and $m < n$ implies $m \in P$). Then ^0H is torsion-free.

PROOF.   Suppose P is a segment. Let $b \neq e$ be an element of ^0H. Then $a^n \in \mu(b)$ for some n such that $|n| > m$ for all $m \in P$. Let k be any finite positive integer. Then $a^{nk} \in \mu(b^k)$. But $|nk| \geq |n| > m$ for all $m \in P$. Hence, $a^{nk} \notin \mu(e)$, $b^k \neq e$. This proves the theorem.

It seems intuitive that the powers of an infinitesimal element in some sense make up a one-parametric group. We shall make this idea precise under the following conditions.

Suppose that an open neighborhood U of the identity e in G is mapped by a topological mapping ϕ on an open neighborhood V of the origin 0 in B such that $\phi(e)=0$, as in section 8.2, and let $f(x,y)$ be defined again by 8.2.6, $f(x,y)=\phi(\phi^{-1}(x)\phi^{-1}(y))$. Suppose more particularly that B is the n-dimensional real Euclidean space, $B=R^n$, $n\geq 1$, so that $x=(x^1,x^2,...,x^n)$, $y=(y^1,y^2,...,y^n)$, where the norm is defined by $\|x\|=|x^1|+|x^2|+...+|x^n|$. Then the function $f(x,y)$ is given by a set of n numerical functions of $2n$ real numbers, $f^i(x^1,x^2,...,x^n,y^1,y^2,...,y^n)$. Let α be a positive real number such that the closed cube C which is given by $|x^j|\leq\alpha$, $j=1,2,...,n$ is contained in V. (Observe that we are discussing certain standard structures so that in the present context we take it for granted that α is standard.) We shall suppose that the following condition is satisfied.

8.3.5 The functions $f^i(x^1,...,x^n,y^1,...,y^n)$ possess first and second derivatives with respect to the y^j, and these derivatives are continuous functions of their $2n$ variables for $|x_j|\leq\alpha$, $|y_j|\leq\alpha$, $j=1,...,n$.

8.3.5 is satisfied by any n-parametric Lie group.

The condition implies that there exists a (standard) positive β such that

$$\left|\frac{\partial^2 f^i}{\partial y^j\partial y^k}\right|\leq\beta, \qquad i,j,k=1,...,n,$$

for all x and y in C, i.e., for

$$|x_j|\leq\alpha, |y_j|\leq\alpha, \qquad j=1,...,n.$$

By Taylor's theorem we have, for x and y in C,

8.3.6 $$f^i(x^1,...,x^n,y^1,...,y^n)=f^i(x^1,...,x^n,0,...,0)+\sum_{j=1}^{n}y^j\left(\frac{\partial f^i}{\partial y^j}\right)_{(x,0)}$$

$$+\frac{1}{2}\sum_{j=1}^{n}(y^j)^2\left(\frac{\partial^2 f^i}{(\partial y^j)^2}\right)_{(x,\theta y)}+\sum_{j\neq k}y^jy^k\left(\frac{\partial^2 f^i}{\partial y^j\partial y^k}\right)_{(x,\theta y)}$$

where the subscript $(x,0)$ indicates that the derivatives in question are evaluated at $(x^1,...,x^n,0,0,...,0)$, while the subscripts $(x,\theta y)$ stand for $(x^1,...,x^n,\theta y^1,...,\theta y^n)$ where $0\leq\theta\leq 1$ and θ depends on x and y.

Putting $y^1 = y^2 = \ldots = y^n = 0$ in 8.3.6, we find that $f^i(x^1, \ldots, x^n, 0, \ldots, 0) = x^i$. Also, $f^i(0, \ldots, 0, y^1, \ldots, y^n) = y^i$. This yields

$$y^i = \sum_{j=1}^{n} y^j \left(\frac{\partial f^i}{\partial y^j}\right)_{(0,0)} + \tfrac{1}{2} \sum_{j=1}^{n} (y^j)^2 \left(\frac{\partial^2 f^i}{(\partial y^j)^2}\right)_{(0,\theta y)}$$
$$+ \sum_{j \neq k}^{n} y^j y^k \left(\frac{\partial^2 f^i}{\partial y^j \partial y^k}\right)_{(0,\theta y)}, \qquad i = 1, \ldots, n.$$

Taking y^i infinitesimal, but different from zero, in these equations, while $y^j = 0$ for $j \neq i$, in turn for $i = 1, \ldots, n$, (or else, by a standard procedure) we conclude without difficulty that

8.3.7 $$\left(\frac{\partial f^i}{\partial y^j}\right)_{(0,0)} = \delta^i_j = \begin{cases} 0, & i \neq j \\ 1, & i = j \end{cases}, \quad i, j = 1, \ldots, n.$$

From these relations we derive the following basic auxiliary result.

8.3.8 THEOREM. Suppose condition 8.3.5 is satisfied. Then for all x and y in $\mu(0)$,

$$f(x, y) = x + y + z,$$

where $\|z\| = r\|y\|$ for some non-negative infinitesimal r which depends on x and y.

Thus, 8.3.5 implies condition 8.2.14, and its consequences, e.g., 8.2.25, whose assumptions are now applicable.

PROOF. We have, for all x and y in the monad of 0,

8.3.9 $$f^i(x^1, \ldots, x^n, y^1, \ldots, y^n) = x^i + y^i + \sum_{j=1}^{n} y^j \left(\left(\frac{\partial f^i}{\partial y^j}\right)_{(x,0)} - \delta^i_j\right)$$
$$+ \tfrac{1}{2} \sum_{j=1}^{n} (y^j)^2 \left(\frac{\partial^2 f^i}{(\partial y^j)^2}\right)_{(x,\theta y)} + \sum_{j \neq k}^{n} y^j y^k \left(\frac{\partial^2 f^i}{\partial y^j \partial y^k}\right)_{(x,\theta y)}, \qquad i = 1, \ldots, n.$$

Now,

$$\left| \tfrac{1}{2} \sum_{j=1}^{n} (y^j)^2 \left(\frac{\partial^2 f^i}{(\partial y^j)^2}\right)_{(x,\theta y)} + \sum_{j \neq k}^{n} y^j y^k \left(\frac{\partial^2 f^i}{\partial y^j \partial y^k}\right)_{(x,\theta y)} \right| \leq$$
$$\beta \left(\tfrac{1}{2} \sum_{j=1}^{n} |y^j|^2 + \sum_{j \neq k}^{n} |y^j| |y^k| \right) = \beta \|y\|^2.$$

At the same time,

$$\left| \sum_{i=1}^{n} \sum_{j=1}^{n} y^j \left(\left(\frac{\partial f^i}{\partial y^j} \right)_{(x,0)} - \delta^i_j \right) \right| \le n \max_{i,j} \left| \left(\frac{\partial f^i}{\partial y^j} \right)_{(x,0)} - \delta^i_j \right| \sum_{j=1}^{n} |y^j| = s \|y\| ,$$

where s is infinitesimal, since the $\partial f^i / \partial y^j$ are continuous functions of x. Hence,

$$\|z\| = \|f(x,y) - x - y\| = \sum_{i=1}^{n} |f^i(x^1,...,x^n, y^1,...,y^n) - x^i - y^i|$$
$$\le (s + n\beta \|y\|) \|y\| .$$

Since $s + n\beta \|y\|$ is infinitesimal, this proves 8.3.8.

Within this framework, we return to the consideration of the subgroup $H = \{a^m\}$ of $*G$, a infinitesimal, $a \ne e$, and of the associated groups H' and ^0H. Putting $x = \phi(a^m)$ for any m such that a^m is infinitesimal and hence $x \simeq 0$, and putting $y = \phi(a)$, we obtain from 8.3.8,

$$\phi(a^{m+1}) = f(\phi(a^m), \phi(a)) = \phi(a^m) + \phi(a) + z_m ,$$

where $\|z_m\| = r_m \|\phi(a)\|$, r_m non-negative and infinitesimal. Hence,

8.3.10 $$\phi(a^{m+1}) - \phi(a^m) - \phi(a) = z_m .$$

Now let k be any positive integer, finite or infinite, such that $a^m \simeq e$ for $m = 0, 1, ..., k-1$. Adding the resulting equations 8.3.10 for these values of n and bearing in mind that $\phi(a^0) = \phi(e) = 0$, we obtain

$$\phi(a^k) - k\phi(a) = \sum_{m=0}^{k-1} z_m = w_k ,$$

where

$$\|w_k\| = \left\| \sum_{m=0}^{k-1} z_m \right\| \le \sum_{m=0}^{k-1} \|z_m\| = \left(\sum_{m=0}^{k-1} r_m \right) \|\phi(a)\| .$$

But the set $\{r_0, r_1, ..., r_{k-1}\}$ is Q-finite and so it attains its maximum for some j, $0 \le j < m-1$, which maximum is therefore infinitesimal. We conclude that

$$\frac{\|w_k\|}{\|k\phi(a)\|} = \frac{1}{k} \sum_{m=0}^{k-1} r_m$$

also is infinitesimal, in other words,

8.3.11
$$\frac{\|\phi(a^k) - k\phi(a)\|}{\|k\phi(a)\|} \simeq 0.$$

Now let k be a negative integer such that $a^{-1}, a^{-2}, \ldots, a^k$ are all infinitesimal. By adding the equations 8.3.10 for $m = -1, -2, \ldots k$, we see without difficulty that 8.3.11 holds for such negative k as well.

Let k be any integer different from zero such that 8.3.11 is satisfied, and suppose that $\phi(a^k) \simeq 0$. Then $k\phi(a) \simeq 0$. For if $\|k\phi(a)\|$ is not infinitesimal then $(1/\|k\phi(a)\|) \, \phi(a^k) \simeq 0$, hence

8.3.12
$$\frac{1}{\|k\phi(a)\|} k\phi(a) \simeq \frac{1}{\|k\phi(a)\|} k\phi(a) - \frac{1}{\|k\phi(a)\|} \phi(a^k).$$

But according to 8.3.11 the right hand side of 8.3.12 belongs to $\mu(0)$ while the norm of the left hand side is

$$\left\| \frac{1}{\|k\phi(a)\|} k\phi(a) \right\| = 1.$$

This contradiction proves $k\phi(a) \simeq 0$.

Conversely, if 8.3.11 holds for an integer $k \neq 0$ and $k\phi(a) \simeq 0$ then $\phi(a^k) \simeq 0$. For 8.3.11 shows that $\|\phi(a^k) - k\phi(a)\| \simeq \sigma \|k\phi(a)\|$ for infinitesimal σ, hence $\|\phi(a^k) - k\phi(a)\| \simeq 0$, hence $\|\phi(a^k)\| \simeq 0$.

8.3.13 THEOREM. Condition 8.3.11 holds for all integers $k \neq 0$ such that $k\phi(a) \simeq 0$.

PROOF. Let k be a positive integer which satisfies the assumptions of 8.3.13. Then we have to show that for any standard positive ε,

8.3.14
$$\frac{\|\phi(a^k) - k\phi(a)\|}{\|k\phi(a)\|} < \varepsilon.$$

We know already that 8.3.11, and hence 8.3.14 holds for any finite positive integer k. But the set of k satisfying 8.3.14 is internal so if there is any positive integer for which 8.3.14 does not hold for a given

$\varepsilon > 0$, then there must be a smallest positive integer $k = \lambda$ of this kind. We then have

$$\frac{\|\phi(a^{\lambda-1}) - (\lambda-1)\phi(a)\|}{\|(\lambda-1)\phi(a)\|} < \varepsilon$$

and so

8.3.15 $\|\phi(a^{\lambda-1}) - (\lambda-1)\phi(a)\| < \varepsilon \|(\lambda-1)\phi(a)\|.$

Now if $\|\lambda\phi(a)\|$ were infinitesimal then $\|(\lambda-1)\phi(a)\|$ also would be infinitesimal. In that case, the right hand side of 8.3.15 would be infinitesimal, hence $\|\phi(a^{\lambda-1}) - (\lambda-1)\phi(a)\| \simeq 0$, hence $\|\phi(a^{\lambda-1})\| \simeq 0$. Hence, 8.3.10 would apply, for $m = \lambda - 1$, entailing

$$\|\phi(a^{\lambda}) - \phi(a^{\lambda-1}) - \phi(a)\| = \varrho\|\phi(a)\|,$$

for infinitesimal ϱ, and hence

8.3.16 $\|\phi(a^{\lambda}) - \phi(a^{\lambda-1}) - \phi(a)\| < \varepsilon \|\phi(a)\|.$

But then, taking into account 8.3.15,

$$\|\phi(a^{\lambda}) - \lambda\phi(a)\| = \|\phi(a^{\lambda}) - \phi(a^{\lambda-1}) - \phi(a) + \phi(a^{\lambda-1}) - (\lambda-1)\phi(a)\| \leq$$
$$\|\phi(a^{\lambda}) - \phi(a^{\lambda-1}) - \phi(a)\| + \|\phi(a^{\lambda-1}) - (\lambda-1)\phi(a)\| < \varepsilon\|\phi(a)\| +$$
$$\varepsilon\|(\lambda-1)\phi(a)\| = \varepsilon\|\lambda\phi(a)\|.$$

Hence, $\|\phi(a^{\lambda}) - \lambda\phi(a)\| < \varepsilon\|\lambda\phi(a)\|$ which shows that 8.3.14 holds also for $k = \lambda$, contrary to the definition of λ. This proves our theorem for positive k. A similar argument is available for negative k.

Let ε be a standard positive number smaller than 1. Then 8.3.13 shows that 8.3.14 applies for all positive integers k such that $\|k\phi(a)\| \simeq 0$. Hence, by a familiar argument there exists a positive integer λ such that $\|\lambda\phi(a)\|$ is not infinitesimal, and hence is greater than some positive standard γ and such that 8.3.14 holds for all positive integers $k \leq \lambda$. We observe that the existence of such λ and γ can also be inferred more directly from the argument used in the proof of 8.3.13. We may choose γ so that $\gamma \leq \frac{1}{2}\alpha$, where α was introduced on p. 217.

Let t be any standard number in the interval $0 \leq t \leq \gamma$. Since $\|\phi(a)\|$ is infinitesimal, and $\|\lambda\phi(a)\| = \lambda\|\phi(a)\| > \gamma$, there exists a natural number k,

$0 \le k < \lambda$ such that $^0\|k\phi(a)\| = t$. We now define a mapping $\psi(t)$ from the interval $0 \le t \le \gamma$ into the group G in the following way. For any t in the interval, we choose a natural number k, $0 \le k < \lambda$ such that $^0\|k\phi(a)\| = t$. And we then put $\psi(t) = {}^0(a^k)$. In order to show that this is a good definition we have to verify (i) that $^0(a^k)$ exists, i.e., that a^k is near-standard, and (ii) that $^0(a^k)$ is independent of the particular choice of k subject only to the condition $^0\|k\phi(a)\| = t$.

Evidently, (i) holds for $k = 0$ for in that case $^0(a^0) = {}^0(e) = e$. For $k \ne 0$, we have, by 8.3.14,

$$\|\phi(a^k)\| \le \|\phi(a^k) - k\phi(a)\| + \|k\phi(a)\| \le (\varepsilon + 1)\|k\phi(a)\| \le 2\gamma \le \alpha.$$

Hence, $\|\phi(a^k)\|$ is finite, and $\phi(a^k)$ is a finite point in n-dimensional Euclidean space and accordingly, possesses a standard part $^0(\phi(a^k))$. Also, $\phi(a^k)$ is in the S-interior of the range of the topological mapping ϕ, so

$$^0(\phi(a^k)) = \phi(^0(a^k)) \qquad \text{and} \qquad \psi(t) = \phi^{-1}(^0(\phi(a^k))).$$

As for (ii), let j and h be two natural numbers such that $0 \le j < h < \lambda$, $^0\|j\phi(a)\| = {}^0\|h\phi(a)\| = t$, $0 \le t \le \gamma$. Then $\|(h - j)\phi(a)\| \simeq 0$. Hence, by 8.3.13, $\|\phi(a^{h-j}) - (h - j)\phi(a)\| \simeq 0$, and so $\|\phi(a^{h-j})\| \simeq 0$. This shows that $a^{h-j} \simeq e$, $a^{h-j} \in \mu(e)$, and hence $a^h = a^{h-j} a^j \in \mu(e) \cdot \mu(^0(a^j)) = \mu(^0(a^j))$, and hence $^0(a^h) = {}^0(a^j)$, as asserted.

We observe that for $t \ne 0$ in the interval of definition of $\psi(t)$, we have $\psi(t) \ne e$. For let $^0\|k\phi(a)\| = t$, then 8.3.14 implies

$$\|\phi(a^k)\| = \|k\phi(a) - (k\phi(a) - \phi(a^k))\| \ge \|k\phi(a)\| - \|\phi(a^k) - k\phi(a)\|$$
$$\ge (1 - \varepsilon)\|k\phi(a)\| > \tfrac{1}{2}(1 - \varepsilon)t.$$

This shows that $\|\phi(a^k)\|$ is not infinitesimal and hence, that a^k does not belong to the monad of e, $\psi(t) = {}^0(a^k) \ne e$.

8.3.17 THEOREM.   Let s and t be two non-negative real numbers such that $s + t \le \gamma$. Then

$$\psi(s + t) = \psi(s)\psi(t).$$

PROOF.   Let j and h be natural numbers such that $^0\|j\phi(a)\| = s$, $^0\|h\phi(a)\| = t$, and so $^0\|(j + h)\phi(a)\| = s + t$. Then $\psi(s) = {}^0(a^j)$, $\psi(t) = {}^0(a^h)$, $\psi(s + t) = {}^0(a^{(j+h)}) = {}^0(a^j){}^0(a^h) = \psi(s)\psi(t)$. This proves our assertion.

As an immediate consequence of 8.3.17 we obtain

8.3.18 COROLLARY. Let t be a standard real number, n a finite natural number, such that $0 \le t \le \gamma$ and $nt \le \gamma$. Then $\psi(nt) = (\psi(t))^n$.

Another consequence of 8.3.17 is

8.3.19 THEOREM. Let s and t be standard real numbers such that $0 \le s < t \le \gamma$. Then $\psi(s) \ne \psi(t)$.

PROOF. Put $r = t - s$, then $\psi(t) = \psi(r)\psi(s)$ by 8.3.17. Assuming $\psi(t) = \psi(s)$ we should then obtain $\psi(r) = e$, $0 < r \le \gamma$ and we showed previously that this is impossible. This proves 8.3.19.

We now define a function $\Psi(t)$ whose domain is the set of standard real numbers, and whose range is in G, in the following way.

For $t \ge 0$, we take any finite positive integer n such that $0 \le (1/n)t \le \gamma$, and we then define $\Psi(t) = (\psi(\frac{1}{n}t))^n$. And for $t < 0$, we put $\Psi(t) = (\Psi(-t))^{-1}$.

In order to show that this is a valid definition, we have to verify that for $t \ge 0$, the value of $(\psi(\frac{1}{n}t))^n$ is independent of the particular choice of n. This is evidently correct for $t = 0$.

Let k and n be finite positive integers such that for a given standard positive t, $0 < \frac{1}{n}t \le \gamma$, $0 < \frac{1}{k}t \le \gamma$. Put $s = \frac{1}{kn}t$. By 8.3.18, $\psi(\frac{1}{k}t) = (\psi(s))^n$ and $\psi(\frac{1}{n}t) = (\psi(s))^k$. Hence $(\psi(\frac{1}{k}t))^k = (\psi(s))^{kn} = (\psi(\frac{1}{n}t))^n$, as required. Observe that, for $0 \le t \le \gamma$, $\Psi(t) = \psi(t)$.

8.3.20 THEOREM. $\Psi(t)$ is a homomorphism.

PROOF. We have to show that, for any standard real s and t, $\Psi(s+t) = \Psi(s)\Psi(t)$. Suppose first that s and t are both non-negative, and choose a finite positive integer n such that $0 \le \frac{1}{n}(s+t) \le \gamma$. Then

$$\psi\left(\frac{1}{n}(s+t)\right) = \psi\left(\frac{1}{n}s\right)\psi\left(\frac{1}{n}t\right),$$

by 8.3.17. Hence

$$\psi\left(\frac{1}{n}(s+t)\right)^n = \left(\psi\left(\frac{1}{n}s\right)\psi\left(\frac{1}{n}t\right)\right)^n = \left(\psi\left(\frac{1}{n}s\right)\right)^n\left(\psi\left(\frac{1}{n}t\right)\right)^n,$$

since $\psi(\frac{1}{n}s)$ and $\psi(\frac{1}{n}t)$ commute, their product in either order being equal to $\psi(\frac{1}{n}(s+t))$. This shows that $\Psi(s+t) = \Psi(s)\Psi(t)$, proving the assertion in this case.

If both s and t are negative, then, by definition, $\Psi(s)=(\Psi(-s))^{-1}$, $\Psi(t)=(\Psi(-t))^{-1}$, $\Psi(s+t)=(\Psi(-s-t))^{-1}=(\Psi(-t-s))^{-1}$ and, as we have shown already, $\Psi(-t-s)=\Psi(-t)\Psi(-s)$. This shows that

$$\Psi(s+t)=(\Psi(-t-s))^{-1}=(\Psi(-s))^{-1}(\Psi(-t))^{-1}=\Psi(s)\Psi(t),$$

and proves the assertion for negative s and t. The remaining possibilities can be disposed of in a similar way.

8.3.20 shows that $\Psi(t)$ represents a homomorphism from the standard real numbers onto a commutative subgroup of G, to be denoted by H_a.

H_a is a subgroup of ^0H. For any t in the interval $0 \le t \le \gamma$ it follows in fact immediately from the definition of $\psi(t)$ that $\psi(t)$ belongs to ^0H. And any $\Psi(t)$ is a finite (positive, negative, or zeroth) power of such an element.

Let K_a be the kernel of the homomorphism $\Psi(t)$. We have shown that K_a cannot contain any numbers t in the interval $0 < t \le \gamma$. It follows that K_a either reduces to $\{0\}$ or that it consists of the integral multiples of a standard positive number τ. In the former case, H_a is isomorphic to the additive group of standard real numbers; in the latter case, it is isomorphic to the additive group of standard real numbers modulo τ (or, which is the same up to isomorphism, modulo 1).

8.3.21 THEOREM. The mapping $\Psi(t)$ is continuous.

PROOF. In view of the fact that $\Psi(t)$ is a homomorphism, we only have to prove that for every standard open neighborhood Q of e there exists a standard $\delta > 0$ such that $\Psi(t) \in Q$ for $|t| < \delta$. Since $\Psi(-t)=(\Psi(t))^{-1}$, we may confine ourselves to positive t (for the general case may be obtained from this by considering $Q \cap Q^{-1}$). At the same time we may restrict our attention to sets Q which are contained in the domain U of the mapping ϕ into B, and on this assumption, we only have to verify that for any standard $\eta > 0$, $\|\phi(\Psi(t))\| < \eta$ for $0 < t < \delta$. Supposing (as we may) that $\delta \le \gamma$, we see that this is equivalent to showing that $\|\phi(\psi(t))\| < \eta$ for $0 < t < \delta$. But for t in the range under consideration, $\psi(t) = {}^0(a^k)$ for a positive integer k such that $t = {}^0\|k\phi(a)\|$ and at the same time 8.3.14 holds for the selected ε, $0 < \varepsilon < 1$. Then 8.3.14 implies

$$\|\phi(a^k)\| \le \|\phi(a^k)-k\phi(a)\| + \|k\phi(a)\| < (1+\varepsilon)\|k\phi(a)\| < 2\delta.$$

Hence,

$$\|\phi(\psi(t))\| = \|\phi(^0(a^k))\| = ^0\|\phi(a^k)\| < 3\delta,$$

so that $\|\phi(\psi(t))\| < \eta$ is satisfied by putting $\delta = \frac{1}{3}\eta$. This proves the theorem.

8.3.21 shows that the group H_a is connected. It follows that H_a is a subgroup of the component of the identity G_e of G.

By way of example, we may consider the group G of non-singular square matrices of order $m \geq 1$, $A = (a_k^j)$, $j,k = 1,...,m$, with real coefficients. The corresponding linear space B is m^2-dimensional. Denoting the coordinates of a point of B by $x^{(jk)}$, $j,k = 1,...,m$, we see that an appropriate mapping ϕ from a neighborhood of the identity $I = (\delta_k^j)$ in G, into B, is given by $x^{(jk)} = a_k^j - \delta_k^j$. Let $A \in G$ and let η be positive infinitesimal then the matrix $I + A\eta$ is an infinitesimal element of the group G. If we put $\alpha = \sum_{j,k=1}^m |a_k^j|$ then according to our previous definitions, $\|\phi(I+A\eta)\| = \alpha\eta$. Then if $t = ^0(k\|\phi(I+A\eta)\|) = \alpha^0(k\eta)$ where k is a (finite or infinite) natural number, we have for sufficiently small standard positive t, $\psi(t) = ^0((I+A\eta)^k)$.

8.4 The Lie algebra of a group. In the classical theory of Lie groups, infinitesimal elements cannot occur directly. They are replaced, or represented, by certain linear differential operators, the so-called *infinitesimal transformations* of G. Taken as a linear vector space, and with the appropriate definition of the Lie product, the infinitesimal transformations of G constitute the Lie algebra of the group. We are going to show how to correlate the infinitesimal transformations of G with the infinitesimal group element of the present theory. Observe that there is nothing infinitesimal about an infinitesimal transformation, in the original sense of the word 'infinitesimal' (which is the sense in which this word is used in the present book). Rather, the term 'infinitesimal transformation' was chosen precisely for the intuitive reason which we are about to explicate.

We shall suppose that condition 8.3.5 is satisfied. Let $g(x) = g(x^1,...,x^n)$ be a standard function which is defined on the cube C, i.e., for $|x_i| \leq \alpha$, $i = 1,...,n$. Assume that $g(x^1,...,x^n)$ has continuous first and second derivatives on C.

Let b be any infinitesimal element of $^*G$, $y = \phi(b)$, and let x be a standard interior point of C, $x = \phi(a)$. Let η be any infinitesimal number such that

$\|\phi(b)\|/\eta$ is finite. We consider the ratio

8.4.1 $\qquad \dfrac{1}{\eta}(g(\phi(ab))-g(\phi(a)))=\dfrac{1}{\eta}(g(f(x,y))-g(x)).$

By 8.3.6,

$$g(f(x,y))-g(x)=g\left(x^1+\sum_{j=1}^{n}y^j\left(\frac{\partial f^1}{\partial y^i}\right)_{(x,0)}+\beta_1\|y\|^2,...,\right.$$

$$\left.x^n+\sum_{j=1}^{n}y^j\left(\frac{\partial f^n}{\partial y^j}\right)_{(x,0)}+\beta_n\|y\|^2\right)-g(x^1,...,x^n),$$

where $0\le|\beta_i|\le\beta$, $i=1,...,n$, β being the standard positive bound on the functions $|(\partial^2 f^i)/(\partial y^j\partial y^k)|$ which was introduced in section 8.3. Applying Taylor's theorem to g, we obtain further

8.4.2 $\qquad g(f(x,y))-g(x)=\sum_{i=1}^{n}\left(\sum_{j=1}^{n}y^j\left(\frac{\partial f^i}{\partial y^j}\right)_{(x,0)}+\beta_i\|y\|^2\right)\frac{\partial g}{\partial x^i}+r.$

In this equation, the remainder term r is given by

$$r=\left(\sum_{i=1}^{n}\left(\sum_{j=1}^{n}y^j\left(\frac{\partial f^i}{\partial y^j}\right)_{(x,0)}+\beta_i\|y\|^2\right)\frac{\partial}{\partial x^i}\right)^2 g,$$

where the second derivatives of g which appear upon evaluating the right hand side are taken at a point

$$\left(x^1+\tau\left(\sum_{j=1}^{n}y^j\left(\frac{\partial f^i}{\partial y^j}\right)_{(x,0)}\right),...,x^n+\tau\left(\sum_{j=1}^{n}y^j\left(\frac{\partial f^i}{\partial y^j}\right)_{(x,0)}\right)\right),$$

for a number τ satisfying $0\le\tau\le1$. Since the second derivatives of g are continuous, hence bounded, on C we conclude that there exists a standard positive number γ such that $|r|\le\gamma\|y\|^2$ for all x and y under consideration. Going back to 8.4.2, taking into account that the first derivatives of g are continuous, hence bounded, on C and putting

$$\psi^i_j(x)=\left(\frac{\partial f^i}{\partial y^j}\right)_{(x,0)}, \qquad i,j=1,...,n,$$

we obtain

8.4.3 $\qquad g(f(x,y))-g(x)=\sum_{i=1}^{n}\left(\sum_{j=1}^{n}y^j\psi^i_j(x)\right)\frac{\partial g}{\partial x^i}+s.$

where $|s| \leq \delta \| y \|^2$ for a standard positive δ which is the same for all x and y under consideration. Now $\| \phi(b) \| / \eta = \| y \| / \eta$ is finite, by assumption. Hence, from 8.4.1 and 8.4.3,

8.4.4 $$\,^0\!\left(\frac{1}{\eta} (g(\phi(ab)) - g(\phi(a))) \right) = \sum_{i=1}^{n} \left(\sum_{j=1}^{n} \lambda^j \psi_j^i(x) \right) \frac{\partial g}{\partial x^i},$$

where $\lambda^j = \,^0(y^j/\eta)$, $j = 1, \dots, n$. Defining the infinitesimal transformation Y by

8.4.5 $$Y = \sum_{i=1}^{n} \left(\sum_{j=1}^{n} \lambda^j \psi_j^i(x) \right) \frac{\partial}{\partial x_i}$$

we therefore see from 8.4.4 that Y can be expressed in terms of the infinitesimal group element b by

8.4.6 $$Yg = \,^0\!\left(\frac{1}{\eta} (g(\phi(ab)) - g(\phi(a))) \right).$$

Since Y is determined by the vector $(\lambda^1, \lambda^2, \dots, \lambda^n)$ we can evidently get the same Y for numerous choices of $\eta, y^1, y^2, \dots, y^n$. If we choose η as a positive infinitesimal number, which is kept fixed throughout the discussion, then the infinitesimal elements of $*G$ which give rise to an infinitesimal transformation as in 8.4.6 are precisely those for which $\| y \| / \eta = \| \phi(b) \| / \eta = \varrho(b,e)/\eta$ is finite, i.e., they are the elements of the group $J_0(\eta)$ introduced in 8.2 above. Two elements of $J_0(\eta)$, b_1 and b_2 determine the same infinitesimal transformation Y if $\phi(b_1)/\eta \simeq \phi(b_2)/\eta$ or, which is the same, if b_1 and b_2 belong to the same coset with respect to $J_1(\eta)$. Again, if $b \in J_0(\eta)$ determines Y then b^{-1} determines $-Y$, and if $b_1 \in J_0(\eta)$ and $b_2 \in J_0(\eta)$ determine the infinitesimal transformations Y_1 and Y_2 respectively, then $b_1 b_2$ determines $Y_1 + Y_2$.

Finally, we have to consider how to obtain the Lie product of two infinitesimal transformations within this setting. For this purpose, we strengthen our differentiability conditions on the composition functions by assuming that the f^i have first, second, and third continuous partial derivatives on the cube C. It then follows from Taylor's theorem that these functions can be written in the form

8.4.7 $$f^i(x^1, \dots, x^n, y^1, \dots, y^n) = x^i + y^i + \sum_{j,k=1}^{n} a_{jk}^i x^j x^k + r_i \| x \|^2 \| y \|$$
$$+ s_i \| x \| \| y \|^2,$$

where r_i and s_i are uniformly bounded by a standard positive number, for all points in the cube.

Now let b_1 and b_2 be two infinitesimal elements of $*G$ and let $y_1 = (y_1^1, \ldots, y_1^n) = \phi(b_1)$, $y_2 = (y_2^1, \ldots, y_2^n) = \phi(b_2)$. Then a straightforward computation (which occurs also in the classical theory) shows that

$$
\textbf{8.4.8} \qquad \phi(b_1 b_2 b_1^{-1} b_2^{-1}) = \left(\sum_{j,k=1}^{n} c_{jk}^1 y_1^j y_2^k + \varrho_1 \|y_1\|^2 \|y_2\| \right.
$$
$$
+ \sigma_1 \|y_1\| \|y_2\|^2, \ldots, \sum_{j,k=1}^{n} c_{jk}^n y_1^j y_2^k + \varrho_n \|y_1\|^2 \|y_2\|
$$
$$
\left. + \sigma_n \|y_1\| \|y_2\|^2 \right),
$$

where the ϱ_i and σ_i are finite and $c_{jk}^i = a_{jk}^i - a_{kj}^i$. This shows that if $\|y_1\|/\eta$ and $\|y_2\|/\eta$ are finite then

$$
\frac{1}{\eta^2} \phi(b_1 b_2 b_1^{-1} b_2^{-1}) \simeq \frac{1}{\eta^2} \left(\sum_{j,k=1}^{n} c_{jk}^1 y_1^j y_2^k, \ldots, \sum_{j,k=1}^{n} c_{jk}^n y_1^j y_2^k \right).
$$

Put $c = b_1 b_2 b_1^{-1} b_2^{-1}$ and let a be any element of the group G such that $\phi(a) = (x^1, \ldots, x^n)$ belongs to the interior of the cube C. For any function $g(x^1, \ldots, x^n)$ as above, we consider $(1/\eta^2)(g(\phi(ac)) - g(\phi(a)))$. Still assuming that $\|y_1\|/\eta$ and $\|y_2\|/\eta$ are finite and taking into account 8.4.8, we obtain (compare the evaluation of 8.4.1)

$$
\frac{1}{\eta^2}(g(\phi(ac)) - g(\phi(a))) \simeq \frac{1}{\eta^2} \sum_{i=1}^{n} \sum_{m=1}^{n} \sum_{j,k=1}^{n} c_{jk}^m y_1^j y_2^k \psi_m^i(x) \frac{\partial g}{\partial x^i}.
$$

Hence, putting $\lambda_i^j = {}^0(y_i^j/\eta)$, $j = 1, \ldots, n$, $i = 1, 2$,

$$
\textbf{8.4.9} \qquad {}^0\!\left(\frac{1}{\eta^2}(g(\phi(ac)) - g(\phi(a))) \right) \simeq \sum_{i=1}^{n} \sum_{m=1}^{n} \sum_{j,k=1}^{n} c_{jk}^m \lambda_1^j \lambda_2^k \psi_m^i(x) \frac{\partial g}{\partial x^i}.
$$

We recall that at the same time

$$
\textbf{8.4.10} \qquad {}^0\!\left(\frac{1}{\eta}(g(\phi(ab_l)) - g(\phi(a))) \right) = \sum_{i=1}^{n} \sum_{j=1}^{n} \lambda_l^j \psi_j^i(x) \frac{\partial g}{\partial x^i}, \qquad l = 1, 2.
$$

Reference to 8.4.5 and 8.4.6 now shows that 8.4.10 and 8.4.9 are equivalent to

8.4.11 $Y_l g = {}^0\!\left(\dfrac{1}{\eta}\left(g\left(\phi\left(ab_l\right)\right)-g\left(\phi\left(a\right)\right)\right)\right), \qquad l=1,2.$

and

8.4.12 $[Y_1,Y_2]g = {}^0\!\left(\dfrac{1}{\eta^2}\left(g\left(\phi\left(ab_1 b_2 b_1^{-1} b_2^{-1}\right)\right)-g\left(\phi\left(a\right)\right)\right)\right),$

respectively. We have thus correlated the commutator of two infinitesimal elements of the group $*G$ directly with the Lie product of the corresponding infinitesimal transformations.

8.5   Remarks and references.   For the classical background to this chapter, the reader may consult PONTRJAGIN [1946] and COHN [1957].

CHAPTER IX

SELECTED TOPICS

9.1 Variations. In the first three sections of this chapter, we shall consider several variational problems and methods from the point of view of Non-standard Analysis. We begin with an informal discussion of the basic notions of the Calculus of Variations. Working at first in the standard system of real numbers R, let F be a family of curves in n-dimensional Euclidean space R^n, $n \geq 1$, all given in parametric form $x^i = x^i(t)$, $i = 1, \ldots, n$, $a \leq t \leq b$. The functions $x^i(t)$ shall be continuous with continuous first derivatives in the closed interval $a \leq t \leq b$, except that the derivatives may have discontinuities of the first kind at a finite number of points. The *distance* between two curves may be given either by

9.1.1
$$\varrho(x_1, x_2) = \sup_{a \leq t \leq b} \|x_1(t) - x_2(t)\| = \sup_{a \leq t \leq b} \sqrt{\{(x_1^1(t) - x_2^1(t))^2 + \cdots + (x_1^n(t) - x_2^n(t))^2\}}$$

or by

9.1.2
$$\bar{\varrho}(x_1, x_2) = \varrho(x_1, x_2) + \varrho(x_1', x_2'),$$

where ϱ is defined by 9.1.1 and $x_i' = (x_i^{1\prime}(t), \ldots, x_i^{n\prime}(t))$, $i = 1, 2$. Let $*R$ be a higher order non-standard model of Analysis so that, in particular, $*R$ may be an enlargement of R.

Let $J[x]$ be a functional which is defined on F, that is to say, a function from F into the real numbers. As we pass from R to $*R$, F is extended to a set $*F$ and $J[x]$ is then defined on $*F$. Let x and y be two curves which belong to $*F$ such that x is standard and $\varrho(x, y) \simeq 0$, or, alternatively, $\bar{\varrho}(x, y) \simeq 0$. We put $\delta x = y - x = (y^1(t) - x^1(t), \ldots, y^n(t) - x^n(t))$ and $\delta J = J[y] - J[x] = J[x + \delta x] - J[x]$. δJ will be called the *first variation* of J. The vector function $(0, \ldots, 0)$ will be denoted simply by 0.

Suppose in particular that $J[x]$ is given as an integral of the form

$$J[x] = \int_a^b g(t, x(t), x'(t)) \, dt$$

such that $x(a)$ and $x(b)$ are the same for all elements x of the family F. Suppose further that g possesses continuous and uniformly bounded first and second derivatives for all sets of arguments $(t, x(t), x'(t))$ which occur for curves of the family F and let $z(t) = (z^1(t), \ldots, z^n(t))$ be an internal function such that $x(t) + \lambda z(t)$ belongs to $*F$ for all λ in the interval $0 \leq \lambda \leq 1$, so that $z(a) = z(b) = 0$. Following the standard procedure of the calculus of variations, we compute $J[x+z] - J[x]$. Supposing in particular that $\bar{\varrho}(x+z, x) = \bar{\varrho}(z, 0)$ is infinitesimal, we obtain

$$J[x+z] - J[x] = \int_a^b \sum_{i=1}^n \left(\frac{\partial g}{\partial x^i} - \frac{d}{dt}\left(\frac{\partial g}{\partial x^{i\prime}}\right) \right) z^i \, dt + m(\bar{\varrho}(z,0))^2$$

where m is finite. Thus, writing δx for z,

9.1.3 $$\delta J \simeq \int_a^b \sum_{i=1}^n \left(\frac{\partial g}{\partial x^i} - \frac{d}{dt}\left(\frac{\partial g}{\partial x^{i\prime}}\right) \right) \delta x^i \, dt \quad \mathrm{mod}\ O((\bar{\varrho}(z,0))^2),$$

where the symbol O is as defined at the beginning of section 3.6. The symbol o also was defined there.

We shall say that J is *stationary* at x if

9.1.4 $$\delta J \simeq 0 \quad \mathrm{mod}\ o(\bar{\varrho}(\delta x, o)).$$

Then 9.1.3 shows that in that case

9.1.5 $$\int_a^b \sum_{i=1}^n \left(\frac{\partial g}{\partial x^i} - \frac{d}{dt}\left(\frac{\partial g}{\partial x^{i\prime}}\right) \right) \delta x^i \, dt \simeq 0 \quad \mathrm{mod}\ o(\bar{\varrho}(\delta x, 0)).$$

It is now not difficult to obtain the equations of Euler–Lagrange from 9.1.5. Generally speaking, our method parallels the classical procedure quite closely. The main difference compared with the standard theory is that here we define the notion of the first variation directly and that the stationary values of J are then given by 9.1.4. Classically, the first variation is defined by an expression which coincides formally with 9.1.5 and whose *vanishing* determines the stationary values of J.

Our definition can be generalized immediately so as to refer to any real valued function x which is defined in a metric space T. $f(x)$ will be said to be *stationary* at a point x_0 which belongs to the S-interior of the domain of definition of $f(x)$ if

$$f(x) \simeq f(x_0) \quad \mod o(\varrho(x,x_0))$$

for all points x in the monad of x_0.

9.2 Riemann's mapping theorem.

The use of Non-standard Analysis in connection with direct variational methods can be of considerable interest. As a first example we consider Riemann's fundamental theorem in conformal mapping.

Let D be a simply connected domain in the standard plane of complex numbers such that D possesses at least two boundary points. Riemann's theorem states that there exists an analytic function $w = f(z)$ which is univalent (schlicht) on D and which maps D on the interior of a circle.

A standard proof of the theorem proceeds as follows.

(i) One chooses an arbitrary point z_0 in D and one shows by means of a simple irrational function that there exist functions $g(z)$ which are analytic, bounded, and univalent on D such that $g(z_0) = 0$ and $g'(z_0) = 1$. Let the class of these functions be denoted by Γ.

(ii) For any $g \in \Gamma$ let $m[g]$ be the least upper bound of the values of $|g(z)|$ in D. m is finite since g is bounded. Let ϱ be the greatest lower bound of all $m[g]$, $g \in \Gamma$. Thus, $\varrho \geq 0$. One proves that this bound is attained in Γ, in more detail,

9.2.1 THEOREM (Standard). There exists a function $f(z) \in \Gamma$ such that $m[f] = \varrho$.

(iii) Finally, one proves that $f(z)$ maps D on a circle. If this were not the case then (as can be shown by another construction) there would exist an $h \in \Gamma$ such that $m[h] < m[f] = \varrho$, and this contradicts the definition of ϱ.

9.2.1 is usually proved by means of Vitali's theorem or by some related method. Using internal analytic functions in $*R$ we may proceed as follows. Since ϱ is the greatest lower bound of the set of numbers $m[g]$, $g \in \Gamma$, the following statement holds in R.

9.2.2 'For every positive δ there exists a function $G(z)$ in Γ such that $m[G] < \varrho + \delta$'.

9.2.2 can be formalized as a sentence X of K where ϱ appears in X explicitly as a constant while δ and G are represented by quantified variables. Let δ' be any positive infinitesimal number. Interpreting X in $*R$, we find that there exists a function $F(z) \in *\Gamma$ such that $m[F] < \varrho + \delta'$. Since D is open, it is a subset of the S-interior B of $*D$. $F(z)$ is defined and analytic in $*D$ and $m[F]$ is the lowest upper bound of the values of $F(a)$ for $z \in *D$. Thus, $F(z)$ is bounded in $*D$ by $\varrho + \delta'$ and hence, is S-continuous on B, by 6.2.1. Let $f(z) = {}^0F(z)$, then $f(z)$ is analytic in D by 6.2.3 and is bounded in D, e.g., by $\varrho + 1$. Also, $f(z_0) = {}^0F(z_0) = 0$ and $f'(z_0) = ({}^0F(z))'_{z=z_0} = 1$ since $F \in *\Gamma$ and so $f(z)$ is not constant. Hence, $f(z)$ is univalent in D, by 6.2.10, and, accordingly, is an element of Γ. Finally, for any $z_1 \in D$, $|F(z_1)| < \varrho + \delta'$ and $|f(z_1) - F(z_1)|$ is infinitesimal. It follows that $|f(z_1)| \leq \varrho + \eta$ for some infinitesimal η. But $f(z_1)$ is a standard number and so, more precisely, $|f(z_1)| \leq \varrho$. This implies that $m[f] \leq \varrho$ and so $m[f] = \varrho$, since $f \in \Gamma$. The proof of 9.2.1 is complete.

9.3 Dirichlet's principle. As a second example for the use of Non-standard Analysis in connection with a direct variational method we consider the solution of the first boundary value problem of Potential Theory (Dirichlet's problem) by means of Dirichlet's principle. We shall require the following result which is of independent interest.

Let D be a domain (open connected set) in R^2 and let $\phi(x,y)$ be a function which is defined on $*D$.

9.3.1 THEOREM. Suppose $\phi(x,y)$ is finite and harmonic (possesses continuous second derivatives and satisfies Laplace's equation) everywhere in $*D$. Then the function ${}^0\phi(x,y)$ exists and is harmonic everywhere in D.

We prove this theorem in several stages. As a first step, we wish to show that $\phi(x,y)$ is S-continuous in the S-interior of $*D$. Let $p = (x_0, y_0)$ be a point in $*D$ and let C be a circle with center p and standard positive radius R_0 such that C and its interior belong to $*D$. Since ϕ is harmonic in $*D$, we have Poisson's integral

9.3.2 $$\phi(x,y) = \frac{1}{2\pi}\int_{-\pi}^{\pi} \frac{(R_0^2 - r^2)\phi(x_0 + R_0 \cos \omega, y_0 + R_0 \sin \omega)}{R_0^2 + r^2 - 2R_0 r \cos(\omega - \theta)}\, d\omega$$

where $q = (x,y)$ is a point in the interior of the circular disk bounded by C, $(x,y) = (x_0 + r \cos\theta, y_0 + r \sin\theta)$, $r < R_0$. Now

9.3.3 $$\left| \frac{R_0^2 - r^2}{R_0^2 + r^2 - 2R_0 r \cos(\omega - \theta)} - 1 \right| = \left| \frac{r(2R_0 \cos(\omega - \theta) - 2r)}{R_0^2 + r^2 - 2R_0 r \cos(\omega - \theta)} \right|$$

Suppose $r < \frac{1}{2}R$ then the denominator on the right hand side of 9.3.3 cannot be smaller than $\frac{1}{4}R_0^2$. It follows that 9.3.3 is smaller than, or equal to, $4r|2 R_0 \cos(\omega - \theta) - 2 r|/R_0^2$, which in turn is smaller than or equal to $16\, r/R_0$. On the other hand, $\phi(x_0 + R_0 \cos\omega, y_0 + R_0 \sin\omega)$ attains its maximum somewhere on the circle C, for $\omega = \omega_0$ say, and since $|\phi(x_0 + R_0 \cos\omega_0, y_0 + R_0 \sin\omega_0)| < m$ for some standard positive m we may write

$$|\phi(q) - \phi(p)| = \frac{1}{2\pi}\left| \int_{-\pi}^{\pi} \phi(x_0 + R_0 \cos\omega, \right.$$

$$\left. y_0 + R_0 \sin\omega)\left(\frac{R_0 - r^2}{R_0^2 + r^2 - 2R_0 r \cos(\omega - \theta)} - 1\right) d\omega \right| \leq \frac{1}{2\pi}\int_{-\pi}^{\pi} m\, \frac{16r}{R_0}\, d\omega = \frac{16mr}{R_0}$$

where $\phi(q)$ and $\phi(p)$ stand for $\phi(x,y)$ and $\phi(x_0,y_0)$ respectively. It follows that if q is in the monad of p so that r is infinitesimal, then $|\phi(q) - \phi(p)| \simeq 0$ and hence $\phi(q) \simeq \phi(p)$. This shows that $\phi(x,y)$ is S-continuous at p. But p may be any point in the S-interior B of $*D$ and so ϕ is S-continuous at all points of B.

Accordingly, we may define ${}^0\phi(x,y) = {}^0\phi(p)$ for all standard points $p = (x,y)$ which belong to ${}^0B$ and in particular for all points of D (which is subset of ${}^0B$) as the joint value of ${}^0(\phi(q))$ at points q which belong to the intersection of the monad of p, $\mu(p)$, with B. Moreover (in agreement with a general conclusion reached earlier) $^0\phi$ is continuous in D. For, given $p = (x,y) \in D$, and given a standard $\varepsilon > 0$, there is a standard $\delta > 0$ such that any standard q which satisfies $\varrho(p,q) < \delta$ is in the interior of D and satisfies also $|\phi(q) - \phi(p)| < \frac{1}{2}\varepsilon$. But $|^0\phi(p) - \phi(p)|$ and $|^0\phi(q) - \phi(q)|$ are infinitesimal and so

$$|^0\phi(q) - {}^0\phi(p)| \le |^0\phi(q) - \phi(q)| + |\phi(q) - \phi(p)| + |\phi(p) - {}^0\phi(p)| < \varepsilon.$$

This shows that $^0\phi$ is continuous in D.

Let C be a circle which is contained in D, with center (x_0, y_0) and radius R_0. We may now conclude that $^0\phi - \phi$ is infinitesimal on C. We may conclude further that there exists a positive infinitesimal ε such that

9.3.4 $|^0\phi(x_0 + R_0 \cos \omega, y_0 + R_0 \sin \omega) - \phi(x_0 + R_0 \cos \omega,$

$$y_0 + R_0 \sin \omega)| < \varepsilon$$

for $-\pi \le \omega \le \pi$, i.e., everywhere on C. This is true since $|^0\phi - \phi|$ is continuous on C and so possesses an absolute maximum on the circle. However, we may deduce the existence of such an ε also without taking into account that $|^0\phi - \phi|$ is continuous. For since $|^0\phi - \phi|$ is an internal function which is infinitesimal everywhere on C, we have $|^0\phi - \phi| < 1/n$ on C, for all finite positive integers n, and we may then conclude in the usual way that this inequality is true also for some infinite n.

In order to prove that $^0\phi(x,y)$ is harmonic in D, it is sufficient to show that for every point p of D there is an open circular disk E which is contained in D, and bounded by a circle C about p as center, such that Poisson's formulae holds for $^0\phi$ in E. That is to say, we have to prove that 9.3.2 holds in E with $^0\phi$ for ϕ, where $p = (x_0, y_0)$ and R_0 is the radius of the circle C. Now for any point $q = (x,y)$ in E, the difference $^0\phi(x,y) - \phi(x,y)$ is infinitesimal. Thus, in order to show that $^0\phi$ satisfies Poisson's formulae of q we only have to verify that the difference

9.3.5 $\Delta = \dfrac{1}{2\pi} \displaystyle\int_{-\pi}^{\pi} H(R_0, \omega, r, \theta)\, {}^0\phi(x_0 + R_0 \cos \omega, y_0 + R_0 \sin \omega)\, d\omega$

$$- \frac{1}{2\pi} \int_{-\pi}^{\pi} H(R_0, \omega, r, \theta)\, \phi(x_0 + R_0 \cos \omega, y_0 + R_0 \sin \omega)\, d\omega$$

also is infinitesimal, where H is the Poisson Kernel, $H(R_0,\omega,r,\theta) = (R_0^2 - r^2)/(R_0^2 + r^2 - 2R_0 r\cos(\omega - \theta))$, and where we observe that $H(R_0,\omega,r,\theta)$ is strictly positive in the range of integration. Then

$$|\Delta| \le \frac{1}{2\pi} \int_{-\pi}^{\pi} |H(R_0,\omega,r,\theta)| \ |^0\phi(x_0 + R_0 \cos\omega, y_0 + R_0 \sin\omega) -$$

$$\phi(x_0 + R_0 \cos\omega, y_0 + R_0 \sin\omega)|\,d\omega \le \frac{\varepsilon}{2\pi} \int_{-\pi}^{\pi} H(R_0,\omega,r,\theta)\,d\omega = \varepsilon,$$

where we have taken into account 9.3.4. This completes the proof of 9.3.1, since ε is infinitesimal.

Going back to the standard plane R^2 for the time being, let D be a bounded domain (open and connected set) whose boundary C consists of a finite number of Jordan arcs, and consider Dirichlet's integral

9.3.6
$$\Delta[\phi] = \iint_D \left(\left(\frac{\partial\phi}{\partial x}\right)^2 + \left(\frac{\partial\phi}{\partial y}\right)^2 \right) dx\,dy$$

for functions $\phi(x,y)$ which are defined on D and satisfy the following conditions. (i) $\phi(x,y)$ is continuous on $D \cup C$, and takes *specified* boundary values on C. (ii) $\phi(x,y)$ possesses continuous first derivatives on D except possibly at a finite number of circular arcs and straight segments C_i where it may have discontinuities of the first kind. That is, if P is a point on one of the C_i then the limits of $\partial\phi/\partial x$ and $\partial\phi/\partial y$ exist on approaching C_i from either side.

Let Φ be a set of functions $\phi(x,y)$ for which $\Delta[\phi]$ exists either as a proper or as an improper integral. We shall suppose that Φ is *not empty*. Then the set of numbers $\Delta[\phi]$ for $\phi \in \Phi$ possesses a greatest lower bound (infimum) $d \ge 0$. Given any positive ε, there exists a function $\phi \in \Phi$ such that $d \le \Delta[\Phi] < d + \varepsilon$.

We pass to $*R$, take ε positive infinitesimal, and choose a function $\phi \in {}^*\Phi$, $\phi_0(x,y)$ say, such that $d \le \Delta[\phi_0] < d + \varepsilon$. Then ϕ_0 satisfies conditions (i) and (ii) above *in the sense of* $*R$. Thus the number of circular arcs at which the derivatives of ϕ_0 are discontinuous may now be infinite in the absolute sense although it is Q-finite.

Now let D' be any closed subdomain of D in R^2. Since D is bounded, D' is bounded. Hence, by the theorem of Heine–Borel, we may cover D' with a finite number of circular disks all of which are contained in D, together with their circumferences. Let γ be one of these disks and let Γ be its circumference.

Passing to $^*R$, let ϕ_γ be the harmonic function in $^*\gamma$ which takes the same values as ϕ_0 on $^*\Gamma$. ϕ_γ may be determined by means of Poisson's integral. Putting

$$\Delta_\gamma[\phi]=\iint_\gamma\left(\left(\frac{\partial\phi}{\partial x}\right)^2+\left(\frac{\partial\phi}{\partial y}\right)^2\right)dx\,dy$$

we claim that

9.3.7 $\Delta_\gamma[\phi_\gamma]\le\Delta_\gamma[\phi_0],$

and we prove 9.3.7 by a familiar application of Green's theorem.

Now let ϕ_1 be the function which is obtained from ϕ_0 by setting $\phi_1=\phi_0$ on $^*D\cup{}^*C-{}^*\gamma$ and $\phi_1=\phi_\gamma$ on $^*\gamma$. Then ϕ_1 also belongs to $^*\Phi$ and $\Delta[\phi_1]\le\Delta[\phi_0]$. Hence, $d\le\Delta[\phi_1]<d+\varepsilon$. Beginning with ϕ_0, we may apply this procedure to all circular disks of the selected finite covering of D' in turn. In this way, we finally obtain a function $\bar\phi$ which has a Dirichlet integral satisfying $d\le\Delta[\bar\phi]<d+\varepsilon$ and which is harmonic in D' except possibly at a finite (*finite* in the absolute sense) number of circular arcs across which the function $\bar\phi$ is still continuous, but its first derivatives may have discontinuities of the first kind, in the sense explained. Such discontinuities may occur only across the circumferences of the disks which constitute the selected finite covering of D'. For although the derivatives of the original ϕ_0 may have had discontinuities of this kind also across additional circular arcs or straight segments, all these were eliminated in the course of the successive modifications of the function.

One shows next that $\bar\phi$ is finite in D. It then follows from 9.3.1 that $^0\bar\phi(x,y)$ exists and is harmonic in the interior of D' except possibly on the circumferences of the disks just mentioned. However, adapting classical methods, one can show that in actual fact no discontinuities occur on these circumferences within D'. Finally, one approximates D by a sequence of closed subdomains $D'_1\subset D'_2\subset D'_3\subset\dots$ such that $\bigcup D'_n=D$ and one shows that the functions $\phi_n(x,y)={}^0\phi(x,y)$ determined in the way just mentioned in the interior of the respective regions D'_n coincide in their joint domains of definition (i.e., $\phi_j(x,y)=\phi_n(x,y)$ in the interior of D_n for $j\geq n$). The function $\psi(x,y)$ which is defined at any point $p=(x,y)\in D$ as the joint value of the $\phi_n(x,y)$ for the D_n' which contain p in their interior, is then harmonic in D. $\psi(x,y)$ can be shown to satisfy the specified boundary conditions.

We have detailed only the steps which mark the salient difference between the approach of Non-standard Analysis and the classical method. Whereas in the classical procedure $\psi(x,y)$ is obtain in terms of limits of minimizing sequences, we arrive at $\psi(x,y)$ by taking the standard parts of functions whose Dirichlet integral is infinitely close to the greatest lower bound d.

9.4 Sources and doublets. In the theory of continuous media, the language of infinitesimals has a long tradition. This tradition remained unbroken even though infinitesimals had been abandoned in pure Analysis. For example, the derivation of Euler's equations of motion is frequently based on the consideration of the balance of forces acting on a *very small* cube and several laws of Fluid Mechanics are expressed in terms of *fluid particles* whose dimensions are supposed to be infinitely small. Again, a doublet is defined by letting two sources of equal and opposite strength move infinitely close to each other while the strength of the sources varies in a suitable manner. It is usually taken for granted, and sometimes shown explicitly, that the use of infinitesimal language can be avoided. Naturally, this is not what we shall do here.

Consider, for example, the derivation of the formula for a doublet in an incompressible fluid. The velocity potential of a source located at a point (x_0,y_0,z_0) is

9.4.1
$$\phi(x,y,z)=\frac{\sigma}{4\pi}\cdot\frac{1}{\sqrt{\{(x-x_0)^2+(y-y_0)^2+(z-z_0)^2\}}}=$$
$$\frac{\sigma}{4\pi}\Phi(x,y,z\,;x_0,y_0,z_0),$$

where σ is the strength of the source. The velocity potential due to a source of strength $-\sigma$ at (x_0,y_0,z_0) together with a source of strength σ located at $(x_0+\lambda r,y_0+\mu r,z_0+\nu r)$, where (λ,μ,ν) is a unit vector, is

9.4.2
$$\psi(x,y,z) = \frac{\sigma}{4\pi}(\Phi(x,y,z;x_0+\lambda r,y_0+\mu r,z_0+vr) - \Phi(x,y,z;x_0,y_0,z_0)).$$

Working in $*R$ and supposing r positive but infinitesimal, we obtain for (x,y,z) outside the monad of (x_0,y_0,z_0),

9.4.3
$$\Phi(x,y,z;x_0+\lambda r,y_0+\mu r,z_0+vr) - \Phi(x,y,z;x_0,y_0,z_0) \simeq$$
$$r\left(\lambda\frac{\partial\Phi}{\partial x_0}+\mu\frac{\partial\Phi}{\partial y_0}+v\frac{\partial\Phi}{\partial z_0}\right) \quad \text{mod } o(r).$$

Multiplying 9.4.3 by $\sigma/4\pi$, where σ may be finite or infinite and putting $\sigma r = \tau$ we obtain, taking into account 9.4.2,

$$\psi(x,y,z) \simeq \frac{\tau}{4\pi}\left(\lambda\frac{\partial\Phi}{\partial x_0}+\mu\frac{\partial\Phi}{\partial y_0}+v\frac{\partial\Phi}{\partial z_0}\right) \quad \text{mod } o(\tau).$$

Assuming τ to be standard, we conclude that $^0\psi(x,y,z)$ exists for all standard points (x,y,z) other than (x_0,y_0,z_0) and

9.4.4
$$^0\psi(x,y,z) = \frac{\tau}{4\pi}\left(\lambda\frac{\partial\Phi}{\partial x_0}+\frac{\partial\Phi}{\partial y_0}+\frac{\partial\Phi}{\partial z_0}\right).$$

Thus, $^0\psi(x,y,z)$ is the velocity potential of a doublet.

So far, we have been dealing with sources and doublets in incompressible flow, i.e., with solutions of Laplace's equation. Next, we consider the corresponding notions for linearized supersonic flow. The appropriate partial differential equation now is, formally, a wave equation,

9.4.5
$$-\beta^2\frac{\partial^2\phi}{\partial x^2}+\frac{\partial^2\phi}{\partial y^2}+\frac{\partial^2\phi}{\partial z^2}=0,$$

and the corresponding (induced) velocity potential of a source located at a point (x_0,y_0,z_0) is $\phi(x,y,z)=(\sigma/2\pi)\Phi(x,y,z;x_0,y_0,z_0)$ where

9.4.6 $\Phi(x,y,z;x_0,y_0,z_0)=$
$$\begin{cases} \dfrac{1}{\sqrt{[(x-x_0)^2-\beta^2\{(y-y_0)^2+(z-z_0)^2\}]}} \\ \qquad\qquad \text{for } (x-x_0)^2 > \beta^2\{(y-y_0)^2+(z-z_0)^2\}, \ x>x_0 \\ 0 \quad \text{elsewhere.} \end{cases}$$

Except for the fact that this makes $\Phi=0$ in the fore-cone of (x_0,y_0,z_0), Φ is a fundamental solution of 9.4.5.

From 9.4.6, we may now obtain the equation of a doublet as before. In particular, if the axis of the doublet is parallel to the x-axis, so that $\lambda=1$, $\mu=\nu=0$, the velocity potential of the doublet is given by

9.4.7
$$\zeta(x,y,z)={}^0\psi(x,y,z)=\overset{0}{\left(\frac{\sigma}{2\pi}\,\Phi(x,y,z;x_0+r,y_0,z_0)\right.}$$
$$\left.-\Phi(x,y,z;x_0,y_0,z_0)\right),$$

where Φ is as in 9.4.6 and $\tau=\sigma r$ is standard, while r is positive infinitesimal. 9.4.7 applies everywhere except for points of the Mach cone
$$(x-x_0)^2-\beta^2\{(y-y_0)^2+(z-z_0)^2\}=0,\qquad x\geq x_0.$$

Consider the surface integral $\iint_S\zeta\,dy\,dz$ where S is an area in the plane $x=\alpha$, α any constant $>x_0$, and consists of the points of that plane which satisfy $\alpha-x_0>\beta\sqrt{\{(y-y_0)^2+(z-z_0)^2\}}$. Thus S is the intersection of the plane $x=\alpha$ with the Mach cone emanating from (x_0,y_0,z_0). Putting $y=y_0+\varrho\cos\theta$, $z=z_0+\varrho\sin\theta$, we then have, with $\varrho_0=(\alpha-x_0)/\beta$,

$$\int_S\Phi(x,y,z;x_0,y_0,z_0)dy\,dz\;=\;\int_0^{\varrho_0}d\varrho\int_0^{2\pi}\frac{\varrho\,d\theta}{\sqrt{\{(\alpha-x_0)^2-\beta^2\varrho^2\}}}$$

$$=2\pi\int_0^{\varrho_0}\frac{\varrho}{\sqrt{\{(\alpha-x_0)^2-\beta^2\varrho^2\}}}$$

$$=-2\pi[\sqrt{\{(\alpha-x_0)^2-\beta^2\varrho^2\}}]_0^{(\alpha-x_0)/\beta}=2\pi(\alpha-x_0).$$

It follows that

9.4.8
$$\overset{0}{\iint_S}\psi\,dy\,dz=\overset{0}{\left[\frac{\sigma}{2\pi}\iint_S(\Phi(x,y,z;x_0+r,y_0,z_0)\right.}$$
$$\left.-\Phi(x,y,z;x_0,y_0,z_0))dy\,dz\right]=\sigma((\alpha-(x_0+r))-(\alpha-x_0))=-\sigma r=-\tau.$$

On the other hand, the surface integral that we set out to calculate originally, i.e., $\iint_S\zeta\,dy\,dz=\iint_S{}^0\psi\,dy\,dz$ actually diverges since

$$^0\psi(x,y,z)=\frac{\tau}{2\pi}\frac{\partial}{\partial x_0}\Phi=\frac{\tau}{2\pi}\frac{x-x_0}{[(x-x_0)^2-\beta^2\{(y-y_0)^2+(z-z_0)^2\}]^{\frac{3}{2}}}.$$

Hence,

9.4.9
$$\iint_S \zeta \, dy \, dz = \frac{\tau}{2\pi} \int_0^{\varrho_0} d\varrho \int_0^{2\pi} \frac{(\alpha - x_0)\varrho \, d\theta}{[(\alpha - x_0)^2 - \beta^2 \varrho^2)]^{\frac{3}{2}}} =$$

$$= \tau \int_0^{\varrho_0} \frac{(\alpha - x_0)\varrho \, d\varrho}{[(\alpha - x_0)^2 - \beta^2 \varrho^2)]^{\frac{3}{2}}}$$

$$= \frac{\tau}{\beta} \left[\frac{\alpha - x_0}{\sqrt{(x_0^2 - \varrho^2)}} \right]_0^{x_0},$$

and this expression is properly divergent. However, 9.4.8 yields precisely the *finite part* of this divergent integral, i.e., $-\tau$, which is obtained classically by neglecting the contribution of the upper limit of integration on the right hand side of 9.4.9.

9.5 Local perturbations.

Let W be a large vessel whose solid walls are at rest and which is provided with a number of small orifices, $Q_1, \dots, Q_k$. The vessel is filled with an irrotational and incompressible fluid (water, say) which flows in and out of the orifices. Then it appears intuitively likely that the field of flow of the fluid inside the vessel in regions far removed from the orifices depends essentially only on the total flow across each of the orifices and is sensibly independent of the distribution of the flow across each orifice. The same idea may be expressed by the assertion that any modification of the flow across that orifice can produce only *local* effects in the sense that the effects produced by it in regions far removed from the orifice are negligible.

It is our purpose in the present section to make this intuitive principle precise.

For a given system of rectangular coordinates relative to which the vessel is at rest, let u, v, w be the velocity components of a specified field of flow within the vessel, as usual, and let $\phi(x, y, z)$ be the corresponding velocity potential, so that

9.5.1
$$u = -\frac{\partial \phi}{\partial x}, \qquad v = -\frac{\partial \phi}{\partial y}, \qquad w = -\frac{\partial \phi}{\partial z},$$

or, in vector notation, $q = -\text{grad } \phi$ where $q = (u, v, w)$. If we suppose, for convenience, that the density of the fluid, ϱ, is equal to 2, then the kinetic energy of the fluid in a given volume V coincides with the Dirichlet integral of ϕ over that volume, $D_V[\phi] = \int_V (\text{grad}\,\phi)^2 \, d\tau$.

We shall require some simple relations involving Dirichlet integrals. Let V be a finite volume which is bounded by a simply connected surface S. Suppose that V is divided into two volumes V_1 and V_2 by a surface S_3. Suppose further that S_3 is bounded by a simple curve C on S which divides S into two surfaces S_1 and S_2, such that S_1 together with S bounds V_1 and S_2 together with S bounds V_2. Let ϕ, ϕ_1, ϕ_2 be three functions which are defined and harmonic in V, V_1, V_2 respectively, and which are continuous in the closures of the respective volumes. Suppose that $(\text{grad}\,\phi_1)_n = (\text{grad}\,\phi)_n$ on S_1 and $(\text{grad}\,\phi_2)_n$ on S_2, and $(\text{grad}\,\phi_1)_n = (\text{grad}\,\phi_2)_n$ on S_3 where the subscript n indicates the component normal to the surface under consideration (the same direction being taken for grad ϕ_1, and grad ϕ_2 in the case of S_3). Using the notation

$$D_V[\phi] = \int_V (\text{grad}\,\phi)^2 \, d\tau, \qquad D_{V_1}[\phi_1] = \int_{V_1} (\text{grad}\,\phi_1)^2 \, d\tau,$$

$$D_{V_2}[\phi_2] = \int_{V_2} (\text{grad}\,\phi_2)^2 \, d\tau,$$

as above, we propose to show that

9.5.2 $$D_V[\phi] \le D_{V_1}[\phi_1] + D_{V_2}[\phi_2].$$

PROOF of 9.5.2. Put $\psi_1 = \phi_1 - \phi$ in V_1, $\psi_2 = \phi_2 - \phi$ in V_2. Applying the divergence theorem to the function ϕ grad ψ_1 in V_1 and similarly to ϕ grad ψ_2 in V_2 we obtain

9.5.3 $$\int_{S_3 \cup S_1} \phi \, \text{grad}\,\psi_1 \, d\sigma = \int_{V_1} \text{div}(\phi \, \text{grad}\,\psi_1) \, d\tau$$

and

9.5.4 $$-\int_{S_3 \cup S_2} \phi \, \text{grad}\,\psi_2 \, d\sigma = \int_{V_2} \text{div}(\phi \, \text{grad}\,\psi_2) \, d\tau,$$

where we assume that the normal to S_3 points from V_1 to V_2. Adding 9.5.3 and 9.5.4 and bearing in mind that grad $\psi_1 n = \text{grad}\,\psi_2 n$ where n is the unit normal to S_3 at any point of this surface, we obtain

9.5.5 $$D_{V_1}[\phi,\psi_1]+D_{V_2}[\phi,\psi_2]=0$$

In this equation, we have made use of the general notation $D_V[\phi,\psi]=\int_V \text{grad } \phi \text{ grad } \psi\, d\tau$.

Now $\phi_1=\phi+\psi_1, \phi_2=\phi+\psi_2$. Hence

$$D_{V_1}[\phi_1]=D_{V_1}[\phi]+D_{V_1}[\psi_1]+2D_{V_1}[\phi,\psi_1],$$
$$D_{V_2}[\phi_2]=D_{V_2}[\phi]+D_{V_2}[\psi_2]+2D_{V_2}[\phi,\psi_2].$$

Adding the last two equations and taking into account 9.5.5, we obtain

$$D_{V_1}[\phi_1]+D_{V_2}[\phi_2]=D_V[\phi]+D_{V_1}[\psi_1]+D_{V_2}[\psi_2].$$

This implies 9.5.2 since $D_{V_1}[\psi_1]\geq0,\ D_{V_2}[\psi_2]\geq0$.

Suppose now that S_1 is divided into two parts, S_0 and Q. Let q_n be a real valued function which is defined on Q such that $\int_Q q_n d\sigma=0$. Consider the problem of determining a harmonic function ϕ in V (which is continuous in V up to and at the boundary of V) and such that the outward normal derivative $\partial\phi/\partial n$ satisfies

9.5.6 $$\frac{\partial\phi}{\partial n}=q_n \text{ on } Q, \qquad \frac{\partial\phi}{\partial n}=0 \text{ on } S-Q.$$

This is a boundary value problem of potential theory, of the second kind, and subject to certain conditions of regularity, which we do not discuss here, it possesses a solution which is uniquely determined except for the possible addition of an arbitrary constant. We select such a solution, $\phi=\phi(x,y,z)$. Since ϕ is harmonic in V_2 and $\partial\phi/\partial n=0$ at all points of S_2 we must have $\int_{S_3}(\partial\phi/\partial n)d\sigma=0$ in other words the flow due to ϕ across S_3 vanishes.

Furthermore, let ψ_1 be defined and harmonic in V_1 such that $(\partial\psi_1/\partial n)=q_n$ on Q and $(\partial\psi_1/\partial n)=0$ over the remainder of the boundary of V_1; and let ψ_2 be defined and harmonic in V_1 such that $(\partial\psi_2/\partial n)=0$ on S_1 and $(\partial\psi_2/\partial n)=(\partial\phi/\partial n)$ on S_3. The problem of finding such a ψ_2 has a solution since $\int_{S_3}(\partial\phi/\partial n)d\sigma=0$, as shown above. Then $\psi_1+\psi_2$ has the same normal derivatives on the boundary of V_1 as ϕ and so $\text{grad }\phi=\text{grad }\psi_1+\text{grad }\psi_2$ throughout V_1. Hence,

$$D_{V_1}[\phi]=D_{V_1}[\psi_1+\psi_2]=D_{V_1}[\psi_1]+D_{V_1}[\psi_2]+2D_{V_1}[\psi_1,\psi_2].$$

Now let λ be an arbitrary constant, and consider the function $(1+\lambda)\psi_2+\psi_1$ which is defined and harmonic in V_1, together with the function $(1+\lambda)\phi$ as defined in V_2. Then $(1+\lambda)\phi$ has the same normal derivatives on S_2 as ϕ (i.e., both vanish at points of S_2) and $(1+\lambda)\psi_2+\psi_1$ has the same normal derivatives on S_1 as ϕ. At the same time, $(1+\lambda)\phi$ and $(1+\lambda)\psi_2+\psi_1$ have the same normal derivatives across S_3. Accordingly, the inequality 9.5.2 applies, with ϕ, $(1+\lambda)\phi$, $(1+\lambda)\psi_2+\psi_1$ for ϕ, ϕ_1, ϕ_2 respectively, so –

9.5.7 $$D_V[\phi]\le D_{V_1}[(1+\lambda)\psi_2+\psi_1]+D_{V_2}[(1+\lambda)\phi].$$

Expanding on the right hand side, we obtain

$$D_V[\phi]\le D_{V_1}[\psi_1+\psi_2]+2\lambda D_{V_1}[\psi_1+\psi_2,\psi_2]+\lambda^2 D_{V_1}[\psi_2]$$
$$+D_{V_2}[\phi]+2\lambda D_{V_2}[\phi]+\lambda^2 D_{V_2}[\phi].$$

Hence, taking into account that $D_{V_1}[\psi_1+\psi_2]=D_{V_1}[\phi]$,

9.5.8 $$0\le 2\lambda(D_{V_1}[\psi_1+\psi_2,\psi_2]+D_{V_2}[\phi])+\lambda^2(D_{V_1}[\psi_2]+D_{V_2}[\phi]).$$

But this inequality can hold for all real λ only if the coefficient of λ on the right hand side vanishes, i.e.,

9.5.9 $$D_{V_1}[\psi_1+\psi_2,\psi_2]+D_{V_2}[\phi]=0.$$

Hence,

9.5.10 $$D_V[\phi]=D_{V_1}[\psi_1+\psi_2,\psi_1+\psi_2]+D_{V_2}[\phi]=D_{V_1}[\psi_1+\psi_2,\psi_1]$$
$$=D_{V_1}[\psi_1]+D_{V_1}[\psi_1,\psi_2]=D_{V_1}[\psi_1]-D_{V_1}[\psi_2]-D_{V_2}[\phi].$$

This shows that

9.5.11 $$D_V[\phi]\le D_{V_1}[\psi_1].$$

At the same time, from 9.5.9 and 9.5.10,

$$D_{V_2}[\phi]=-D_{V_1}[\psi_1,\psi_2]-D_{V_1}[\psi_2]=D_{V_1}[\psi_1]-D_V[\phi]-D_{V_1}[\psi_2],$$

and so

9.5.12 $$D_{V_2}[\phi]\le D_{V_1}[\psi_1]-D_V[\phi].$$

Observe that the right hand side of 9.5.12 is non-negative, in view of 9.5.11.

Consider now a one-parametric family of vessels $W(t)$, defined for $t>0$, such that $W(t_1)$ is part of $W(t_2)$ for $t_1 \leq t_2$, and such that all the vessels have an orifice, Q, in common. For example, the $W(t)$ may be cylindrical, with generators parallel to the x-axis, extending from $x=0$ to $x=t$, open at $x=0$ and closed at $x=t$. Thus, Q is precisely the base of the vessel at $x=0$, and the normal component of the flow at Q is, except for the sign, the x-component u. Supposing that u is specified on Q, so that $\int u\,dy\,dz=0$, we denote the resulting velocity potential in the vessel $>(t)$ by ϕ_t. Then the kinetic energy of the fluid within the vessel, assuming $\varrho=2$ for the density, is precisely $D_{W(t)}[\phi_t]$, i.e., the Dirichlet integral of ϕ taken over the volume of the vessel. Putting $D(t)=D_{W(t)}[\phi_t]$, we find that $D(t)$ is a decreasing (i.e., non-increasing) function of t. For if $t_1 < t_2$ then by 9.5.11, with ϕ_{t_1}, ϕ_{t_2} for ψ_1 and ϕ respectively, $D(t_1) \geq D(t_2)$. Also Denoting by $D(t_1, t_2)$ the Dirichlet integral of $\phi(t_2)$ over the excess volume of $W(t_2)$ over $W(t_1)$, so that $D(t_1, t_2)$ is defined for $0 \leq t_1 \leq t_2$, we obtain from 9.5.12,

9.5.13 $D(t_1, t_2) \leq D(t_1) - D(t_2)$ for $0 < t_1 \leq t_2$.

But $D(t) \geq 0$, and so $\lim_{t \to \infty} D(t) = d$ exists, and $d \geq 0$. Passing to *R, we conclude that $D(t) \simeq d$ for all infinite t, and so

9.5.14 $D(t_1, t_2) \simeq 0$

for all infinite t_1 and $t_2 \geq t_1$.

From this result, we may draw the following conclusions. Suppose that the vessel W is a cylindrical pipe with generators parallel to the x-axis, with a base which is given by a standard equation in the yz plane (e.g., circular, with standard radius r), and extending from $x=0$ to $x=t$, t positive infinite. Suppose that W is open at both ends, and denote the two bases by Q_0 and Q_t respectively. Suppose further that the x-components of the normal velocity across Q_0 and Q_t are specified as standard functions of y and z, so that $\int_{Q_0} u\,dx\,dy = \int_{Q_t} u\,dx\,dy = UA$, say, where A is the cross sectional area of the cylinder, and hence $\int_{Q_0} (u-U)\,dx\,dy = \int_{Q_t} (u-U)\,dx\,dy = 0$. Then the integral $\int ((u-U)^2 + v^2 + w^2)\,dx\,dy\,dz$ is infinitesimal when taken over any volume of fluid in the pipe which is infinitely far removed from both ends.

To take another simple example, $W(t)$ may have the shape of a cone whose base is in the y, z-plane while its apex is on the positive x-axis. Suppose that the vessel is open at the base of the cone. Then for any standard distribution of normal velocities across the base, $D(t_1, t_2) \simeq 0$ for infinite $t_1 < t_2$. We conclude that if the height of the cone, t_2, is infinite, then the kinetic energy of the fluid at the part of the vessel for which $t \geq t_1, t_1$ infinite, is infinitesimal.

We now wish to estimate the magnitude of the velocity at a point of the vessel which is far removed from the specified orifice or orifices. For this purpose, we make use of the mean value theorem of potential theory which states that if a sphere S together with its interior belong to a domain in which a function ϕ is harmonic, then

9.5.15
$$\phi_0 = \frac{1}{4\pi R_0^2} \int_S \phi \, d\sigma .$$

In this formula, R_0 is the radius of the sphere and ϕ_0 is the value of ϕ at its center. From 9.5.15 one obtains without difficulty the less familiar

9.5.16
$$\phi_0 = \frac{3}{4\pi R_0^3} \int_V \phi \, d\tau ,$$

where the volume of integration V is the interior of the sphere. Hence, by the Cauchy–Schwarz inequality,

9.5.17
$$\phi_0^2 = \frac{9}{16\pi^2 R_0^6} \left(\int_V \phi \, d\tau \right)^2 \leq \frac{9}{16\pi^2 R_0^6} \int_V d\tau \int_V \phi^2 \, d\tau = \frac{3}{4\pi R_0^3} \int_V \phi^2 \, d\tau .$$

Replacing ϕ in 9.5.17 in turn by the partial derivatives $\partial\phi/\partial x$, $\partial\phi/\partial y$, $\partial\phi/\partial z$, and adding, we obtain

$$\left(\left(\frac{\partial\phi}{\partial x}\right)^2 + \left(\frac{\partial\phi}{\partial y}\right)^2 + \left(\frac{\partial\phi}{\partial z}\right)^2 \right)_0 \leq \frac{3}{4\pi R_0^3} \int_V \left(\left(\frac{\partial\phi}{\partial x}\right)^2 + \left(\frac{\partial\phi}{\partial y}\right)^2 + \left(\frac{\partial\phi}{\partial z}\right)^2 \right) d\tau ,$$

where the subscript 0 indicates again that the value of the expression in parentheses is taken at the center of the sphere. Thus, briefly,

9.5.18 $$((\text{grad } \phi)^2)_0 \leq \frac{3}{4\pi R^3} D_V[\phi].$$

We conclude that grad ϕ is infinitesimal at any point P which belongs
to the S-interior of a volume for which $D[\phi]$ is infinitesimal. For in this
case, we may surround P with a sphere S of standard radius $R_0 > 0$ such
that S and its interior belong to the volume in question. Then the right
hand side of 9.5.18 is infinitesimal and the same therefore applies to the
left hand side. Thus, if q is the velocity vector of the flow through a long
cylindrical pipe considered above then $q \simeq (U,0,0)$ at all points which
belong to the S-interior of the pipe and are infinitely far from both ends;
while in the case of a conical vessel considered later, q is infinitesimal in
the S-interior of the vessel at all points which are infinitely far from its
open base.

The restriction of our conclusion to the S-interior of the vessel is un-
avoidable. At a sharp corner which projects into the vessel, the flow
through a single orifice, however far, may produce an infinite velocity.
On the other hand, it should be possible to establish our results even for
infinite vessels which do not belong to a standard one-parametric family
of vessels, although our present method depended on that condition.

9.6 Boundary layer theory.   In this section we propose to show how our
notions can be used in order to discuss boundary layers in viscous fluid
flow.

Consider two-dimensional flow along a semi-infinite flat plate which
extends from $x=0$ along the positive x-axis. The external flow may be
non-uniform. Let it be given by a velocity vector $(\bar{u}, \bar{v})$ which, on the
positive x-axis, reduces to $\bar{u} = U(x)$, $\bar{v} = 0$.

Passing to $*R$, we now suppose that the field of flow (u, v) is given by
$u = \bar{u}$, $v = \bar{v}$ at all points (x, y) except those for which $x > 0$, $^0y = 0$. Thus, we
still have to determine (u, v) at all points which are infinitely close to the
positive x-axis. For this purpose, we proceed as follows. We suppose that
the dynamic viscosity, $v = \mu/\varrho$, is a definite positive infinitesimal number.
We introduce another positive infinitesimal number δ, which will be
circumscribed more closely in due course.

The flow is governed by the equations of Navier–Stokes,

9.6.1 $$u\frac{\partial u}{\partial x}+v\frac{\partial u}{\partial y}=-\frac{1}{\varrho}\frac{\partial p}{\partial x}+v\left(\frac{\partial^2 u}{\partial x^2}+\frac{\partial^2 u}{\partial y^2}\right),$$

$$u\frac{\partial v}{\partial x}+v\frac{\partial v}{\partial y}=-\frac{1}{\varrho}\frac{\partial p}{\partial y}+v\left(\frac{\partial^2 v}{\partial x^2}+\frac{\partial^2 v}{\partial y^2}\right),$$

together with the equation of continuity,

9.6.2 $$\frac{\partial u}{\partial x}+\frac{\partial v}{\partial y}=0.$$

The boundary conditions at the plate are

9.6.3 $$u=v=0 \qquad \text{for } x>0, y=0.$$

Putting $v'=v/\delta$ and taking x and $y'=y/\delta$ as independent variables in place of x and y we obtain from 9.6.1, 9.6.2, and 9.6.3,

9.6.4 $$u\frac{\partial u}{\partial x}+v'\frac{\partial u}{\partial y'}=-\frac{1}{\varrho}\frac{\partial p}{\partial x}+v\left(\frac{\partial^2 u}{\partial x^2}+\frac{1}{\delta^2}\frac{\partial^2 u}{\partial y'^2}\right)$$

$$\delta\left(u\frac{\partial v'}{\partial x}+v'\frac{\partial v'}{\partial y'}\right)=-\frac{1}{\varrho\delta}\frac{\partial p}{\partial y'}+v\left(\delta\frac{\partial^2 v'}{\partial x^2}+\frac{1}{\delta}\frac{\partial^2 v'}{\partial y'^2}\right), \qquad \frac{\partial u}{\partial x}+\frac{\partial v'}{\partial y'}=0,$$

and

9.6.5 $$u=v'=0 \qquad \text{for} \qquad x>0, \ y'=0.$$

We now suppose that p, μ and, v' are standard functions of x and y' for finite positive x and for finite y' and we try to satisfy 9.6.4 for such x and y' modulo infinitesimal quantities. Thus, we neglect $v(\partial^2 u/\partial x^2)$ on the right hand side of the first equation of 9.6.4. Next, we introduce the usual assumption that the inertia forces which are given by the left hand side of the first equation of 9.6.4 multiplied by ϱ are *of the same order of magnitude* as the viscous forces which, after the discarding of $\varrho v(\partial^2 u/\partial x^2)$, are given by $(\varrho v/\delta^2)(\partial^2 u/\partial y'^2)$. The appropriate interpretation of this assumption within our framework leads us to postulate that the ratio v/δ^2 is a finite and non-infinitesimal number v', so that $v=v'\delta^2$. Making use of this relation in the second equation of 9.6.4, we find that

9.6.6
$$\frac{1}{\varrho\delta}\frac{\partial p}{\partial y'}\simeq 0.$$

Since p is a standard function of x and y', 9.6.6 implies that $\partial p/\partial y'=0$, p is independent of y'. Thus, the equations governing the flow in the region under consideration reduce to

9.6.7
$$u\frac{\partial u}{\partial x}+v'\frac{\partial u}{\partial y'}=-\frac{1}{\varrho}\frac{\partial p}{\partial x}+v'\frac{\partial^2 u}{\partial y'^2}$$

$$\frac{\partial p}{\partial y'}=0$$

$$\frac{\partial u}{\partial x}+\frac{\partial v'}{\partial y'}=0,$$

in the second system of coordinates, i.e., to

9.6.8
$$u\frac{\partial u}{\partial x}+v\frac{\partial u}{\partial y}=-\frac{1}{\varrho}\frac{\partial p}{\partial x}+v\frac{\partial^2 u}{\partial y^2},$$

$$\frac{\partial p}{\partial y}=0,$$

$$\frac{\partial u}{\partial x}+\frac{\partial v}{\partial y}=0,$$

in the original system. These are the boundary layer equations. Since all other quantities in the first equation of 9.6.7 are standard, we conclude that v' must be assumed standard. (Strictly speaking, this argument depends on the assumption that $\partial^2 u/\partial y'^2$ is not identically zero.)

At the plate, $u=v'=0$. In addition we have to link u',v with the external flow components $\bar{u}$ and $\bar{v}$. We shall consider conditions above the plate. Corresponding arguments hold below the plate.

As y' tends to infinity through the standard real positive numbers, $y=y'\delta$ remains infinitesimal and bounded, for example by the infinitesimal number $\sqrt{\delta}$. Nevertheless, we shall assume that

9.6.9
$$\lim_{y'\to\infty} u(x,y')=\bar{u}(x,0).$$

That is to say, we assume that $u(x,y)$ realizes its external flow value already in the interior of the monad of $y=0$. Correspondingly, we may amend our earlier definition of the domain of validity of the expressions for the external flow by supposing that

9.6.10 $u(x,y) = \bar{u}(x,y), \qquad v(x,y) = \bar{v}(x,y)$

above the plate for all y such that y/δ is infinite.

In order to complete the set of boundary conditions, we add

9.6.11 $\lim_{x \downarrow 0} u(x,y') = \bar{u}(0,0).$

In terms of the original system, 9.6.9 and 9.6.11 are equivalent to

9.6.12 $\lim_{y/\delta \to \infty} u(x,y) = \bar{u}(x,0)$

and

9.6.13 $\lim_{x \downarrow 0} u(x,y) = \bar{u}(0,0).$

Although 9.6.13 is a commonly accepted condition for the leading edge, the validity of 9.6.13, and hence 9.6.12 warrants further investigation.

Suppose the functions $u, v'p$ provide a solution for 9.6.7 with the boundary conditions 9.6.9 and 9.6.11. Since the functions on the left hand side of the equations in 9.6.7 are standard, we conclude that, for $x > 0$,

$$u(x,y') \simeq \bar{u}(x,0)$$

for all infinite y'. Transforming back to the original system, we conclude that

9.6.14 $u(x,y) \simeq \bar{u}(x,0)$

for infinite y/δ.

In the standard procedure, one replaces 9.6.12 by

9.6.15 $\lim_{y \to \infty} u(x,y) = \bar{u}(x,0)$

and one may then expect 9.6.14 to be satisfied approximately for infinite y/δ. There is a conceptual difficulty in using boundary conditions such as 9.6.15, which refer to very large y, in order to determine the field of flow as a solution of equations which apply only for very small y. Since it is known that in practice acceptable solutions are obtained by the use of such boundary conditions, the difficulty has long since become a grain de beauté of boundary layer theory.

However, we may equally well continue to work with 9.6.7 where this difficulty does not occur since y remains infinitesimal although y' becomes infinite. Actually, the problem of solving 9.6.7 with the boundary condition 9.6.9 and $u=v'=0$ at the plate is formally equivalent to the problem of solving 9.6.8 for the boundary conditions 9.6.9 and $u=v=0$ at the plate (for v read v'). For example, for the case of zero external pressure gradient

9.6.16 $p=\text{const.}, \qquad \bar{u}=\text{const}=U$

we may find the solution by a well known procedure, which depends on the introduction of the variable

9.6.17 $$\eta=\tfrac{1}{2}\sqrt{\frac{U}{v'x}}\,y'.$$

We introduce a modified stream function ψ' which is linked to u and v' by $u=\partial\psi'/\partial y'$, $v'=-(\partial\psi'/\partial x)$. Assuming $\psi'=\sqrt{(Uv'x)}f(\eta)$ we find from the first equation of 9.6.7 that

9.6.18 $$\frac{d^3f}{d\eta^3}+f\frac{d^2f}{d\eta^2}=0, \qquad 0<y<\infty.$$

The appropriate boundary conditions are $f=df/d\eta=0$ for $\eta=0$ and $df/d\eta=2$ for $\eta\to\infty$, and several methods of integration are available for this problem.

The expression for the shear stress at the plate is

9.6.19 $$\tau=\mu\left(\frac{\partial u}{\partial y}+\frac{\partial v}{\partial x}\right)=\left[\varrho v\left(\frac{1}{\delta}\frac{\partial u}{\partial y'}+\delta\frac{\partial v'}{\partial x}\right)\right]_{y'=0}$$

Hence

9.6.20
$$\frac{\tau}{\sqrt{v}} = \varrho \sqrt{v'} \left[\frac{\partial u}{\partial y'} + \delta^2 \frac{\partial v'}{\partial x} \right]_{y'=0}$$

$$\simeq \varrho \sqrt{v'} \left(\frac{\partial u}{\partial y'} \right)_{y'=0}$$

But, for the case under consideration,

$$\frac{\partial u}{\partial y'} = \frac{\partial^2 \psi'}{\partial y'^2} = \tfrac{1}{4} U \sqrt{\frac{U}{v'x}} f''(\eta),$$

and so

$$\varrho \sqrt{v'} \left(\frac{\partial u}{\partial y'} \right)_{y'=0} = \tfrac{1}{4} \varrho U \sqrt{\frac{U}{x}} f''(0).$$

Hence, from 9.6.20,

$$\sqrt{\frac{U}{v}} \tau \simeq \tfrac{1}{4} \varrho U^2 \frac{f''(0)}{\sqrt{x}}.$$

For a plate of length c, this leads to the following expression for the drag D, taking into account both top and bottom surfaces,

9.6.21
$$\sqrt{\frac{Uc}{v}} D = 2 \sqrt{\frac{Uc}{v}} \int_0^c \tau \, dx \simeq \sqrt{c} \int_0^c \tfrac{1}{2} \varrho U^2 \frac{f''(0)}{\sqrt{x}} \, dx = \varrho U^2 c f''(0).$$

Introducing the Reynolds number $R_e = Uc/v$ on the left hand side of 9.6.21, so that R_e is infinite, we obtain

9.6.22
$$\sqrt{R_e} D \simeq \varrho U^2 c f''(0)$$

For practical purposes, 9.6.22 may be interpreted as stating that for large Reynold's numbers, $\sqrt{R_e} D$ is approximately equal to $\varrho U^2 c f''(0)$ and this is in fact a well-known conclusion of boundary layer theory. We have gone through the calculation in order to show that it can be carried out consistently within our framework. It is significant that the particular choice of δ, which serves as a sort of unit of measurement for the infinitesi-

mal region in which we are interested, does not affect the final result
obtained for the drag. Nor does it affect the results obtained for other
physical quantities such as the displacement thickness.

This conclusion applies also in the general case. For let $\delta = \sqrt{v}$, which is
a permissible choice for δ since $\sqrt{v}$ is infinitesimal. Then $v' = v/\delta^2 = 1$. Let
the corresponding solution of 9.6.7 for the stated boundary conditions be
given by $u_0(x,y')$, $v_0'(x,y')$. A simple argument shows that for any other
choice of δ, a solution is then provided by $u(x,y') = u_0(x, y'/\sqrt{v'})$,
$v'(x,y') = v_0'(x, y'/\sqrt{v'})$. The equations for the shear stress, 9.6.19 and
9.6.20 are still applicable and so

$$\frac{\tau}{\sqrt{v}} \simeq \varrho \sqrt{v'} \left(\frac{\partial u}{\partial y'}\right)_{y'=0} = \varrho \sqrt{v'} \left(\frac{\partial u_0}{\partial y'}\right)_{y'=0} = \varrho \left(\frac{\partial u_0}{\partial y''}\right)_{y''=0}.$$

where $u_0 = u_0(x, y'')$, $y'' = y'/\sqrt{v'} = y/\sqrt{v}$. But $\varrho(\partial u_0/\partial y'')_{y''=0}$ is precisely the
corresponding expression obtained for $\delta = \sqrt{v}$, $v' = 1$. This shows that our
particular choice of δ does not affect the result, so long as δ is positive
infinitesimal and v' is finite and, moreover, standard.

9.7 Saint-Venant's principle. The hydrodynamic problem of section 9.5
has an older counterpart in the Theory of Elasticity. We recall the following facts and common notation of that theory.

The components of the *strain tensor* will be denoted by e_{xx}, e_{xy}, e_{xz}, etc.
so that $e_{xy} = e_{yx}$, etc. The *strain energy function* W is a positive quadratic
form of the six distinct components of the strain tensor. The components
of the *stress tensor*, X_r, X_y, X_z, etc., are given by

9.7.1 $$X_x = \frac{\partial W}{\partial e_{xx}}, \qquad X_y = Y_x = \frac{\partial W}{\partial e_{xy}}, \quad \text{etc.}$$

Also, in the absence of body forces,

9.7.2 $$\frac{\partial X_x}{\partial x} + \frac{\partial X_y}{\partial y} + \frac{\partial X_z}{\partial z} = 0,$$

$$\frac{\partial Y_x}{\partial x} + \frac{\partial Y_y}{\partial y} + \frac{\partial Y_z}{\partial z} = 0,$$

$$\frac{\partial Z_x}{\partial x} + \frac{\partial Z_y}{\partial y} + \frac{\partial Z_z}{\partial z} = 0.$$

Throughout the discussion, we shall adopt the assumptions of the classical Theory of Elasticity, and we shall confine ourselves to isotropic and homogeneous solids.

Let the volumes V, V_1, V_2 and the surfaces S, S_1, S_2, S_3 be subject to the conditions stated at the beginning of section 9.5, so that V is divided by S_3 into V_1 and V_2, etc. For any volume A, we define

$$9.7.3 \qquad D_A[F] = \int_A W \, d\tau,$$

where F is a symbol which stands for a particular field of strain in A or, if we wish, for the strain tensor, and $W = W[F]$ is the corresponding strain energy function. Thus, $D_A[F]$ is the strain energy stored in a material volume A which is subject to a strain F. Given two fields of strain in A, F and F', we define a scalar function $U[F, F']$ with domain A by

$$U[F, F'] = \tfrac{1}{2}[X_x e'_{xx} + Y_x e'_{xy} + Z_x e'_{xz} + \cdots],$$

where $X_x, \ldots$ are the stress components of the field F and $e'_{xx}, \ldots$ are the components of strain given by F'. Then $U[F', F] = U[F, F']$ and $U[F, F] = W[F]$. Put

$$9.7.4 \qquad D_A[F, F'] = \int_A U(F, F') \, d\tau.$$

Now let F, F_1, F_2 be three fields of strain which are defined in V, V_1, V_2, respectively, and are continuous in the closures of the respective volumes. Suppose also that the stresses given by F_1 and F respectively are the same across S_1; that the stresses given by F_2 and F respectively, are the same across S_2 and that the stresses given by F_2 and F_3 respectively, are the same across S_3. By this we mean, for example, that if $d\sigma$ is an element of S_3, then the stresses on $d\sigma$ which are due to F_2 and F_3, coincide. Under these conditions, we propose to show (compare 9.5.2) that

$$9.7.5 \qquad D_V[F] \leq D_{V_1}[F_1] + D_{V_2}[F_2].$$

For this purpose, we verify first that

$$9.7.6 \qquad D_{V_1}[F, F_1 - F] + D_{V_2}[F, F_2 - F] = 0,$$

where $F_1 - F$ and $F_2 - F$ are the fields obtained by subtracting the corresponding strain tensors. Thus, the stresses due to $F_1 - F$ and $F_2 - F$ vanish on S_1 and S_2 respectively, and are the same across S_3. Hence, if (u, v, w) is the displacement vector due to F, and (X_n, Y_n, Z_n) is the stress vector on S_3 due to both $F_1 - F$ and $F_2 - F$, then the first and second terms on the left hand side of 9.7.6 are different in sign but are both equal in magnitude to

$$\int_{S_3} (X_n u + Y_n v + Z_n w)\, d\sigma.$$

This proves 9.7.6. Referring back to 9.5 we see that 9.7.6 is analogous to 9.5.5. Proceeding as in that section we arrive at

$$D_{V_1}[F_1] + D_{V_2}[F_2] = D_V[\phi] + D_{V_1}[F_1 - F] + D_{V_2}[F_2 - F].$$

This proves 9.7.5 since $D_{V_1}[F_1 - F] \geq 0$, $D_{V_2}[F_2 - F] \geq 0$.

Suppose that S_1 is divided into two parts, S_0 and Q and let Φ be a system of external forces, or stresses, acting on Q and *in static equilibrium* (i.e., with zero resultant force and moment). We make the assumption (which is usually taken for granted in the Theory of Elasticity) that Φ determines a unique field of strain in any material volume which will be considered here. Thus, let F be the field of strain produced by Φ in V, and G_1 the field of strain produced by Φ in V_1, no other tractions across the boundaries of V and V_1 being present, respectively.

The stresses due to F across S_3 must be in static equilibrium, since there are no other surface tractions on the boundary of V_2. Thus, there exists a field of strain G_2 in V_1 which is free of stresses across S_1 and whose stresses across S_3 coincide with those due to F. Then $F = G_1 + G_2$ in V_1 since the stresses due to F across the boundary of V_1 coincide with those due to $G_1 + G_2$. Hence,

$$D_{V_1}[F] = D_{V_1}[G_1 + G_2] = D_{V_1}[G_1] + D_{V_1}[G_2] + 2D_{V_1}[G_1, G_2].$$

Introducing the fields $(1 + \lambda)G_2 + G_1$ where λ is a real variable and taking into account 9.7.5 we obtain, exactly as in section 9.5 (see 9.5.11 and 9.5.12),

9.7.7 $$D_V[F] \leq D_{V_1}[G_1]$$

and

9.7.8 $$D_{V_2}[F] \leq D_{V_1}[G_1] - D_V[F],$$

where the right hand side of 9.7.8 is evidently non-negative.

Continuing as in 9.5, we consider a one-parametric family of finite elastic bodies $B(t)$ defined for $t > 0$ such that $B(t_1)$ is part of $B(t_2)$ for $t_1 \leq t_2$ and such that there is an area, Q, which is common to the surfaces of all the $B(t)$. For example, $B(t)$ may be a prismatic bar whose generators are parallel to the x-axis with bases at $x = 0$ and $x = t$, while Q is the base at $x = 0$. Let Φ be a system of forces, or stresses, acting on Q and in static equilibrium. Φ will be kept fixed throughout the discussion. Let F_t be the field of strain produced by the given loading in $B(t)$ so that $D(t) = D_{B(t)}[F_t]$ is the corresponding strain energy in $B(t)$. For $t_2 \geq t_1$ let $D(t_1, t_2)$ be the strain energy due to the loading Φ of $B(t_2)$, and stored in the excess volume of $B(t_2)$ over $B(t_1)$. Then (compare 9.5.13),

9.7.9 $$D(t_1, t_2) \leq D(t_1) - D(t_2) \quad \text{for} \quad 0 < t_1 \leq t_2.$$

Passing to $*R$, we conclude, as in 9.5, that

9.7.10 $$D(t_1, t_2) \simeq 0$$

for all infinite t_1 and $t_2 \geq t_1$.

Again, we wish to deduce from 9.7.10 that the strain components themselves are infinitesimal at all points which belong to the S-interior of $B(t_2) - B(t_1)$. This can be proved by elementary methods and it will be sufficient to indicate the main steps.

We first write the strain energy function W in the form

9.7.11 $$W = \tfrac{1}{2}\lambda(e_{xx} + e_{yy} + e_{zz})^2 + \mu(e_{xx}^2 + e_{yy}^2 + e_{zz}^2 + \tfrac{1}{2}e_{yz}^2 + \tfrac{1}{2}e_{zx}^2 + \tfrac{1}{2}e_{xy}^2),$$

where λ and μ now are the elastic constants of the material. 9.7.11 shows that the dilatation $\Delta = e_{xx} + e_{yy} + e_{zz}$ satisfies the inequality $\Delta^2 \leq 4W^2/\lambda^2$. Hence, for any volume A,

9.7.12
$$\int_A \Delta^2 \, d\tau \le \frac{4}{\lambda^2} D_A[F],$$

where F is the field of flow whose strain energy function is W. Thus, $\int_A \Delta^2 \, d\tau$ is infinitesimal if $D_A[F]$ is infinitesimal. But Δ is known to be a solution of Laplace's equation. Using the mean value theorem of potential theory we conclude as in 9.5 that Δ is infinitesimal at any point which belongs to the S-interior of $B(t_2) - B(t_1)$, i.e., which can be surrounded by a sphere of standard radius whose interior belongs entirely to $B(t_2) - B(t_1)$. Moreover, using Poisson's integral, we may then show that the first and second (and higher) derivatives of Δ also are infinitesimal in the S-interior of $B(t_2) - B(t_1)$.

Now the equations of elastic equilibrium,

9.7.13
$$(\lambda + \mu) \frac{\partial \Delta}{\partial x} + \mu \nabla^2 u = 0, \quad \text{etc.,}$$

show that the components of strain, $e_{xx} = \partial u/\partial x$, $e_{xy} = \partial v/\partial x + \partial u/\partial y$, etc., are solutions of Poisson's equation.

9.7.14
$$\nabla^2 f(x,y,z) = -\sigma,$$

where

$$\sigma = \left(1 + \frac{\lambda}{\mu}\right) \frac{\partial^2 \Delta}{\partial x^2} \qquad \text{for} \qquad f = e_{xx}$$

and

$$\sigma = 2\left(1 + \frac{\lambda}{\mu}\right) \frac{\partial^2 \Delta}{\partial x \partial y} \qquad \text{for} \qquad f = e_{xy},$$

with similar expressions for the remaining components. Thus, σ is infinitesimal in the S-interior of $B(t_2) - B(t_1)$. At the same time, 9.7.12 shows that the integral $\int_A f^2 \, d\tau$ is infinitesimal for $A = B(t_2) - B(t_1)$. Using particular solutions of 9.7.14, of the form $\int (\sigma/4\pi r) \, d\tau$, it is then not difficult to prove that $f(x,y,z)$ is itself infinitesimal in the S-interior of $B(t_1) - B(t_1)$. Hence, finally, *the components of strain and hence, the components of stress, are infinitesimal in the S-interior of $B(t_2) - B(t_1)$. In* particular, if P is an infinite point which belongs to the S-interior of $B(t_2)$, t_2 infinite, then we can find an infinite $t_1 < t_2$ such that P belongs to the S-interior of $B(t_2) - B(t_1)$. Hence,

9.7.15 THEOREM.   If $B(t)$ defines a family of bodies as detailed, and P is any infinite point in the S-interior of some body $B(t_0)$ for infinite t_0 then the strains and stresses at P due to a standard self-equilibrating load distribution on Q, are infinitesimal.

We may 'translate' 9.7.15 into a standard theorem by a procedure which has been used many times previously. The result is

9.7.16 THEOREM (Standard).   Let $B(t)$ be a family of elastic bodies as detailed, and let Φ be a specified load distribution on Q, which is in static equilibrium. For any $\varepsilon > 0$ and $\delta > 0$ there exists a positive $R_0 = R_0(\varepsilon, \delta)$ such that for any member $B(t_0)$ of the family, and any point P whose distance from the surface of $B(t_0)$ is greater than ε, and whose distance from Q is greater than R_0, the stresses at P are smaller in magnitude than δ. There is a corresponding result for strains.

Saint-Venant's original principle, which was verified by him experimentally in certain cases, asserts that the stresses due to a local distribution of loads which are in static equilibrium are negligible at any distance from the loaded region which is sufficiently large compared with the dimensions of the region. Theorem 9.7.16 establishes this principle not only for cylindrical bars, but also e.g., for bars with variable cross section. The restriction to regions which are at some distance ε from the surface of the body is inevitable, since the principle may actually fail at the surface. On the other hand, 9.7.16 applies to loads which are distributed over the entire base of a bar, as required, for example, for the classical applications of the principle to problems of torsions and flexure. As proved here, there is a certain uniformity in the conclusion since $R_0(\varepsilon, \delta)$ is actually independent of the particular choice of $B(t_0)$. However, it should be possible to go further and to prove that the same R_0 is suitable for a wide variety of load distributions and of elastic bodies.

9.8 Remarks and references.   For the classical proof of Riemann's mapping theorem whose adaptation to Non-standard Analysis is sketched in 9.2, compare DIENES [1931]. A related method is given in HILLE [1962].

A standard text on Dirichlet's principle is COURANT [1950]. The hydro-dynamical principle of section 9.5 has been formulated by analogy with Saint-Venant's principle (9.7). I have not seen it stated elsewhere. The methods used in 9.5 and 9.7 were inspired in part by energy arguments due to O. Zanaboni (ZANABONI [1937]), but our results are more precise. Another relatively recent attempt to endow Saint-Venant's principle with a definite meaning is due to E. Sternberg, who used a Green's function method (STERNBERG [1954]). However, it appears that Sternberg's results are not applicable to external forces which are distributed over a finite area, although it is for such cases that Saint-Venant's principle is most important in practice. While carrying out the analysis of 9.7, I had the benefit of stimulating discussions with R. A. Toupin. More recently, Toupin has obtained some interesting quantitative formulations of Saint-Venant's principle for cylindrical bars.

The reader is referred to SCHLICHTING [1955] or THWAITES [1960] for information on boundary layer theory; and to WARD [1955] or ROBINSON and LAURMANN [1956] for an introduction to the theory of linearized supersonic flow.

CHAPTER X

CONCERNING THE HISTORY OF THE CALCULUS

10.1 Introduction. The history of a subject usually is written in the light of later developments. For over half a century now, accounts of the history of the Differential and Integral Calculus have been based on the belief that even though the idea of a number system containing infinitely small and infinitely large elements might be consistent, it is useless for the development of Mathematical Analysis. In consequence, there is in the writings of this period a noticeable contrast between the severity with which the ideas of Leibniz and his successors are treated and the leniency accorded to the lapses of the early proponents of the doctrine of limits. We do not propose here to subject any of these works to a detailed criticism. However, it will serve as a starting point for our discussion to try and give a fair summary of the contemporary impression of the history of the Calculus, as follows.

'After a long development which involved the determination of areas, volumes, and tangents in special cases, the general theory of differentiation and integration was put forward in the second half of the seventeenth century by Newton and, somewhat later but independently, by Leibniz. As far as the foundations of the new subject were concerned, Newton vacillated, referring sometimes to infinitesimals, sometimes to limits, and sometimes to basic physical intuition, and his immediate successors gave preference to the last-mentioned approach. On the other hand, Leibniz and his followers based the development of the theory on infinitely small differentials of first and higher orders. The technical advantages of the differential notation at first helped to produce rapid progress in the Calculus and its applications on the continent of Europe, where it was adopted. However, after a while the glaring internal contradictions of the theory led to the realization that alternative foundations were required. Lagrange believed that he had discovered a suitable approach by taking the Taylor expansion of a function as fundamental. However the satisfactory solution to the problem was provided by Cauchy who gave the first rigorous development of Mathematical Analysis. Cauchy based his theory on the concept of the limit, which, after Newton, had been advocated again by d'Alembert. Cauchy's method was subsequently given a more formal guise by Weierstrass, who had been anticipated to some extent by Bolzano. As the theory of limits became firmly established, the use of infinitely small and infinitely large quantities in Analysis became discredited and survived only as a manner of speaking, e.g., in the statement that a variable tends to infinity. The significance of subsequent developments in the theory of non-archimedean fields was confined entirely to Algebra.'

Since (as we believe) we have shown that the theory of certain types of non-archimedean fields can indeed make a positive contribution to classical Analysis, it seems appropriate to conclude our work with a number of remarks which attempt to indicate in what ways the picture sketched above should be supplemented or even redrawn. Our comments will by necessity be fragmentary since a complete history of the Calculus would be beyond the scope of this book.

10.2 Leibniz. In spite of occasional slight inconsistencies, Leibniz' attitude towards infinitely small and infinitely large quantities in the Calculus remained basically unchanged during the last two decades of his life. He approved entirely of their introduction, but thought of them as ideal elements, rather like the imaginary numbers. These ideal elements are governed by the same laws as the ordinary numbers. However, they are only a useful fiction, adopted in order to shorten the argument and to facilitate mathematical invention (or, discovery). And if one so desires, one can always dispense with infinitely small or infinitely large numbers and revert to the style of the mathematicians of antiquity by arguing in terms of quantities which are large enough or small enough to make the error smaller than any given number.

All this is stated clearly and repeatedly in Leibniz' writings. We proceed to quote some passages which are representative of his point of view.

'... on n'a pas besoin de prendre l'infini ici à la rigueur, mais seulement comme lorsqu'on dit dans l'optique, que les rayons du soleil viennent d'un point infiniment éloigné et ainsi sont estimés parallèles. Et quant il y a plusieurs degrés d'infini ou infiniment petits, c'est comme le globe de la terre est estimé un point à l'égard de la distance des fixes, et une boule que nous manions est encore un point en comparaison du semidiamètre du globe de la terre, de sorte que la distance des fixes est un infiniment infini ou infini de l'infini par rapport au diamètre de la boule. Car au lieu de l'infini ou de l'infiniment petit, on prend des quantités aussi grandes et aussi petites qu'il faut pour que l'erreur soit moindre que l'erreur donnée, de sorte qu'on ne diffère du style d'Archimède que dans les expressions, qui sont plus directes dans notre méthode et plus conformes à l'art d'inventer.' (LEIBNIZ [1701]).

'... D'où il s'ensuit, que si quelqu'un n'admet point des lignes infinies et infiniment petites à la rigueur métaphysique et comme des choses réelles, il peut s'en servir sûrement comme des notions idéales qui abrègent le raisonnement, semblable à ce qu'on appelle racines imaginaires dans l'analyse commune (comme par exemple $\sqrt{-2}$)... C'est encore de la même façon qu'on conçoit des dimensions au delà de trois ..., le tout pour établir des idées propres à abréger les raisonnements et fondées en réalités.

Cependant il ne faut point s'imaginer que la science de l'infini est dégradée par cette explication et réduite à des fictions; car il reste toujours un infini syncategorématique, comme parle l'école, et il demeure vrai par exemple que 2 est autant que $\frac{1}{1}+\frac{1}{2}+\frac{1}{4}+\frac{1}{8}+\frac{1}{16}+\frac{1}{32}$, etc., ce qui est une série infinie dans laquelle toute les fractions dans les numérateurs sont 1 et les dénominateurs de progression géometrique double, sont comprises á la fois, quoiqu'on n'y emploie toujours que des nombres ordinaires et quoiqu'on n'y fasse point entrer aucune mention infiniment petite, ou dont le dénominateur soit un nombre infini ...

... et il se trouve que les règles du fini réussissent dans l'infini comme s'il y avait des atomes (c'est à dire des élements assignables de la nature), quoiqu'il n'y en ait point la matière étant actuellement sousdivisée sans fin; et que vice versa les règles de l'infini réussissent dans le fini, comme s'il y'avait des infiniment petits métaphysiques, quoiqu'on n'en n'ait point besoin; et que la division de la matière ne parvienne jamais à des parcelles infiniment petites: c'est parce que tout se gouverne par raison, et qu'-autrement il n'aurait point de science ni règle, ce qui ne serait point conforme avec la nature du souverain principe.' (LEIBNIZ [1701]).

'Entre nous je crois que M. de Fontenelle, qui a l'esprit galant et beau, en a voulu railler, lorsqu'il a dit qu'il voulait faire des éléments métaphysiques de notre calcul. Pour dire le vrai, je ne suis pas trop persuadé moi-même qu'il faut considérer nos infinis et infiniment petits autrement que comme des choses idéales et comme des fictions bien fondées. Je crois qu'il n'y a point de créature au dessous de laquelle n'ait une infinité de créatures, cependant je ne crois point qu'il y en ait, ni même qu'il en puisse avoir d'infiniment petites et c'est ce que je crois pouvoir démontrer.' (LEIBNIZ [1702a]).

Finally we quote from a letter which was written towards the end of Leibniz' life.

'II. Pour ce qui est du calcul des infinitésimales, je ne suis pas tout à fait content des expressions de Monsieur Herman dans sa réponse à Monsieur Nieuwentijt, ni de nos autres amis. Et M. Naudé a raison d'y faire des oppositions. Quand ils se disputèrent en France avec l'Abbé Gallois, le Père Gouge et d'autres, je leur témoignai, que je ne croyais point qu'il y eût des grandeurs véritablement infinies ni véritablement infinitésimales, que ce n'étaient que des fictions mais des fictions utiles pour abréger et pour parler universellement... Mais comme M. le Marquis de l'Hospital croyait que par là je trahissais la cause, ils me prièrent de n'en rien dire, outre ce que j'en avais dit dans un endroit des actes de Leipsic, et il me fut aisé de déférer à leur prière.' (LEIBNIZ [1716]).

The last sentence provides a reason for a certain lack of decisiveness in some of the statements quoted earlier. (For the passage mentioned as having been published in the Actes de Leipsic, see LEIBNIZ [1689].)

However, although Leibniz assures us that the infinitely small and infinitely large numbers are ideal elements which are only introduced for convenience, and that arguments involving them are equivalent to the method of Archimedes (i.e., the so-called method of exhaustion), there is no attempt to justify this claim except for the appeal to the 'sovereign principle'. The same principle is supposed to serve as a justification for the assumption that the infinitely small and infinitely large numbers obey the same laws as the ordinary (real) numbers. Elsewhere, Leibniz appeals to his principle of continuity to justify this assumption.

We note that although Leibniz wishes to make it clear that he can dispense with infinitely small and infinitely large numbers and at any rate does not ascribe to them any absolute reality, he accepts the idea of potential ('syncategorematic') infinity and finds it in the totality of terms of a geometrical progression.

We shall make no further reference to Nieuwentijt and Fontenelle, beyond mentioning that Nieuwentijt was the author of two works (NIEUWENTIJT [1694, 1695]) in which among other things he rejected the notion of higher order differentials, while Fontenelle became famous by giving full reign to the tendencies criticized here (FONTENELLE [1727]). Leibniz replied to Nieuwentijt in the Acta Eruditorum of Leipzig (LEIBNIZ [1695]).

10.3 De l'Hospital. The marquis de l'Hospital was the author of the first textbook of the Differential Calculus, 'Analyse des infiniment petits pour l'intelligence des lignes courbes', whose first edition was published in 1696. De l'Hospital was a loyal disciple of Leibniz, and adhered consistently to his methods. In the preface to his book de l'Hospital acknowledges his indebtedness to Leibniz and to the Bernoulli brothers without mentioning details. However, in the present discussion we shall not be concerned with the problem of assigning priorities but with the evolution of the fundamental notions of the calculus.

De l'Hospital begins his work with the statement of two definitions and a corollary followed by two axioms. The essential parts of these definitions and axioms are as follows (DE L'HOSPITAL [1715]).

'Définition I. On appelle quantités *variables* celles qui augmentent ou diminuent continuellement; et au contraire quantités *constantes* celles qui demeurent les mêmes pendant que les autres changent. ...'

'Définition II. La portion infiniment petite dont une quantité variable augmente ou diminue continuellement, en est appeleé la *différence* ...'

For *difference* read *differential*. The sentence just quoted is followed by examples of differences (i.e., differentials) of the coordinates and of the arc of a curve, and of the area under a curve.

'I. Demande ou supposition. On demande qu'on puisse prendre indifféremment l'une pour l'autre deux quantités qui ne diffèrent entr'elles que d'une quantité infiniment petite: ou (ce qui est la même chose) qu'une quantité qui n'est augmentée ou diminuée que d'une autre quantité infiniment moindre qu'elle, puisse être considérée comme demeurant la même...'

'II. Demande ou supposition. On demande qu'une ligne courbe puisse être considérée comme l'assemblage d'une infinité de lignes droites, chacune infiniment petite: ou (ce qui est la même chose) comme un polygône d'un nombre infini de côtés, chacun infiniment petit, lesquels déterminent par les angles qu'ils font entr'eux, la courbure...'

Here again, we have omitted the examples, which refer to a diagram.

In laying down definitions and axioms as the point of departure of his theory, de l'Hospital of course, merely followed the classical examples of Euclid and Archimedes. We may recall that a short time earlier, in 1687, Newton had adopted a similar scheme with regard to the physical notions and assumptions of his Mathematical Principles of Natural Philosophy. We may also point out that in the approach of Euclid and Archimedes, which is also the approach of de l'Hospital, a definition frequently is an explication of a previously given and intuitively understood concept, and an axiom is a true statement from which later results are obtained deductively. Thus, the axiomatic approach implied a belief in the reality of the objects mentioned (e.g., differentials), which, as we have seen, Leibniz was unwilling to concede. It is interesting to contrast this with our present attitude according to which an axiom implies no ontological commitment (except possibly in set theory).

Coming next to the details of the two axioms, they were both contained in the works of Leibniz as permissible assumptions. The assumption given as the second axiom, in particular, has a long history and has been traced back to antiquity. The first axiom is of greater importance since it attempts to provide a basis for the justification of the fundamental rules of the differential calculus. Additional interest is lent to the axiom by the fact that it implies, clearly and immediately, a glaring contradiction, i.e., that the (infinitesimal) difference between two quantities may, at the same time, be both equal to, and different from, zero. In this connection it is of interest to read Leibniz' formulation of the same principle in a letter to de l'Hospital.

'... et je compte pour égales les quantités dont la différence leur est incomparable. J'appelle grandeurs incomparables dont l'une multipliée par quelque nombre fini que ce soit, ne saurait excéder l'autre, de la même façon qu'Euclide l'a pris dans sa cinquième définition du cinquième livre.' (LEIBNIZ [1695a]). The definition mentioned here is actually the *fourth* definition of the fifth book of Euclid according to the version presently accepted (EUCLID).

Berkeley pointed out some forty years later (BERKELEY [1734]) that the same weakness was present in the method of fluxions. This is true although the absence of a clear basis for that method made it harder to argue against. It must also be said that the lore of the method of fluxions provided a better starting point for the ideas of d'Alembert and Cauchy (see below). However, there can be little doubt that the inconsistency in question contributed to the eventual eclipse of the method of infinitely small and infinitely large numbers at the beginning of the nineteenth century. In Non-standard Analysis, the inconsistency does not arise since two numbers a and b which differ only by an infinitesimal quantity are equivalent, $a \simeq b$, but not necessarily equal. It is true that this leads to a formalism which is in some ways more complicated than that introduced by Leibniz. Thus, for infinitesimal dx and corresponding dy we only have $dy/dx \simeq f'(x)$, so that $^0(dy/dx) = f'(x)$, while $dy/dx = f'(x)$ directly, according to Leibniz. However, this is a small price to pay for the removal of an inconsistency.

Another weakness of Leibniz' theory was that neither he nor his successors were able to state *with sufficient precision* just what rules were supposed to govern their extended number system. Leibniz did say, in one of the passages quoted above, that what succeeds for the finite numbers succeeds also for the infinite numbers and vice versa, and this is remarkably close to our transfer of statements from R to $*R$ and in the opposite direction. But to what sort of laws was this principle supposed to apply? To take a non-trivial example, why did not Archimedes' axiom apply to the extended number system? The possible answer might have been that the principle was supposed to apply only to the algebraic laws of the system but this permits the application of the idea only in the simplest cases. In fact, what was lacking at the time was a formal language which would have made it possible to give a precise expression of, and delimitation to, the laws which were supposed to apply equally to the finite numbers and to the extended system including infinitely small and infinitely large numbers as well. In our own theory the answer to the question whether Archimedes' axiom is true not only in R but also in $*R$ is, yes and no! If by 'Archimedes' axiom' we mean the sentence of K which formalizes the statement 'For every pair of real numbers a and b such that $0 < a < b$ there exists a natural number n for which $b < na$' then this sentence holds in R and hence also in $*R$. In the latter system n may then be realized by an infinite natural number. However, if we mean the postulate that for any pair of real numbers a and b, $0 < a < b$ there exists a natural number n

in the ordinary sense such that

$$b < a + a + a + \cdots + a \qquad (n \text{ times})$$

then this postulate is not true for $*R$.

10.4 Lagrange and d'Alembert. For over fifty years Leibniz' method ruled supreme on the European continent. However, in the second half of the eighteenth century the doubts regarding the soundness of the theory, or rather its evident lack of soundness, led to the call for alternative foundations. Lagrange proposed to base the entire theory on the supposed existence of a Taylor expansion for any (continuous) function (LAGRANGE [1797]). He thought that in this way the calculus could be developed independently of 'any consideration of infinitely small or vanishing numbers, or of limits, or of fluxions.' One might try to justify the procedure by choosing to regard it as a precursor to the method of formal power series, but this would be erroneous since Lagrange is concerned with the *values* of a function and of its expansion.

Nevertheless the episode is interesting. The fact that a man of Lagrange's stature could propound such a method, and that for a time it was quite widely accepted (compare LACROIX [1797]), serves to illustrate the point that the mathematicians of that age were worried about the basic nature of the concepts of the Calculus and not, primarily, about the deductive inadequacy of their methods (see 10.7 below).

D'Alembert's contribution is more relevant to our discussion. It is expounded in several articles of the Encyclopédie Méthodique (Mathématiques).

D'Alembert based the Calculus on the notion of limit. We quote some passages from the article under that heading.

'On dit qu'une grandeur est la *limite* d'une autre grandeur, quand la seconde peut approcher de la première plus près que d'une grandeur donnée, si petite qu'on la puisse supposer, sans pourtant que la grandeur qui approche, puisse jamais surpasser la grandeur dont elle approche; en sorte que la différence d'une pareille quantité à sa *limite* est absolument inassignable...'

'La théorie des *limites* est à la base de la vraie Métaphysique du calcul différentiel... A proprement parler, la limite ne coincide jamais, ou ne devient jamais égale, à la quantité dont elle est la limite; mais celle-ci s'en approche toujours de plus en plus et peut en différer aussi peu qu'on voudra...'

D'Alembert then proceeds to discuss the sum of a geometrical progression as an example of a limit.

The following passage is taken from the article *Différentiel*.

'... il ne s'agit point, comme on le dit encore ordinairement, de quantités infiniment petites dans le calcul *différentiel*; il s'agit uniquement de limites de quantités finies. Ainsi la métaphysique de l'infini et des quantités infiniment petites plus grandes ou plus petites les unes que les autres est totalement inutile au calcul *différentiel*. On ne se sert du terme d'*infiniment petit*, que pour abréger les expressions. Nous ne dirons donc pas avec bien des géomètres qu'une quantité est infiniment petite non avant qu'elle s'évanouisse, non après qu'elle est évanouie, mais dans l'instant même où elle s'évanouit; car que veut dire une définition si fausse, cent fois plus obscure que ce qu'on veut définir. Nous dirons qu'il n'y a point dans le calcul *différentiel* de quantités infiniment petites...'

We notice that the definition of the limit reads, at first sight, like an acceptable though informal statement of our present viewpoint, except for d'Alembert's insistence on the condition that a quantity which tends to a limit does not ever become equal to that limit. This condition is not only unnecessary but would prove extremely inconvenient in practice. For example, with d'Alembert's definition it would be incorrect to say that $x \sin (1/x)$ tends to zero as x tends to zero, since this function has this value also on a set of argument values different from zero but with zero as limit. The introduction of the restriction may well have been motivated by the feeling that if a quantity *tends* towards a limit it cannot get there beforehand.

The idea of a limit went back to Newton and had since been discussed repeatedly in England. The assumption of 'many geometers' that a quantity is infinitely small neither before nor after it vanishes but just at the moment of its vanishing, which is sharply criticized in the last of the passages quoted above, also is in the Newtonian tradition. On the other hand, Leibniz' remarks on the 'style of Archimedes' which was quoted earlier, while intended as a description of the method of exhaustion, also is fairly close to our contemporary notions on limits, and indeed there are related formulations antedating both Newton and Leibniz. At any rate, from a historical point of view, d'Alembert's intervention was important

because it helped to turn continental Mathematics towards the notion of
limit as the central concept of Analysis.

10.5 Cauchy. The quotations and comments given so far have amended
the picture drawn in section 10.1 chiefly by showing that Leibniz' theory
of infinitely small and infinitely large numbers was a serious attempt to
provide foundations for the Calculus which avoided the complication in-
herent in the Greek method of exhaustion and that, in spite of its inconsistencies, it may be regarded as a genuine precursor of the theory of
the present book. The description in 10.1 would now lead us to expect
that, following the rejection of Leibniz' theory by Lagrange and D'Alembert, infinitely small and infinitely large quantities would have no place
among the ideas of Cauchy, who is generally regarded as the founder of
the modern approach, or that they might, at most, arise as figures of
speech, as in 'x tends to infinity'. However, this expectation is mistaken.
We proceed to give a number of relevant quotations from Cauchy's Cours
d'Analyse (Analyse Algébrique) (CAUCHY [1821]).

'En parlant de la continuité des fonctions je n'ai pu me dispenser de
faire connaître les propriétés principales des quantités infiniment petites,
propriétés qui servent de base au calcul infinitésimal...'

'Quant aux méthodes, j'ai cherché à leur donner toute la rigueur qu'on
exige en géométrie, de manière à ne jamais recourir aux raisons tirées de la
généralité de l'algèbre. Les raisons de cette espèce, quoique assez com-
munément admises surtout dans le passage des séries convergentes aux
séries divergentes, et des quantités réelles aux expressions imaginaires, ne
peuvent être considérés, ce me semble, que comme des inductions propres
à faire pressentir quelquefois la vérité, mais qui s'accordent peu avec
l'exactitude si vantée des sciences mathématiques.'

'On nomme quantité variable celle que l'on considère comme devant
recevoir successivement plusieurs valeurs differentes les unes des autres...
Lorsque les valeurs successivement attribuées à une même variable
s'approchent indéfiniment d'une valeur fixe, de manière à finir par en
différer aussi peu que l'on voudra, cette dernière est appelée la limite de
toutes les autres.'

There follow two familiar examples, one being that of the circle as the
limit of inscribed polygons. Cauchy then continues:

'Lorsque les valeurs numériques successives d'une même variable décroissent indéfiniment, de manière à s'abaisser au'dessous de tout nombre donné, cette variable devient ce qu'on nomme un *infiniment petit* ou une quantité *infiniment petite.* Une variable de cette espèce a zéro pour limite.'

'Lorsque les valeurs numériques d'une même variable croissent de plus en plus, de manière a s'élever au dessus de tout nombre donné, on dit que cette variable a pour limite l'infini positif...'

Chapter II of the same book begins with a detailed discussion of the orders of the infinitely small and infinitely large quantities. Continuity is considered next. We quote,

'... la fonction $f(x)$ sera, entre les deux limites assignées à la variable x, fonction *continue* de cette variable, si, pour chaque valeur de x intermédiaire entre ces limites, la valeur numérique de la différence $f(x+\alpha) - f(x)$ décroît indéfiniment avec celle de α. En d'autres termes, *la fonction $f(x)$ restera continue par rapport à x entre les limites données, si, entre ces limites, un accroissement infiniment petit de la variable produit toujours un accroissement infiniment petit de la fonction elle-même.*'

We gather from the above passages that infinitely small quantities are fundamental in Cauchy's approach to Analysis. However, these quantities are not numbers but *variables*, or rather, states of variables whose limit is zero. As for variables in general they are not symbols (i.e., elements of the formal language which describes the system under consideration) but are regarded as basic entities as a special kind. If one tries to make this notion precise in contemporary terms one might perhaps describe a variable as a function whose range is numerical while its domain may be any ordered set without last element. However, Cauchy evidently does not wish to base the notion of a variable on the notion of a function; indeed, in a passage which is not quoted here he defines the notion of a function in terms which involve the notion of a variable. At the same time it should be mentioned that elsewhere Cauchy treats functions $f(x)$ such that $\lim_{x \to 0} f(x) = 0$ as infinitely small quantities, implying that functions may be regarded also as variables.

Whatever the precise picture of an infinitely small quantity may have been in Cauchy's mind, we may examine his subsequent definitions and see what they amount to if we interpret the infinitely small and infinitely large quantities mentioned in them in the sense of Non-standard Analysis. For the notion of continuity, Cauchy's definition may thus be interpreted as stating that for $f(x)$ defined in the interval $a < x < b$, $f(x)$ is continuous in that interval if, for infinitesimal α, the difference $f(x+\alpha)-f(x)$ is *always (toujours)* infinitesimal. If now we interpret 'always' as meaning 'for all standard x' then we obtain ordinary continuity in the interval but if by 'always' we mean 'for all x' then we obtain uniform continuity.

Cauchy regarded his theory of infinitely small quantities as a satis-factory foundation for the theory of limits and (d'Alembert's suggestion notwithstanding) he did not introduce the latter in order to replace the former. His proof procedures thus involved *both* infinitely small (and in-finitely large) quantities *and* limits. In spite of his own warning against arguments relying on the generality of Algebra, which was quoted above, and which may have been directed against Leibniz' principle of continuity he did not hesitate to add infinitely small quantities to one another and even to divide them by one another. The ε-technique does appear in his writings but we may take it, for example, that the condition that $|f(x+\alpha)-f(x)|<\varepsilon$ for $|\alpha|<\delta$ would have been regarded by him as a criterion for continuity and not as a definition.

We proceed to consider a famous error of Cauchy's, which has been discussed repeatedly in the literature. Cauchy introduces a series of functions

10.5.1 $$u_0(x)+u_1(x)+u_2(x)+\cdots$$

and states the following theorem (CAUCHY [1821]).

10.5.2 'Lorsque les différents termes de la série (1) (i.e., in our numbering 10.5.1) sont des fonctions d'une même variable x, continues par rapport à cette variable dans le voisinage d'une valeur particulière pour laquelle la série est convergente, la somme s de la série est aussi, dans le voisinage de cette valeur particulière, fonction continue de x.'

The assertion is that if all the functions $u_n(x)$ are continuous in an interval $a<x<b$ (which includes a particular x_0) and if the series 10.5.1 converges in that interval to a sum $s(x)$ then $s(x)$ also is continuous. For the proof, Cauchy introduces the partial sums $s_n(x)=u_0(x)+u_1(x)+\ldots+u_n(x)$ and the remainders $r_n(x)=s(x)-s_n(x)$ and then argues as fol-lows.

'Cela posé, considérons les accroissements que reçoivent ces trois fonctions (i.e., r_n, s_n and s), lorsqu'on fait croître x d'une quantité infiniment petite. L'acroissement de s_n sera, pour toutes les valeurs possibles de r_n une quantité infiniment petite; et celui de r_n deviendra insensible en même temps que r_n, si l'on attribue à n une valeur très considérable. Par suite, l'accroissement de la fonction x ne pourra être qu'une quantité infiniment petite.'

According to Cauchy this leads immediately to the theorem stated above. Interpreted in terms of Non-standard Analysis, the argument runs as follows. Let x_1 be a standard number such that $a < x_1 < b$. In order to prove that $s(x)$ is continuous at x_1 we attempt to show that $s(x_1 + \alpha) - s(x_1)$ is infinitesimal for all infinitesimal α. Now

10.5.3 $$s(x_1 + \alpha) - s(x_1) = (s_n(x_1 + \alpha) - s_n(x_1)) + (r_n(x_1 + \alpha) - r_n(x_1))$$

Following Cauchy's argument we might be inclined to claim that the left hand side is infinitesimal since $s_n(x_1 + \alpha) - s_n(x_1)$ is infinitesimal for all n and $r_n(x_1 + \alpha)$ and $r_n(x_1)$ are infinitesimal for all infinite n. However, this argument is erroneous. For although $r_n(x_1)$ is indeed infinitesimal for all infinite n (since x_1 is standard), $r_n(x_1 + \alpha)$ must needs be infinitesimal only for sufficiently *high* infinite n; while $s_n(x_1 + \alpha) - s_n(x_1)$ is infinitesimal for all *finite* n, and hence, by one of our basic lemmas (3.3.20), for sufficiently *small* infinite n. In order to prove that the left hand side of 10.5.3 is infinitesimal we have to ensure that there exists an n for which $r_n(x_1 + \alpha)$ and the difference $s_n(x_1 + \alpha) - s_n(x_1)$ are infinitesimal simultaneously. Two natural alternatives offer themselves, (i) to assume that 10.5.1 is uniformly convergent in the interval $a < x < b$, so that $r_n(x_1 + \alpha)$ is infinitesimal for *all* infinite n, or (ii) to assume that the family $\{s_n(x)\}$ is equicontinuous in the interval, so that $s_n(x_1 + \alpha) - s_n(x_1)$ is infinitesimal for *all* infinite n.

The classical example of a convergent series of continuous functions whose sum is not continuous everywhere is (ABEL [1826])

10.5.4 $$\sin x + \tfrac{1}{2}\sin 2x + \tfrac{1}{3}\sin 3x + \cdots$$

Indeed, 10.5.4. has the sum $-\tfrac{1}{2}(\pi + x)$ for $-\pi < x < 0$, the sum 0 for $x = 0$ and the sum $\tfrac{1}{2}(\pi - x)$ for $0 < x < \pi$. The sum is thus discontinuous at the point $x = 0$.

Over thirty years after the publication of Theorem 10.5.2 above, Cauchy returned to the problem (CAUCHY [1853]). It appears from the text of his paper that he made a careful study of the counterexample 10.5.4. This led him to the following revised version of 10.5.2.

10.5.5 'Si les différents termes de la série

$$(1) \qquad u_0, u_1, u_2, \ldots, u_n, u_{n+1}, \ldots$$

sont des fonctions de la variable réelle x, continue, par rapport à cette variable, entre des limites données; si d'ailleurs, la somme

$$(3) \qquad u_n + u_{n+1} + \cdots + u_{n'-1}$$

devient toujours infiniment petite pour des valeurs infiniment grandes des nombres entiers n et n' > n, la série (1) sera convergente, et la somme s de la série (1) sera entre les limites données, fonction continue de la variable x.'

If we interpret this theorem in the sense of Non-standard Analysis, so that 'infiniment petite' is taken to mean 'infinitesimal' and translate 'toujours' by 'for all x' (and not only 'for all standard x'), then the condition introduced by Cauchy in 10.5.5 amounts precisely to uniform convergence in accordance with (i) above. We should add that the problem had in the meantime been considered by other mathematicians, who introduced similar conditions (Compare HARDY [1914]).

Let us now consider Cauchy's definition of the derivative (Cauchy[1823]).

'Lorsque la fonction $y = f(x)$ reste continue entre deux limites données de la variable x, et que l'on assigne à cette variable une valeur comprise entre les deux limites dont il s'agit, un accroissement infiniment petit, attribué a la variable, produit un accroissement infiniment petit de la fonction elle-même. Par conséquent, si l'on pose alors $\Delta x = i$, les deux termes du *rapport aux différences*

$$(1) \qquad \frac{\Delta y}{\Delta x} = \frac{f(x+i) - f(x)}{i}$$

seront des quantités infiniment petites. Mais tandis que ces deux termes s'approcheront indéfiniment et simultanément de la limite zéro, le rapport lui-même pourra converger vers une autre limite, soit positive, soit négative. Cette limite, lorsqu'elle existe, a une valeur déterminée pour chaque valeur particulière de x; ...'

– and this value is the derivative of the function of the point under consideration.

Later generations have overlooked the fact that in this definition Δx and Δy were explicitly supposed to be infinitely small. Indeed according to our present standard ideas, we take $f'(x)$ to be the limit $\Delta y/\Delta x$ as Δx tends to zero, whenever that limit exists, without any mention of infinitely small quantities. Thus, as soon as we consider limits, the assumption that Δx and Δy are infinitesimal is completely redundant. It is therefore the more interesting that the assumption is there, and, indeed, appears again and again also in Cauchy's later expositions of the same topic (CAUCHY [1829, 1844]). We are forced to conclude that Cauchy's mental picture of the situation was significantly different from the picture adopted today, in the Weierstrass tradition. Thus, for $f(x)=x^m$, Cauchy states that $f'(x)=mx^{m-1}$ is the limit of

$$\frac{(x+i)^m - x^m}{i} = mx^{m-1} + \frac{m(m-1)}{1\cdot 2}x^{m-2}i + \cdots + i^{m-1},$$

where i, and hence $(x+i)^m - x^m$ is infinitely small. He also calls the derivative of a given point 'la dernière raison des différences infiniment petites Δy et Δx', where the term 'dernière raison' (ultimate ratio) is taken from the Newtonian terminology. It seems reasonable to conclude from all this that in Cauchy's mind a function did not approach its limit directly, but only via expressions involving infinitesimals. In fact, a non-standard analyst would be tempted to equate the expression 'limit of $\Delta y/\Delta x$' for infinitely small Δx and Δy with 'standard part of $\Delta y/\Delta x$, i.e., $^0(\Delta y/\Delta x)$' and this would indeed give the right answer, $f'(x)$, whenever $f'(x)$ exists. However, Cauchy would most probably have rejected this suggestion since he regarded Δx and Δy as variables, i.e., as 'quantities taking a succession of values'.

At any rate, Cauchy used infinitely small or large quantities freely in both definitions and proofs. An interesting example is provided by the proof of the theorem that if $y'=f'(x_0)$ is positive then $f(x)$ increases in the neighborhood of x_0 (CAUCHY [1829]). Cauchy argues that since $\Delta y/\Delta x$ has the limit $f'(x_0)$ for infinitesimal Δx and $\Delta y = f(x_0+\Delta x) - f(x_0)$, and since $f'(x_0)$ is positive, it follows that $\Delta y/\Delta x$ must be positive also for sufficiently small 'numerical' values of Δx, where the adjective 'numerical' means that Δx is an ordinary number, i.e., in the terminology of Non-standard Analysis, that Δx is standard. This passage from a ratio of infinitely small numbers to a ratio of sufficiently small standard numbers is legitimate in Non-standard Analysis, because it can be justified there by a rigorous proof. In Cauchy's theory however, it is a logical non sequitur and may be said to involve an unconscious use of the 'principle of continuity', whose application had been criticized by Cauchy himself.

What has been said so far concerning differentiation applies equally well to Cauchy's theory of integration. The following quotation is typical (CAUCHY [1823]).

'D'après ce qui a été dit dans la dernière leçon, si l'on divise $X - x_0$ en éléments infiniment petits $x_1 - x_0$, $x_2 - x_1$, ..., $X - x_{n-1}$ la somme

$$(1) \qquad S = (x_1 - x_0) f(x_0) + (x_2 - x_1) f(x_1) + \cdots + (X - x_{n-1}) f(x_{n-1})$$

convergera vers une limite représentée par l'intégrale définie

$$(2) \qquad \int_{x_0}^{X} f(x) \, dx \, .\,'$$

Even questions concerning the (Cauchy) principal value of a singular integral are discussed in the same setting.

On the other hand, although his definition of the derivative involved infinitely small increments, Cauchy introduced *differentials* in the way generally accepted in classical analysis. That is to say, for $dx = b$ where $b \neq 0$, otherwise arbitrary, he defined dy by $dy = f'(x)b = f'(x) \, dx$ so that the relation $dy/dx = f'(x)$ is exact.

It appears that Cauchy never abandoned the techniques described above although the following quotation, which is taken from a paper published in 1844 may indicate a slight shift in his position.

'Pour écarter complètement l'idée que les formules employées dans le calcul différentiel sont des formules approximatives, et non des formules rigoureusement exactes, il me paraît important de considérer les différentielles comme des quantités finies, en les distinguant soigneusement des accroissements infiniment petits des variables. La considération de ces derniers accroissements peut et doit être employée comme moyen de découverte ou de démonstration dans la recherche des formules ou dans l'établissement des théorèmes. Mais alors le calculateur se sert des infiniment petits comme d'intermédiaires qui doivent le conduire à la connaissance des relations qui subsistent entre des quantités finies; et jamais, à mon avis, des quantités infiniment petites ne doivent être admises dans les équations finales, où leur présence deviendrait sans objet et sans utilité...' (CAUCHY [1844]).

The remark in this passage that it is important to consider differentials as finite quantities, may be contrasted with the fact that, fifteen years earlier, Cauchy had stated that $dx = b$ is arbitrary and might even be infinitesimal (CAUCHY [1829]).

Thus, Cauchy stands in the history of the Calculus not as a man who broke with tradition and swept away old and rotten foundations to make room for new and sound ones but rather as a link between the past and the future. The elements of his theory can be traced back to Newton and to Leibniz (and beyond) but he provided a synthesis of the doctrine of limits on one hand and of the doctrine of infinitely small and large quantities on the other by assigning a central role to the notion of a variable which tends to a limit, in particular to the limit zero. And although his basic approach was actually superseded half a century later, the immense care with which he handled his tools and the great scope and detail of his work in Analysis constituted a giant step in the direction of the final (or, presently accepted) solution.

10.6   Bolzano, Weierstrass, and after.   '...if, for every $\varepsilon > 0$, there exists a $\delta > 0$...', '...if, for every $\varepsilon > 0$ there exists a natural number n...' – it is by means of these phrases that we now define limits, continuity, differentiation, and related notions, in an approach which is universally accepted and which needs no introduction here. The credit for the general adoption of this approach goes to Weierstrass, although the absence of adequate records makes it hard to determine when exactly Weierstrass began to make his ideas public. (Compare HEINE [1872]). It is certain that, in essence, the method was conceived already by Bolzano, who was a contemporary of Cauchy's (BOLZANO). Bolzano's lack of standing in the mathematical world may have been one of the reasons why he did not succeed in spreading his ideas. However, as mentioned, there are traces of the 'ε, δ-approach' already in Cauchy and an interesting example of an intermediate attitude is to be found in the following (originally unpublished) remark of Riemann's.

'Unter dem Ausdruck: die Grösse w ändert sich stetig mit z zwischen den Grenzen $z = a$ und $z = b$ verstehen wir: in diesem Intervall entspricht jeder unendlich kleinen Änderung von z eine unendlich kleine Änderung von *w* oder, greiflicher ausgedrückt: für eine beliebig gegebene Grösse ε lässt sich stets die Grösse α so annehmen, dass innerhalb eines Intervalls für *z*, welches kleiner als α ist, der Unterschied zweier Werthe von *w* nie grösser als ε ist. Die Stetigkeit einer Function führt hiernach, auch wenn dies nicht besonders hervorgehoben ist, ihre beständige Endlichkeit mit sich' (RIEMANN).

Thus, Riemann asserted the identity of Cauchy's notion of continuity in an interval and of the 'Weierstrassian' notion of uniform continuity, but stated that the Weierstrassian notion is more concrete (greiflicher). It is hard to say in what sense this could be regarded as a precise statement at the time when it was made. However, with the spread of Weierstrass' ideas, arguments involving infinitesimal increments, which survived, particularly in differential geometry and in several branches of applied mathematics began to be taken automatically as a kind of shorthand for corresponding developments by means of the ε, δ-approach (or, later on, for some more sophisticated method). Usually, this assumption has turned out to be correct although in several cases its justification was complicated and hard to achieve. However, it may well have been this attitude which gave rise to the impression that Cauchy's theory was already basically the same as Weierstrass', and that the former used infinitesimals really as a kind of shorthand. The question remains whether Cauchy himself also held this opinion. As we have seen he did in fact regard them (at least toward the end of his life) as in some way provisional but there seems to be no indication that he wished to, or would have been able to, eliminate them altogether. At the same time, we may recall that Leibniz already claimed that his mention of infinitely small or infinitely large numbers merely provided an abbreviation of 'Archimedes' method'. And it must be said that in certain respects the Greek method of exhaustion is much closer to the method of Weierstrass than to those of Newton, Leibniz, or Cauchy.

The introduction of the Weierstrass approach in Analysis, the arithmetization of the system of real numbers, and the discovery of set theory by Cantor all occurred within the same short time span and were in fact interrelated. The same period also saw the first attempts to develop a theory of infinitely small and infinitely large quantities in accordance with the newly adopted standards of rigor. Following Cauchy's idea that an infinitely small or infinitely large quantity is associated with the behavior of a function $f(x)$, as x tends to a finite value or to infinity, du Bois-Reymond produced an elaborate theory of orders of magnitude for the asymptotic behavior of functions (DU BOIS-REYMOND [1875]). Stolz tried to develop also a theory of arithmetical operations for such entities, but conceded that his results were inadequate as far as their practical applicability is concerned (STOLZ [1885]). However, the custom of regarding functions $f(x)$ such that $\lim f(x) = 0$ as infinitely small quantities and to make correct, though modest, use of this idea in Analysis has been retained in several standard texts (e.g., KHINCHIN [1953]).

Beginning with the turn of the century, interest in non-archimedean fields was restricted more and more to their algebraic aspects. This period culminated in the development of the beautiful theory of formally-real fields of Artin and Schreier. Although purely algebraic, this concept plays a part in the more recent theory of hyperreal fields (HEWITT [1948]), which, as we have observed, can serve as non-standard models of analysis. It seems likely that Skolem's idea to represent infinitely large natural numbers by number-theoretic functions which tend to infinity (SKOLEM [1934]), also is related to the earlier ideas of Cauchy and du Bois-Reymond. It is a fact that much of Skolem's work in Number Theory was concerned with rational points on algebraic curves and involved a detailed discussion of the asymptotic properties of such curves. Finally, we should record that, as far as these matters can be ascertained at all, Skolem's works on non-standard models of Arithmetic was the greatest single factor in the creation of Non-standard Analysis.

We may complete the picture of recent developments by mentioning that during the period under consideration attempts were still made to define or justify the use of infinitesimals in Analysis (e.g., GEISSLER [1904], NATORP [1923]). The most successful among these is the theory of Schmieden and Laugwitz (SCHMIEDEN and LAUGWITZ [1958], LAUGWITZ [1961, 1961a]). The starting point of the Schmieden–Laugwitz theory is, once again, the identification of infinitely small or infinitely large numbers with *functions*, with particular reference to their asymptotic behavior, except that these functions now are mappings from the natural into the rational or real numbers, i.e., sequences. The resulting system is not an ordered field but a ring with zero divisors. Nevertheless, the reader who consults the original papers will find that the theory is of considerable interest. Among other things, it includes a substitute for the theory of distributions.

It might be thought that the introduction and development of the theory of transfinite cardinal and ordinal numbers also led to an upswing in the theory of infinitesimals. This is true indirectly since abstract set theory forms a historical background to the free and easy handling of infinite sets that is required in Non-standard Analysis. However, the discoverer of Set Theory, G. Cantor, actually went so far as to claim that he could prove the impossibility of infinitely small numbers by means of that theory (CANTOR [1887/1888]). And we may quote, without comment, the words of one of the foremost authorities on Set Theory, A. H. Fraenkel, in a classical textbook on the subject (FRAENKEL [1928]). Fraenkel mentions, with approval, the opinion that the test for the efficacity of infinitely small numbers is their applicability to the Differential and Integral Calculus, and then goes on:

'*Bei dieser Probe hat aber das Unendlichkleine restlos versagt.* Die bisher in Betracht gezogenen und teilweise sorgfältig begründeten Arten unendlichkleiner Grössen haben sich zur Bewältigung auch nur der einfachsten und grundlegendsten Probleme der Infinitesimalrechnung (etwa zum Beweis des Mittelwertsatzes der Differentialrechnung oder zur Definition des bestimmten Integrals) als völlig unbrauchbar erwiesen... es besteht auch kein Grund zu der Erwartung, dass sich hierin künftig etwas ändern werde. Gewiss wäre es an sich *denkbar* (wenn auch aus guten Gründen äusserst unwahrscheinlich und jedenfalls beim heutigen Stand der Wissenschaft in ungreifbarer Ferne liegend), dass ein zweiter Cantor dereinst eine einwandfreie arithmetische Begründung neuer unendlichkleiner Zahlen gäbe, die sich als mathematisch brauchbar erweisen und ihrerseits vielleicht einen einfachen Zugang zur Infinitesimalrechnung eröffnen könnten. Solange das aber nicht der Fall ist, wird man weder die (in vieler Hinsicht interessanten) Veroneseschen noch andere unendlichkleine Zahlen in Parallele zu den Cantorschen setzen dürfen, sondern sich auf den Standpunkt stellen müssen, dass von der mathematischen und damit logischen Existenz des *Unendlichkleinen* in einem gleichen oder ähnlichen Sinn wie beim Unendlichgrossen in keiner Weise gesprochen werden kann.'

10.7   Infinitely small and large numbers and the infinite.   So far, we have concentrated on the purely mathematical aspects of the development of Analysis. However, the history of the subject cannot be fully understood if we disregard its philosophical background. Nowadays the Philosophy of Mathematics is considered in connection with more general notions such as consistency, predicativity, or constructivity. So it might seem to us that the evolution of the foundations of the calculus amounts merely to the development of certain mathematical techniques, and the rejection or modification of such techniques on grounds of obvious inconsistency. This picture is incomplete. It ignores the fact that, from the seventeenth to the nineteenth century, the history of the Philosophy of Mathematics is largely identical with the history of the foundations of the Calculus.

The gingerly way in which the Greek mathematicians, Eudoxus, Euclid, and Archimedes, handled problems of quadrature and related questions, was not a sign of the uncertainty of a nascent science, but was due to a profound and abiding distrust of the notion of infinity. The most famous expression to this uneasiness is given by some of Zeno's paradoxes. Aristotle discussed the matter in several of his writings and rejected the actual infinite while accepting the notion of a potential, or growing infinite (ARISTOTLE). The discussion of the problem continued in the Middle Ages, but we shall not deal with this phase. Writing in 1689, Locke stated a position on infinity which is close to Aristotle's (LOCKE [1690]).

Against this background, we have to understand Leibniz' position in the following way. Leibniz did not reject the actual infinite in general, but, at least in his later years, he considered that it has no place in the Calculus. Moreover, while in his mind the terms of an infinite series constituted only a potentially ('syncategorematically') infinite assemblage, the infinitely large and small numbers were thought to belong to the actual infinite. He therefore regarded the latter as ideal or fictitious while accepting the former as real. The opinion that infinitely large or small numbers were forms of the actual infinite seems to have been common ground at that time and for some time after. Thus the curiously ambivalent and shifting explanations of Newton and his followers can be explained, at least in part, by their reluctance to admit the existence of such entities. The vigorous attack directed by Berkeley against the foundations of the Calculus in the forms then proposed is, in the first place, a brilliant exposure of their logical inconsistencies. But in criticizing infinitesimals of all kinds, English or continental, Berkeley also quotes with approval a passage in which Locke rejects the actual infinite (BERKELEY [1734, 1735, −]). It is in fact not surprising that a philosopher in whose system perception plays the central role, should have been unwilling to accept infinitary entities (compare STRONG [1957]).

As the Calculus continued to develop and thrive, the logical and philosophical weaknesses of the system were overlooked for a period. Indeed, the great success of the theory was ascribed by some to the acceptance of the reality of the infinitely small and large, by contrast with its rejection by Archimedes and his predecessors. But when Cauchy constructed his edifice in response to renewed doubts he quite probably chose the notion of a variable as basic since, by its very nature, the idea of a variable seems to express potentiality and not actuality. (Compare CANTOR [1885], CARRUCIO [1957].) In this respect the shift from Cauchy's outlook to the views commonly held at present is crucial, since the ε, δ-conditions are interpreted naturally in terms of actual infinite totalities, e.g., the real numbers.

Thus, the problem of infinity which, in the view of many, is still one of the chief concerns of the Philosophy of Mathematics, although with a widened scope and with changed emphasis, is basically the same problem that worried the founders of mathematical Analysis. Having produced a theory of infinite sets of remarkable coherence and beauty, Cantor and his followers thought that they had finally caught the actual infinite, just as de l'Hospital had believed, nearly two hundred years earlier, to have found it in the Calculus of Infinitesimals. On the other hand, the outlook of the intuitionists and other constructivists may be compared with that of Cauchy; while the spirit of Formalism, or at least of one version of it, in its assessment of wide portions of contemporary Mathematics (e.g., the theory of transfinite cardinal and ordinal numbers) is close to Leibniz' as expressed in his pronouncement (on infinitely small and large numbers), 'que ce n'étaient que des fictions, mais des fictions utiles...'

At the moment Cantor's point of view is that held by the majority of mathematicians. But perhaps our historical review suggests that just as the Calculus of infinitesimals, which was triumphant in the middle of the eighteenth century was still put on completely different foundations within the next one hundred years, so future generations of mathematicians, while accepting the formal results of Set Theory, may reject the platonistic claims commonly associated with it.

We should add that, to a logical positivist, a discussion of the ontological significance of infinitary notions of any kind is meaningless. However, presumably even a positivist would concede the historical importance of expressions involving the term 'infinity' and of the (possibly, subjective) ideas associated with such terms.

Returning now to the theory of this book, we observe that it is presented, naturally, within the framework of contemporary Mathematics, and thus appears to affirm the existence of all sorts of infinitary entities. However, from a formalist point of view we may look at our theory syntactically and may consider that what we have done is to introduce *new deductive procedures* rather than new mathematical entities. Whatever our outlook and in spite of Leibniz' position, it appears to us today that the infinitely small and infinitely large numbers of a non-standard model of Analysis are neither more nor less real than, for example, the standard irrational numbers. This is obvious if we introduce such numbers axiomatically; while in the genetic approach both standard irrational numbers and non-standard numbers are introduced by certain infinitary processes. This remark is equally true if we approach the problem from the point of view of the empirical scientist. For all measurements are recorded in terms of integers or rational numbers, and if our theoretical framework goes beyond these then there is no compelling reason why we should stay within an Archimedean number system. For, repeating a quotation given earlier in this chapter, 'on ne diffère du style d'Archimède que dans les expressions qui sont plus directes dans notre méthode et plus conformes à l'art d'inventer'.

BIBLIOGRAPHY

ABEL, N. H.
  1826. Untersuchungen über die Reihe $1 + \frac{m}{1}x + \frac{m(m-1)}{1 \cdot 2}x^2 + \ldots$ u.s.w., Journal für die reine und angewandte Mathematik, **1**, 311–339.

ALEMBERT, J. LE R. D'.
  1784–1789. Articles in 'Encyclopédie méthodique ou par ordre de matières (Mathématiques)', 3 vols., Paris–Liège.

ARISTOTLE,
  ——. Physics (Loeb Classical Library, vols. 228, 255), Book III.

ARONSZAJN, N. and SMITH, K. T.
  1954. Invariant subspaces of completely continuous operators, Annals of Mathematics, **60**, 345–350.

BERKELEY, G.
  1734. The Analyst, Collected Works, vol. 4, (ed. A. A. Luce and T. E. Jessop) London 1951.
  1735. A defence of freethinking in Mathematics, Collected works, vol. 4 (ed. A. A. Luce and T. E. Jessop) London 1951.
  ——. Of infinites (first published 1901), Collected Works, vol. 4, (ed. A. A. Luce and T. E. Jessop) London 1951.

BERNSTEIN, A. R. and ROBINSON, A.
  1966. Solution of an invariant subspace problem of K. T. Smith and P. R. Halmos, Pacific Journal of Mathematics, 16, 421–431.

BETH, E. W.
  1955. The Foundations of Mathematics, Studies in Logic and the Foundations of Mathematics, Amsterdam.

BOIS-REYMOND, P. DU.
  1875. Über asymptotische Werte, infinitäre Approximationen und infinitäre Auflösung von Gleichungen, Mathematische Annalen, **8**, 363–414.

BOLZANO, B.
  ——. Functionenlehre, Schriften, vol. 1, (ed. K. Rychlik) Prague 1930.

BOYER, C. B.
  1949/1959. The History of the Calculus and its Conceptual Development, New York.

CANTOR, G.
  1885. Über die verschiedenen Standpunkte in bezug auf das aktuelle Unendliche, Gesammelte Abhandlungen, (ed. E. Zermelo) Berlin 1932, 370–377.

1887–1888. Mitteilungen zur Lehre vom Transfiniten, Gesammelte Abhandlungen, (ed. E. Zermelo) Berlin 1932, 378–439.

CARRUCCIO, E.

1957. I fondamenti dell'Analisi matematica nel pensiero di Agostino Cauchy, Bolletino dell' Unione Matematica Italiana, ser. 3, **12**, 298–307.

CARTWRIGHT, M. L.

1957. Integral Functions, Cambridge Tract no. 44, Cambridge.

CAUCHY, A. L.

1821. Cours d'Analyse de l'Ecole Royale Polytechnique, 1re partie, Analyse Algébrique, Oeuvres complètes, ser. 2, **3**.

1823. Résumé des leçons sur le calcul infinitésimal, Oeuvres complètes, ser. 2, **4**.

1829. Leçons sur le Calcul différentiel, Oeuvres complètes, ser. 2, **4**.

1844. Mémoire sur l'Analyse infinitésimale (Exercises d'Analyse et de Physique Mathématique), Oeuvres complètes, ser. 2, **13**, 9–58.

1853. Note sur les séries convergentes dont les divers termes sont des fonctions continues d'une variable réelle ou imaginaires, entre des limites données, Comptes rendus de L'Académie des Sciences, vol. 36, p. 454, March 14, Oeuvres complètes, ser. 1, **12**, 30–36.

CHANG, C. C. and MOREL, A. C.

1958. On closure under direct product, Journal of Symbolic Logic, **23**, 149–154.

COHN, P. M.

1957. Lie Groups, Cambridge Tract No. 46, Cambridge.

COURANT, R.

1950. Dirichlet's Principle, Conformal Mapping, and Minimal Surfaces, New York.

DIENES, P.

1931. The Taylor Series, Oxford.

EUCLID.

——. Elements (Heiberg–Heath edition, 3 vols.), Book V.

FONTENELLE, B. LE B. DE.

1727. Eléments de la géométrie de l'infini, Suite des Mémoires de l'Académie Royale des Sciences (1725).

FRAENKEL, A. H.

1928. Einleitung in die Mengenlehre, Grundlehren der mathematischen Wissenschaften, vol. 9, Berlin.

FRAYNE, T., MOREL, D. C. and SCOTT, D. S.

1962. Reduced direct products, Fundamenta Mathematicae, **51**, 195–227.

GEISSLER, K.

1904. Grundgedanken einer übereuklidischen Geometrie durch die Weitenbehaftungen des Unendlichen, Jahresberichte der Deutschen Mathematiker-Vereinigung, **13**, 233–240.

GILLMAN, L. and JERISON, M.

1960. Rings of Continuous Functions, Princeton.

GILMORE, P. C. and ROBINSON, A.

1955. Metamathematical considerations on the relative irreducibility of polynomials, Canadian Journal of Mathematics, **7**, 483–489.

GÖDEL, K.

1930. Die Vollständigkeit der Axiome des Logischen Funktionenkalküls, Monatshefte für Mathematik und Physik, **31**, 349–360.

HALMOS, P. R.

1963. A glimpse into Hilbert space. Lectures on Modern Mathematics (ed. T. L. Saaty) vol. 1, 1–22.

HALPERN, J. D.

1964. The independence of the axiom of choice from the Boolean prime ideal theorem, Fundamenta Mathematicae, **55**, 57–66.

HARDY, G. H.

1914/1919. Sir George Stokes and the concept of uniform convergence, Proceedings of the Cambridge Philosophical Society, **18/19**, 148–156.

HEINE, E.

1872. Die Elemente der Functionenlehre, Journal für die reine und angewandte Mathematik, **74**, 172–188.

HENKIN, L.

1949. The completeness of the first-order functional calculus, Journal of Symbolic Logic, **14**, 159–166.

1950. Completeness in the theory of types, Journal of Symbolic Logic, **15**, 81–91.

HEWITT, E.

1948. Rings of real-valued continuous functions I, Transactions of the American Mathematical Society, **64**, 54–99.

HILLE, E.

1962. Analytic Function Theory, vol. 2, Boston.

HOSPITAL, G. F. A. DE L'.

1715. Analyse des infiniment petits pour l'intelligence des lignes courbes, 2nd ed. (1st ed. 1696) Paris.

JERISON, M. and GILLMAN L.

see L. Gillman and M. Jerison.

JULIA, G.

1923. Leçons sur les fonctions uniformes à point singulier essentiel isolé, Paris.

KAKEYA, S.

1917. On zeros of a polynomial and its derivatives, Tohoku Mathematical Journal, **11**, 5–16.

KEISLER, H. J.

1962. Ultraproducts and elementary classes, Proceedings of the Royal Academy of Sciences, Amsterdam, Ser. A, **64**, 477–495.

KEMENY, J. G.

1958. Undecidable problems of elementary number theory, Mathematische Annalen, **135**, 160–169.

KHINCHIN, A. Y.

1953. Kratkii Kurs Matematicheskovo Analisa, Moscow.

KOCHEN, S.
> 1961. Ultraproducts in the theory of models, Annals of Mathematics, ser. 2, **79**, 221–261.

LACROIX, S. F.
> 1797/1798. Traité du calcul différentiel et du calcul intégral, 2 vols., Paris.

LAGRANGE, J. L.
> 1797. Théorie des fonctions analytiques, Paris.

LAUGWITZ, D.
> 1961. Anwendungen unendlichkleiner Zahlen, I. Zur Theorie der Distributionen. Journal für die reine und angewandte Mathematik, **207**, 53–60.

> 1961a. Anwendungen unendlich kleiner Zahlen, II. Ein Zugang zur Operatorenrechnung von Mikušinski, Journal für die reine und angewandte Mathematik, **208**, 22–34.

LAUGWITZ, D. and SCHMIEDEN, C.
> see C. Schmieden and D. Laugwitz.

LAURMANN, J. A. and ROBINSON, A.
> see A. Robinson and J. A. Laurmann.

LEIBNIZ, G. W.
> 1689. Tentamen de motuum coelestium causis, Acta Eruditorum (Mathematische Schriften, ed. C. I. Gerhardt, vol. 6, 1860, 144–161).

> 1695. Responsio ad nonnullas difficultates a Dn. Bernardo Nieuwentijt circa methodum differentialem seu infinitesimalem motas. Acta Eruditorum (Mathematische Schriften, ed. C. I. Gerhardt, vol. 5, 1858, 320–328).

> 1695a. Leibniz to de l'Hospital, June 14/24. (Mathematische Schriften, ed. C. I. Gerhardt, vol. 2, 1850, 287–289).

> 1701. Mémoire de M. G. G. Leibniz touchant son sentiment sur le calcul différentiel, Journal de Trévoux (Mathematische Schriften, ed. C. I. Gerhardt, vol. 5, 1858, p. 350).

> 1702. Leibniz to Varignon, February 2 (Mathematische Schriften, ed. C. I. Erhardt, vol. 4, 1859, 91–95).

> 1702a. Leibniz to Varignon, June 20 (Mathematische Schriften, ed. C. I. Gerhardt, vol. 4, 1859, 106–110).

> 1716. Letter to Dangicourt, sur les monades et le calcul infinitésimal etc., September 11 (Opera Omnia, ed. L. Dutens, vol. 3, 1789, 499–502).

LOCKE, J.
> 1690. Essay Concerning Humane Understanding.

ŁOŚ, J.
> 1955. Quelques remarques, théorèmes, et problèmes sur les classes définissables d'algèbres, Mathematical Interpretation of Formal Systems (Symposium, Amsterdam 1954), Studies in Logic and the Foundations of Mathematics, Amsterdam, 98–113.

LUXEMBURG, W. A. J.
> 1962. Non-Standard Analysis. Lectures on A. Robinson's theory of infinitesimals and infinitely large numbers, Pasadena.

1962. Two applications of the method of construction by ultrapowers to Analysis, Bulletin of the American Mathematical Society, ser. 2, **68**, 416–419.

1963. Addendum to 'On the measurability of a function which occurs in a paper by A. C. Zaanen', Proceedings of the Royal Academy of Sciences, Amsterdam, **65**, 587–590.

MacDowell, R. and Specker, E.

1961. Modelle der Arithmetik, Infinitistic Methods (Symposium on Foundations of Mathematics, Warsaw 1959) Warsaw, 257–263.

Malcev, A. I.

1936. Untersuchungen aus dem Gebiete der Mathematischen Logik, Matematicheskii Sbornik, vol. 1 (43), 323–335.

Marden, M.

1949. The Geometry of the Zeros of a Polynomial in a Complex Variable, Mathematical Surveys, **3**, New York.

Mendelson, E.

1961. On non-standard models for Number Theory, Essays on the Foundations of Mathematics (Fraenkel anniversary volume), Jerusalem, 259–268.

Mikusiński, J, and Sikorski, R.

1957. The elementary theory of distributions (I) Rozprawy Matematyczne, **12**.

Milloux, H.

1924. Le théorème de M. Picard, suites de fonctions holomorphes, fonctions méromorphes et fonctions entières, Journal de Mathématiques pures et appliquées, new series, **3**, 345–401.

1928. Les cercles de remplissage des fonctions méromorphes ou entières et le théorème de Picard–Borel, Acta Mathematica, **52**, 189–255.

Montel, P.

1923. Sur les modules des zéros des polynomes, Annales scientifiques de l'Ecole normale supérieure, ser. 3, **40**, 1–34.

1927. Leçons sur les familles normales de fonctions analytiques et leurs applications, Paris.

1933. Leçons sur les fonctions univalentes ou multivalentes, Paris.

Morel, A. C. and Chang, C. C.

see C. C. Chang and A. C. Morel.

Morel, A. C., Frayne, T. and Scott, D. S.

see T. Frayne, A. C. Morel, and D. S. Scott.

Morley, M. and Vaught, R. L.

1962. Homogeneous universal models, Mathematica Scandinavica, **11**, pp. 37–57.

Müller, G. H.

1961. Nicht-Standard Modelle der Zahlentheorie, Math. Z. 77, pp. 414–438.

Natorp, P.

1923. Die Logischen Grundlagen der exakten Wissenschaften, Wissenschaft und Hypothese, **12**, 3rd ed., Leipzig and Berlin.

Nieuwentijt, B.

1694. Considerationes circa analyseos ad quantitates infinite parvas applicatae principia, et calculi differentialis usum in resolvendis problematibus geometricis.

1695. Analysis infinitorum seu curvilineorem proprietates ex polygonum natura deductae, Amsterdam.

OSTROWSKI, A.

1926. Über Folgen analytischer Funktionen und einige Verschärfungen des Picardschen Satzes, Mathematische Zeitschrift, **24**, pp. 215–258.

PONTRJAGIN, L.

1946. Topological Groups (translated by E. Lehmer), Princeton.

RABIN, M. O.

1961. Non standard models and independence of the induction axiom, Essays on the Foundations of Mathematics (Fraenkel anniversary volume), Jerusalem, 287–299.

RIEMANN, B.

——. Gesammelte mathematische Werke, 2nd ed., Leipzig 1892 (New York 1953) p. 46.

ROBINSON, A.

1952. On the application of Symbolic Logic to Algebra, Proceedings of the International Congress of Mathematicians (Cambridge, U.S.A., 1950) vol. 1, 686–694.

1955. Théorie métamathematique des idéaux, Collection de logique mathématique, ser. A, **8**, Paris–Louvain.

1961. Model theory and non-standard Arithmetic, Infinitistic Methods (Symposium on Foundations of Mathematics, Warsaw 1959), Warsaw, 265–302.

1961a. Non-standard Analysis, Proceedings of the Royal Academy of Sciences, Amsterdam, ser. A, **64**, 432–440.

1962. Complex Function Theory over Non-archimedean Fields, Technical-scientific note, No. 30, U.S.A.F. contract No. 61 (052)–187, Jerusalem.

1963. Introduction to Model Theory and to the Metamathematics of Algebra, Studies in Logic and the Foundations of Mathematics, Amsterdam.

1964. On generalized limits and linear functionals, Pacific Journal of Mathematics, **14**, 269–283.

1965. Topics in non-archimedean Mathematics (Symposium on Model Theory, Berkeley 1963).

1965a. On the theory of normal families, Acta Philosophica Fennica, fasc. 18 (Rolf Nevanlinna anniversary volume) 159–184.

ROBINSON, A. and GILMORE, P. C.

see P. C. Gilmore and A. Robinson.

ROBINSON, A. and LAURMANN, J. A. 1956. Wing Theory, Cambridge.

SAINT-VENANT, B. DE

1855. Mémoire sur la torsion des prismes, Mémoires des Savants étrangers, **14**, 233–560.

1856. Mémoire sur la flexion des prismes, Journal de Mathématiques pures et appliquées, second series, **1**, 89–189.

SCHLICHTING,

1955. Boundary Layer Theory (translated by J. Kestin), New York.

SCHMIEDEN, C. and LAUGWITZ, D.

1958. Eine Erweiterung der Infinitesimalrechnung, Mathematische Zeitschrift, **69**, 1–39.

SCHWARTZ, L.
1950–1951. Théorie des distributions, 2 vols., Paris.

SCOTT, D.
1961. On constructing models for arithmetic, Infinitistic Methods (Symposium on Foundations of Mathematics, Warsaw 1959) Warsaw, 235–255.

SCOTT, D., FRAYNE. T. and MOREL, A. C.
see T. Frayne, D. C. Morel and D. S. Scott.

SIKORSKI, R. and MIKUSIŃSKI, J.
see J. Mikusiński and R. Sikorski.

SKOLEM, T.
1934. Über die Nicht-charakterisierbarkeit der Zahlenreihe mittels endlich oder unendlich vieler Aussagen mit ausschliesslich Zahlenvariablen, Fundamenta Mathematicae, **23**, 150–161.

SMITH, K. T. and ARONSZAJN, N.
see N. Aronszajn and K. T. Smith.

SPECKER, E. and MacDOWELL, R.
see R. MacDowell and E. Specker.

STERNBERG, E.
1954. On Saint-Venant's principle, Quarterly of Applied Mathematics, **11**, 393–402.

STOLZ, O.
1885. Vorlesungen über allgemeine Arithmetik, Leipzig.

STRONG, E. W.
1957. Mathematical reasoning and its objects, George Berkeley Lectures, University of California Publications, pp. 65–88.

TAKEUTI, G.
1962. Dirac space, Proceedings of the Japanese Academy, **38**, 414–418.

TARSKI, A.
1952. Some notions and methods on the borderline of Algebra and Metamathematics, Proceedings of the International Congress of Mathematicians, (Cambridge, U.S.A. 1950) vol. I, 705–720.

THWAITES, B. (ed.),
1960. Incompressible Aerodynamics, Oxford.

VALIRON, G.
1923. Lectures on the General Theory of Integral Functions, Toulouse.

VAUGHT, R. L. and MORLEY, M.
see M. Morley and R. L. Vaught.

WARD, G. N.
1955. Linearized Theory of Steady High-Speed Flow, Cambridge.

ZANABONI, O.
1937. Dimostrazione generale del principio del De Saint-Venant, Atti dell'Accademia Nazionale dei Lincei, vol. 25, pp. 117–121.

INDEX OF AUTHORS

SUBJECT INDEX

(Consult also the list of Contents)